TREE FRUIT PRODUCTION
THIRD EDITION

TREE FRUIT PRODUCTION
THIRD EDITION

Benjamin J.E. Teskey

Professor, Horticultural Science
Ontario Agricultural College
University of Guelph
Guelph, Ontario, Canada

James S. Shoemaker

Horticulturist Emeritus
University of Florida
Agricultural Experiment Station
Gainesville, Florida

AVI PUBLISHING COMPANY, INC.
Westport, Connecticut

°Copyright 1978 by
THE AVI PUBLISHING COMPANY, INC.
Westport, Connecticut

Second Printing 1982

Library of Congress Cataloging in Publication Data

Teskey, Benjamin J   E
      Tree fruit production.

      Includes index.
      1. Fruit-culture. I. Shoemaker, James Sheldon,
1898–   joint author. II. Title.
SB355.T34 1978   634   78–13092
ISBN–0–87055–265–1

Printed in the United States of America

Contents

Preface to the Third Edition

This is a revised and updated edition of the book *Tree Fruit Production*, first published in 1959 and extensively revised in the second edition in 1972. Considerable advances have been made in recent years in the scientific production and handling of deciduous tree fruits in North America. This third edition brings together in up-to-date usable textbook form the essence of pertinent research and practical experience on the subject. Although the principles involved in the different operations of orchard management, such as pruning, soil management, fruit thinning, and harvesting remain constant, practices and techniques have been undergoing considerable change. Economic and social changes have been brought to bear in altering the approach to such aspects of pomology as tree size, plant density, mechanical harvesting, pest control and irrigation.

Greatly increased costs of production have swung the emphasis of attention toward the wider use of organic chemicals in the orchard. Growth regulating substances are finding a place in the orchard, not only for fruit thinning, preharvest drop control and weed suppression, but also for other purposes such as promotion of early flowering, tree training, pruning and the advancement and extension of the harvest season.

The trend toward the smaller, more easily and economically managed apple tree which began slowly some three or more decades ago and increased rapidly in subsequent years is now complete. This same trend, for the same reasons, is now beginning to involve the pear, the peach and the cherry. Thus, the constant search for suitable, size-controlling rootstocks is spreading from the apple to the other tree fruits as well.

To comply with the demands of a changing market and with altering marketing methods, new cultivars of tree fruits are being developed. The modern cultivar may have to fit into a processing, farm market or consumer-pick operation. It must also satisfy the demands of new manual, semi-mechanical and mechanical harvesting methods. The cultivar must also be compatible with the dwarfing rootstock used and suited to the environmental conditions prevailing in each specific production area.

Irrigation, which in earliest times was often a vital factor in fruit production, is now becoming a standard practice in many orchards outside of the arid and semiarid areas of the continent. As furrow irrigation has given way to the sprinkler system, so now the overhead sprinkler is being replaced by the more precise drip or trickle method.

The large reference list has been updated where necessary and enriched where possible in order to provide even greater substance and scope to principles and practices presented in the text. Many changes have been made to the nearly 100 photographs and drawings with which the text is illustrated so that the illustrations truly reflect the timely character of the treatise.

It is hoped that the material presented and the way in which it is presented will prove useful as a text and reference book and as a guide for orchard practice.

B.J.E. TESKEY
J.S. SHOEMAKER

March 1978

Preface to the First Edition

This is a book on the culture and handling of apples, pears, peaches, cherries, plums, apricots, nectarines, quinces, and citrus fruits. It has been written to answer the need of students, teachers, and others who require the most up-to-date information on the production of tree fruits. Included in the discussion are subjects which confront every grower of these fruits and every student of pomology.

Orchard operations, such as soil management, use of fertilizers, pruning, thinning, harvesting, storage, and marketing, that are commonly referred to as standard practices, are constantly undergoing change. Research developments resulting in better techniques and materials, together with increased mechanization and the pressure of economic competitition, force the adoption of improved procedures such as are presented in this text.

The text is an outgrowth of experimental work, research, classroom, and a practical orchard experiences over a wide range of territory in North America. These firsthand personal experiences were gained over many years and in widely differing areas, from the colder parts, through the more moderate, to the warmer parts of this continent—from the province of Alberta to the state of Florida, including extensive periods of service in Ontario, Iowa, Minnesota, and Ohio.

Also, in this book a great deal of emphasis has been placed on recent research by horticultural scientists throughout this country and abroad, who have contributed to needed improvements in production and handling methods with tree fruits. To this end, information from more than 800 references has been incorporated into the treatise. All of this has resulted in a text that is adapted to widespread regions in the United States and Canada.

Increasing costs of orchard management and the introduction of automatic spraying equipment have, within the last few years, created a growing interest in smaller fruit trees. A need has thus arisen for the consolidation and presentation of presently available information on this subject. For this reason a separate chapter on dwarf apple and pear trees has been included in the text.

Apricots, nectarines, and quinces have been discussed in one chapter. Each of the other chapters deals with one particular kind of tree fruit and comprises all phases and aspects of its culture, production, and handling. The reasons for certain orchard practices are given, and physiological phenomena involved in tree growth and fruit production are explained.

The headings and subheadings throughout each chapter have been carefully arranged to make specific points stand out clearly for easy reading. The text is illustrated with a total of 103 photographs and drawings. The detailed index given at the end of the book will help the reader find information quickly on any phase of production for any of the specific tree fruit crops discussed.

A chapter is included on citrus fruits even though such fruits are not adapted to regions where the major deciduous fruit trees are at their best. The large citrus industry, however, has many unique features which should be of interest in northern regions. Conversely, many suggestions applying to deciduous fruit trees should be of interest in citrus regions.

Different regions vary in certain practices, such as spraying, and the recommendations are commonly modified from year to year. Therefore, pest control schedules are not presented in this text. Local bulletins, circulars, and other publications can often be used to advantage to supplement the discussion in the text.

It is hoped that the material presented, and the manner in which it is presented, will prove useful as a text and reference work, and as a guide for orchard practice.

J.S. SHOEMAKER
B.J.E. TESKEY

November 1958

Related AVI Books

Apples

Many different fruits and vegetables have been known as "apples." The fleshy fruit referred to in this book is the major deciduous tree fruit of North America and of the world. The term "pomology," the science of fruit production, comes from the apple, a pome.

The commercial cultivars of today are far removed from the wild apples from which they developed over many centuries. Several species may have contributed to the development of the modern apple. For this reason the taxonomy of the apple is somewhat confused. Horticultural authorities generally use *Malus domestica* (Bork) (which is preferred in this text) as the scientific name of the cultivated apple (Zielinski 1955).

The early history of the domestic apple becomes lost in antiquity. The apple probably originated in the region south of the Caucasus, where it appears to be indigenous. It has existed in Europe, both in wild and in cultivated forms, from time immemorial and appears to have been well developed there at the beginning of the Christian era. The westward movement of the apple from Europe was a natural sequence of colonization of North America by the early settlers. Indians, traders, and missionaries first took it westward across this continent. An interesting legend is that of Johnny Appleseed (John Chapman) who assisted in the planting and distribution of apple seeds among the westward moving families. By 1868 the apple had become established from coast to coast.

Botanically, the apple is a pome fruit developed from an inferior ovary, and is derived both from the ovary wall and the floral tube, which is composed of the basal parts of the sepals, petals, and stamens. This tube is fused with the ovary wall, becomes fleshy, and ripens with it. The fleshy mesocarp constitutes the main edible portion. The five cavities may each contain two seeds (Fig. 1.1).

PRODUCTION

Apples are grown in nearly every state of the United States and in most of the provinces of southern Canada, but the three main regions of production in North America are (1) the Eastern section comprising the Northeastern States, the Atlantic States, Ontario, Quebec, New Brunswick and Nova Scotia (about 90 million bushels or 1.5 million tonnes annually), (2) Central United States, including Ohio to Arkansas (about 0.5 million tonnes), and (3) Western section from Colorado to British Columbia (about 1.25 million tonnes). The state of Washington leads

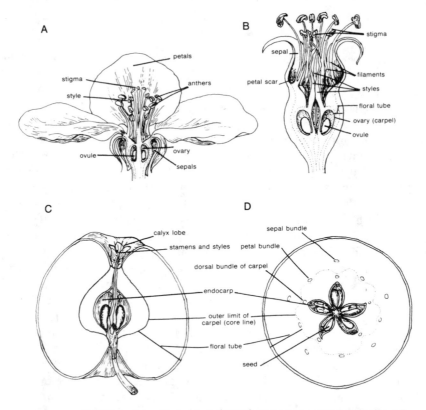

Redrawn from Weir et al. (1970) and used with permission of
John Wiley & Sons

FIG. 1.1.   THE ONTOGENY OF THE FLOWER AND FRUIT (POME)
OF APPLE (*MALUS DOMESTICA*)

(A) Median longitudinal section of flower; (B) median longitudinal section of maturing fruit; (C) median section of fruit; (D) cross-section of fruit.

with more than 0.5 million tonnes, New York is second with less than 0.5 million tonnes, and Michigan with about 0.33 million tonnes is third. The total North American production of about 3.33 million tonnes per year is nearly 25% of the world apple production.

PROPAGATION OF NURSERY STOCK

Apples reproduce themselves by means of seeds, but their offspring differ widely; thus, they must be maintained by vegetative propagation. By means of grafting, two distinct parts (namely, the root system and the stem and branch system), which may have very diverse characteristics, are made to grow together as a single plant.

Seedling Rootstocks

Rootstocks are raised either from seed or by vegetative propagation (for the latter see Dwarfed Apples).

French crab seedlings and domestic seedlings have been commonly used by nurserymen. The French crab seedlings were grown in France and then imported to America, or the seed was imported and the stock grown on this continent.

Seed from Delicious, Yellow Newtown, Wealthy, and Rome gives good germination and vigorous seedlings. Seed from triploid cultivars, e.g., Rhode Island Greening, Gravenstein, and the Winesap group, is unsatisfactory for rootstock production.

Producing Nursery Seedlings

Afterripening.—Seeds of apple normally will not germinate directly after harvest. A period of afterripening is essential for certain chemical and other changes to take place in the seed and start the dormant embryo into growth. The length of the afterripening period of seed of domestic cultivars is usually 70−80 days at 4°−10°C (40°−50°F); the optimum is 4°−5°C (40°−41°F) under moist conditions. Temperatures of 0.6°−3°C (33°−38°F) are more favorable than those above 5°C (41°F). Seed may become dry before afterripening begins without reducing germination, but during the following afterripening it must be kept moist until sown.

For afterripening, soak dry seeds in water for several hours and place them in mouse-proof containers with a moisture-holding material such as damp peat moss. Bury the containers over winter in the ground in a sheltered place or store them in a cool place. Keep the medium, in which the seeds are stored, moist. Polyethylene film envelopes, permitting ready examination of the seed, may be used if they contain some damp peat moss or perlite.

Sowing the Seed.—Seeds sown directly in the nursery row should be evenly distributed in a narrow, shallow trench and covered with soil. Fall sowing of the seed may provide a suitable environment for after-ripening, but apple seed is more commonly afterripened under controlled conditions and sown in the spring.

In soil that tends to form a hard crust, cover the seed first with either moist peat or softwood sawdust and then cap the rows with sandy soil. When seed is sown in the fall, mound up additional soil over the rows to protect the seed over winter, but draw this soil away in early spring, leaving only a thin cover through which the sprouting seed can easily emerge.

Lining Out.—Trim the rootstocks and plant them in early spring. In trimming, cut back the lateral roots, remove all side branches, and cut the top to a total plant length of about 38 cm (15 in.). This trimming expedites planting, since the rootstocks are inserted in a narrow vertical trench opened by a special trenching plow. Many nurseries buy lining-out stock from other nurseries.

Budding

Budding is the usual method of propagating fruit trees in the nursery. Normally only one bud is inserted on each rootstock. A fast budder working under favorable conditions on properly cleaned stock, and with someone else doing the tying, may insert 1200 or more buds in a day. Probably 800–1000 buds per day is a good average.

When to Bud.—Budding is done in the summer when buds of the current season are well formed and the bark slips well. It may take place from June till September, depending on the locality and the fruit. Buds may be too immature, but seldom are too mature, for successful budding.

In New York, budding is done on the following stocks in the order given: mid-July to mid-August, Western sandcherry, European plum, pear, apple, mazzard cherry, and quince; late August to early September, myrobalan plum, mahaleb cherry, and peach (Brase 1956).

Budsticks.—Cut shoots of the current season's growth from either bearing or nursery trees. These shoots which carry the buds for budding are called budsticks.

To check loss of moisture by evaporation from the leaves after the budstick has been cut from the parent tree, promptly remove leaves and keep the budsticks moist. Leaving a short portion of the petiole in place as a "handle," wrap the budsticks in moist cloth, in plastic bags, or place them with the basal end in water in a container. They can thus be stored in a cool place for several days; but it is better to use them soon after cutting. Use the well-developed plump and hard buds from the mid-portion of the shoot, discarding the soft tip buds and basal buds.

Preparing the Seedlings.—Prepare the seedling nursery stock by stripping off the lateral shoots on the lower 15 cm (6 in.) of the stem in early summer. Wipe the stock clean of soil particles near the point of bud insertion.

T-cut.—At budding time make a T-cut in the bark of the stock 5–7.6 cm (2–3 in.) above ground. Make the cut through the bark to cambium depth (not into the wood). Some budders prefer to make the transverse cut first, about 1/3 around the stock. They then make a vertical cut upwards to meet the transverse cut. As it reaches the transverse cut, the knife blade is twisted to raise the edges of the bark just enough, without tearing, so that the bud may be easily inserted. Other budders prefer to make the upward cut first, then the transverse cut.

Cutting the Bud.—Cut the bud with a shield of bark by holding the budstick by the top end, with the lower end away from the body. Place the knife 1.3 cm (1/2 in.) below the first suitable bud and by a shallow slicing movement pass the knife beneath the bud approaching the surface 2.5 cm (1 in.) above it.

Cut the shield bearing the bud fairly thin, but not so thin that the soft growing tissue beneath the bark and wood is injured. Retain the thin strip of wood that was cut with the shield.

Inserting the Bud.—After cutting, hold the bud by the petiole and insert the bud into the T-shaped incision. Fast budders slip the bud on their knife directly to the T-cut. A properly inserted bud is at least 2 cm (3/4 in.) below the transverse cut. Avoid undue manipulation or prying of the bark flaps.

Buds are usually placed on the same side of the stock along the row so that they may be readily inspected and manipulated the following season. The side from which prevailing winds come is the preferable one to prevent subsequent breakage.

Tying.—After inserting the bud wrap it snugly. Be sure to leave the bud exposed. Rubber budding strips have largely replaced raffia or string as binding material. Rubber strips have the merit of expanding with growth of the rootstock and after exposure of a month to the sun rot and fall off. By this time a good- union between bud and stock has taken place.

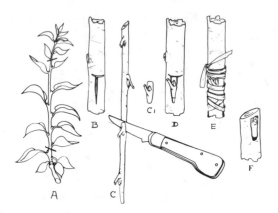

FIG. 1.2. SHIELD BUDDING

(A) Terminal growth of current season, the source of buds; (B)
the T-cut in the stock; (C) the prepared budstick showing
method of cutting the bud; (C-1) the shield bud; (D) the bud in
place; (E) the bud tied tightly against the stock; (F) the branch
of the stock cut off close to the bud the following spring.

When active cambial layers are in close contact they establish an intimate contact of the cambial region, usually by producing thin-walled parenchyma cells. This step is followed by interlocking of parenchyma cells (callus tissue). When parenchyma cells of both stock and bud (scion) are locked, a new or "bridge" cambium is produced which begins typical cambium activity and starts formation of new xylem and phloem. Probably the origin of new xylem tissue is from the bud (or scion) rather than from the stock.

Care After Budding

The first indication that the bud has united with the stock is the dropping off of the leaf stem. In successful budding, the bud usually will have grown to the stock in 2−3 weeks. Shriveled adhering leaf stems often indicate failure; if the bark still separates readily from the wood, a new bud may be inserted in a new position on the stock.

Buds inserted in late July or later remain dormant until the following spring. Buds properly united with the stock do not require any winter protection such as plowing up to budded stock in the fall. Besides eliminating two extra time-consuming operations, namely, covering in the fall and uncovering in the spring, buds that have no soil cover over winter start earlier and more uniformly the following spring. Cut off rootstocks immediately above the grafted bud in early spring.

Rub off all suckers that appear on the rootstocks during spring and early summer; 2–3 "sproutings" may be necessary before growth from the inserted bud is strong enough to be completely dominant. Maintain healthy foliage and good growth by weed and pest control, fertilization, and, if necessary, by irrigation.

Apple and pear trees often are left to grow for a second season rather than dug as 1-year whip (unbranched) trees. In early spring of the second growing season head back the young trees to 60–90 cm (24–36 in.) measured from the bud union. To form the tree head, leave 4–6 buds near the top of the young tree that has been headed back, and rub off all other buds as soon as growth starts.

In southern regions where budding can be done early, it may be necessary to break the seedling a short distance above the bud about two weeks after budding so as to force the bud into growth, thus producing a tree the same year the bud is set. Such a tree is called a "June bud."

Whip-and-tongue Graft

The whip-and-tongue graft is sometimes used in propagating fruit cultivars. Grafting is done in winter and the grafted stock is lined out in the spring.

Use 1-year-old wood that is smooth, straight, and free from branches. The diameters of stock and scion should match. From about 5 cm (2 in.) above the basal end of the scion make a diagonal cut to the base. To make the tongue, a straight cut 3 cm (1-1/4 in.) long is made halfway along the cut surface, toward the apical end of the scion. Similar cuts are made on the rootstock, and the scion and stock are then interlocked so that the cambium layers make as much contact as possible. Secure the graft with grafting tape or waxed string. Waxing is not necessary but is helpful.

Grades for Nursery Trees

All trees should have reasonably straight stems. Grades 1 cm (6/16 in.) and larger should be branched, and well rooted; 1.4–1.7 cm (9/16–11/16 in.) should have 3 or more branches. Caliper is taken 5 cm (2 in.) above the collar or bud; height is taken from the collar, if grafted, or from the union, if budded. Caliper governs; height represents average height; slow growing kinds may fall short of the height specified. Age, in years, refers to the top, not to the rootstock. All nursery trees are tied 10 per bundle or 20 per bundle according to caliper.

In the more northerly regions it is often difficult to obtain well-grown 1-year-old apple trees, so 2-year trees are generally recommended there.

TABLE 1.1
GRADES FOR NURSERY TREES

| Apple, Standard | | Apple, Dwarf | |
Caliper (cm)	Minimum Height (m)	Caliper (cm)	Minimum Height (m)
1.7 to 2.5	1.4 and up	1.4 and up	1.1 and up
1.4 to 1.7	1.2 and up	1.1 to 1.4	0.8 and up
1.1 to 1.4	0.9 and up	0.8 to 1.1	0.6 and up
0.8 to 1.1	0.6 and up		

In most regions, however, well-grown 1-year-old trees have these advantages: their training can commence at an earlier age before many undesirable branches have already formed; younger trees transplant more readily; and the cost of growing, handling, and shipping is less with unbranched trees.

Identification of Nursery Trees

Correct identification of cultivars when the trees are in the nursery is of a great advantage to both nurseryman and fruit grower. Studies of the varietal tree characters have made possible the development of keys for identification of apple cultivars (Shaw 1943). A description of one cultivar is given here as an example of the kind of information available.

Cortland.—Tree is vigorous, diverging, somewhat straggly. Shoots are long, medium size, not much curved, with medium internodes. Bark is dark reddish to dark reddish-olive. Lenticels are medium in number and size, roundish, russet, slightly raised. Leaf blade is medium-to-large, flat, often downfolded, straight or slightly reflexed, even broad oval-to-oblong, medium green, spreading. Serrations are moderately dull, shallow. Surface is dull, rough, with moderate pubescence.

CULTIVARS (VARIETIES)

Apple cultivars have originated as a result of planned breeding, selection, and as chance seedlings. Many of the leading cultivars today were well known to previous generations of growers and consumers. Some of the North American apple cultivars are listed in order of ripening.

Summer

Quinte.—Attractive red; superior quality for season; hardy; ripens unevenly, suited to colder apple regions. Origin: Seedling of Crimson Beauty × Melba, Ottawa, Canada, 1964.

Yellow Transparent.—Yellow; good quality, tree hardy, biennial bearer. Russia, 1870.

Lodi.—Yellow; similar to Yellow Transparent but larger. Tree hardy; annual bearer. Montgomery × Yellow Transparent, New York, 1924.

Melba.—Red; good quality; ripens unevenly. Tree vigorous, hardy, early bearer, suited to colder regions. McIntosh seedling, Ontario, 1909.

Duchess (Oldenburg.)—Red-striped; good cooking variety. Tree very hardy, early, heavy, biennial bearer. Russia, 1815.

Williams.—Red; attractive. Tree rather small and lacking vigor unless topworked. Massachusetts, 1830.

Early Fall

Jerseymac.—Red, large, good quality. McIntosh progeny, New York, 1971.

Milton.—Pinkish-red; large; good quality; matures throughout September. Hardy, early, suited to colder regions. Yellow Transparent × McIntosh seedling, New York, 1923.

Gravenstein.—Red sports are highly colored; good size and quality. Tree vigorous, lacks hardiness, annual bearer. Germany, 1760.

Wealthy.—Red; good cooking quality; very suited for U-pick. Tree hardy, early, biennial bearer. Chance seedling, Minnesota, 1810.

Late Fall

McIntosh.—Red; excellent appearance and quality; stores well; leading cultivar in East. Subject to scab, drops badly. Tree fairly hardy, vigorous, strong, annual bearer. Chance seedling, Ontario, 1796.

Cortland.—Red blush; large; firm white flesh; good quality. Tree heavy feeder, vigorous, smaller than McIntosh. Ben Davis × McIntosh, New York, 1918.

Spartan.—Red; more highly colored, clings better, and freer from scab than McIntosh; good quality. Tree similar to McIntosh. McIntosh × Yellow Newtown, British Columbia, 1936.

Jonagold.—Red-striped; large; stores well. Tree productive; strong; semi-vigorous; triploid. G. Delicious × Jonathan, New York, 1968.

Empire.—McIntosh type; dark red; stores well. Tree spreading; semi-vigorous; semi-spur, McIntosh × (Delicious?), New York. 1966.

Delicious.—Red-striped (several solid red sports); distinctive conical form with five prominent points at calyx end; high dessert quality. Tree upright, narrow angles, sparse foliage; requires deep, fertile soil. Chance seedling, Iowa, 1895.

Jonathan.—Red; medium size; good quality. Tree medium size, moderate vigor reliable cropper. Not suited to colder apple sections. New York, 1800. Blackjohn is one of several sports.

Rhode Island Greening.—Green; good quality for cooking; stores well. Tree vigorous, spreading, moderately hardy. Chance seedling, Rhode Island, 1720.

Stayman.—Red; good color and quality when well grown. Tree moderately vigorous, suited to warmer apple sections. Winesap seedling, Kansas, 1866.

Idared.—Red; attractive, firm, good quality; stores well. Tree medium size, hardy, productive. Wagner × Jonathan, Idaho, 1951.

Golden Delicious.—Yellow; resembles Delicious in form; very good quality. Tree moderately hardy, not suited to most northern apple sections. Fruit requires thinning. Chance seedling, Virginia, 1916.

Mutsu.—Green-yellow; medium large; good quality; triploid; performs well on dwarf stock. Golden Delicious seedling × Indo, Japan, 1948.

Northern Spy.—Red-striped; excellent quality; stores very well; a leading cultivar in the Northeast. Tree large, vigorous, late bearing, relatively hardy. Chance seedling, New York, 1874.

York.—Red; lopsided; good quality; stores very well. Tree vigorous on rich soil, and in warmer apple sections. Seedling, Pennsylvania, 1800.

Winesap.—Red; good quality; medium small. Tree medium size and vigorous in warmer apple sections. Arkansas, Arkansas Black, Paragon, and Stayman Winesap are seedlings of Winesap. Seedling, Virginia, 1800.

Yellow Newtown.—Yellow, often with pinkish blush; good quality; stores well. Tree moderately vigorous. A leading cultivar in western apple sections. Chance seedling, New York, 1730.

Rome.—Red or yellowish-red; firm; fair quality; stores well. Tree medium size and moderate vigor, not suited to northern apple sections. Seedling, Ohio, 1848.

CHOICE OF LOCATION AND SITE

When dealing with a long-term enterprise such as apple growing, where a large capital outlay is involved, a grower cannot afford to make serious mistakes. The commercial life of an apple orchard may be 30–50 years. Careful planning and preparation are time well spent and result in greater returns in the years to follow.

The prospective grower must select with care the district in which he will establish his fruit farm. Certain regions, over a period of years, have become known as particularly well-adapted to this type of farming. Information may be obtained from federal and state authorities, or from successful fruit growers, as to the suitability of any particular district for apple growing. If there are no successful growers in the district there may be a good reason; either the district is not suited to the growing of apples or is better adapted to some other crop.

Modern transportation facilities have done much to extend the fruit growing regions. Today there are small sections, far removed from well-known fruit regions, producing apples within trucking distance of towns and cities that were not previously well supplied with good fruit. Such locations, provided they are not overplanted, may be profitable because of tax differences and better prices.

Well-established fruit regions often are well known to home and foreign markets. There is great interest in improved practices and each grower is a possible source of new information and ideas. In these sections, savings are possible through cooperative buying of supplies and equipment, cooperative selling, by improved storage and transportation facilities, and through better outlets for by-products. Moreover, direct benefit can be obtained from government services which are usually established in such localities.

Climatic Conditions

Precipitation.—In regions with less than 50–60 cm (20–24 in.) of precipitation, well distributed throughout the growing season, chances of success without irrigation are much less than where the rainfall is equal to or greater than this amount. Where irrigation is possible at a reasonable cost, the amount of rainfall is less important, but the grower must be assured of a continuous supply of irrigation water before making his planting.

Humidity.—In districts where the humidity is likely to be high during the fruiting season, the cost of production is usually high because heavy cost is involved in control of fungus diseases, particularly apple scab.

Although there is probably no part of North America where humidity is so high that disease cannot be controlled, nevertheless there is much variation in this respect and spraying costs vary accordingly. Other things equal, districts where fungus disease is easy to control are preferable.

Sunshine.—Abundant sunshine is important in growing apples since it is largely responsible for high color, a characteristic required by the market. Because the ultraviolet rays of sunlight play an important part in the formation of red color, areas where the air is free from dust and smoke are better for apple orchards.

Quality of light as much as quantity is important in fruit production. Length of light period rather than intensity of incident light appears to have the greatest influence on productivity. Within the tree are distinct zones of adequate light which affect the quality of the fruit accordingly (Heinicke 1966).

Temperature.—Late summer and early fall temperatures are important in developing red color. In some districts it is difficult to grow fruit of sufficiently high color, and market prices suffer accordingly. Even red sports may be of a dull rather than bright color.

Winter temperatures directly affect the suitability of a particular district for apple growing. This fact does not necessarily mean that districts suffering occasionally from winter injury are not so profitable as others for they may possess advantages that offset this drawback. However, consideration must be given to winter cold in determining the probable life of the orchard and the success of the investment. Apples succeed best in districts where the trees are uninterrupted in the winter's rest by violent fluctuations in temperature. Even though the weather generally may be relatively mild, sudden temperature changes may result in severe winter injury to the trees. Apple cultivars differ in their resistance to low temperatures, and in their responses to mild winters and hot summers.

Proximity to Large Bodies of Water.—Districts near large, deep, open bodies of water are preferable because the cooling effect of water tends to retard spring growth until danger of frost injury is past. By the end of summer, the water acts as a reservoir of heat, thus extending the period of ripening and development in the fall and reducing temperature fluctuations during the dormant period. Locations adjacent to large open bodies of water usually have good air drainage because of the constant air currents set up between the water and the land.

Marketing Outlets and Facilities.—The prospective grower should also consider whether or not the contemplated location is reasonably assured of a home market or of distant or foreign markets at prices sufficiently remunerative to pay production and shipping charges.

Home Market.—The best markets are often those near at hand, and locating near large consuming centers is likely to be a sound business venture, particularly where early and perishable cultivars are grown. A district with an excellent market for its product has an advantage which may largely, if not wholly, offset its disadvantages in climatic or soil factors. Orchards planted near cities often show an annual net profit in excess of that obtained in more favorable climatic localities. Established growers in these districts are sometimes able to absorb annually a large part of the initial investment. On the other hand, land values and taxes may "squeeze out" the old grower and make it impossible for a new grower to succeed.

FIG. 1.3.   THE FARM AND ROADSIDE MARKET IS ACCOUNTING
FOR A LARGE PROPORTION OF THE FRUIT CROP

Farm Market.—The farm market has gained in importance greatly in recent years. On-the-farm sales and roadside markets are accounting for an ever increasing proportion of the fruit crop. A great deal of variation in size, scope and form exists among sales outlets of this type. Systems where the consumer picks the fruit himself have also become very widespread and popular. The grower may combine two or more marketing methods, but nearness to large populations of consumers largely determines the success of the home market.

Distant or Foreign Markets.—Certain areas grow fruit largely for distant and foreign markets. Before locating in a region dependent on such markets the grower should assure himself that the transportation charges are offset by advantages due to location.

Transportation.—The location should be reasonably near a shipping point or main truck highway. The advantages of a short haul over good roads to a shipping point are obvious.

Size of Enterprise

Before selecting the actual site, the grower must decide what size orchard he wishes to plant. He must consider the market to be served and the extent of other farm enterprises.

The market to be served may be: (1) general, in which case the grower or several cooperating growers, must have hectarage (acreage) large enough for shipment by carlots or transport trucks; (2) local, which may not take so large a quantity at a given time, but responds better to a continuous supply in smaller quantities throughout the year.

Apple growers should grow only other crops that do not conflict seriously in labor and other requirements with those of the orchard. Such crops may be grown on parts of the farm not well adapted to apples. This system enables expenses to be met from year to year until the orchard comes into bearing.

The orchard may be operated as part of a general farm business, along with annual cash crops and livestock. Or, apples may be grown along with other tree fruits or with small fruits and vegetables. In some districts the trend is to produce apples in large units, with few if any supplementary enterprises. In some such cases the best possible use is not being made of the land. Vegetables or small fruits may be grown for a time at least between the rows of trees, providing high levels of organic matter and fertility are maintained.

Site

Great variation often occurs, even within a farm, as to the adaptability of certain fields for successful apple culture. Future success depends largely on the selection of a suitable site.

Elevation and Exposure.—Orchards in a hollow or low area, surrounded by higher land, may suffer not only from late spring frost but from winter injury in severe years. A gentle slope with low lying land at its base, to which the colder air may drain, is desirable. Select the site at an elevation a little higher than surrounding land so as to ensure good air drainage through and away from the orchard. This location not only lessens the risk from late frosts but also helps avoid rapid spread of fungus diseases.

The value of elevation in relation to winter injury may be very marked in certain orchards in the colder regions during a severe or "test" winter. That part located on a hillside with a gentle slope may come through the winter in excellent condition, while part of the same orchard on the low land at the base of the slope may suffer great injury. The difference in elevation between the two parts in one such instance was 1.8−2 m (6−7 ft) in 90 m (300 ft). Temperatures during frosty nights in spring may register −15°C (5°F) lower than at points only 30 m (100 ft) distant but 1 m (4 ft) higher in elevation. Such variations may mean the difference between profit and loss.

Successful apple orchards occur on land that slopes to any point of the compass. If winter injury is a factor, select a slope protected from the prevailing winds. Apple trees growing on a site exposed to the full blast of the prevailing winds are susceptible to winter injury through desiccation. Exposed hilltops, although they are less susceptible to frost injury than slopes or lower lands, may be subjected to strong winds which deform the trees, blow off fruit, and interfere with spraying.

Southern slopes tend to accelerate development of buds in the spring, whereas northern slopes tend to retard development. Winter sunscald is more likely to occur on southwestern than on southeastern slopes.

Drainage.—Apple trees will not tolerate "wet feet." A good orchard site has ample surface drainage to take care of excess water in late winter and early spring when drainage through the soil is difficult. A low spot in an otherwise desirable field may be tile drained, but usually it is better to plant on land that does not require expensive drainage. Excessive drainage also must be avoided.

Soil.—Although some cultivars do better on certain soils, apples generally thrive on a range of soil types. The nature of the subsoil on the proposed site may be even more important than the kind and quality of the surface soil. The subsoil should be well drained so that tree roots at no time stand in water, and it should be loose enough to let the roots make an extended growth. A hard, impervious subsoil restricts root development, with a subsequent effect on the tree's life and vigor. A very open subsoil that holds little or no water is the opposite extreme and should be avoided.

An ideal subsoil is a gravelly loam. Avoid sites on heavy clay or compact subsoil and hardpan formations that are near the surface. A soil 2 m (6 ft) or more deep is desirable, but there are localities where profitable orchards are thriving on shallower soils. These are usually in districts where conditions are very favorable to apple tree growth. If trees are to be thrifty and long-lived, their roots must be able to penetrate the soil readily.

Apple trees are tolerant to a fairly wide range of soil pH; about 6.5−6.8 appears to be optimum.

Windbreaks.—In exposed areas, a well established windbreak affords protection and lessens the velocity of the wind over a considerable area. This protection reduces moisture loss from evaporation, reduces damage to the fruit and to the trees during wind storms and may lessen the danger of winter injury. These effects of the windbreak vary from 3.5−6.5 m (11−22 ft) for every meter (foot) in height of the windbreak. As the apple trees grow in size, they develop their own protection in a great measure, except at the edge of the planting.

Avoid a still-air pocket on the leeward side of the windbreak by keeping the branches of the windbreak trees pruned from the bottom to a height of 1.5−1.8 m (5−6 ft); or else set the windbreak 15−23 m (50−75 ft) from the nearest row of orchard trees. Trees, such as wild apple, hawthorn, and mountain ash, that may harbor pests should be eliminated from the windbreak.

PLANTING THE ORCHARD

Preparation of the Land

Time and money expended in building up the fertility and in establishing a good physical condition of the soil prior to planting is worth many times more than similar efforts after the trees have struggled along indifferently for a period of time. Probably the most important period in the life of an orchard is the first 4−5 years, when the future vigor and shape of the trees are being determined.

It pays to prepare the land a year ahead by cultivating and using a green-manure crop or a row crop such as corn, potatoes, or roots. Soil tests can be made and fertilizer and manure applied accordingly. If the chosen site is already in good tilth, a year's delay is not necessary. Level minor depressions to promote surface drainage. Where necessary, provide surface drains so that excess water flows readily from the orchard area.

Planning the Orchard

Previous to actual planting operations, prepare a plan of the arrangement of the trees. Orchardists may select from several methods.

Square.—This method, which simply consists of placing the trees so that one stands at each corner of a square figure, is the easiest to plant and takes the same number of trees as the alternate plan, but permits easier cross cultivation and spraying than any of the other arrangements. Planting distance depends on the rootstock, soil, and cultivar.

Quincunx.—This method is similar to the square, except that a tree is planted in the center of each square. The tree in the center is usually a filler tree, to be removed later.

Hexagonal or Triangular.—This system, which places a tree at each corner of a hexagon and one in the center, enables the rows to be closer without changing the tree spacing. It permits more trees per hectare (acre) (about 15%) than the square system and makes full use of air and ground space.

Hedgerow.—The rows are placed just far enough apart to allow for adequate light and air and room for orchard operations. The trees are spaced in the row so as to form a continuous hedge when mature but without crowding. This method results in a high density planting with maximum number of trees per hectare (acre).

Contour Method.—On sloping land where erosion and runoff are problems, it is wise to plant on the contour. If cultivation is to be carried on, as with young trees, it is almost imperative that they be planted on the contour where slopes are involved.

In contour planting all trees in the same row are at the same elevation. Cultivation is done on the contour, not up and down the slope. This procedure tends to build up a low terrace at the tree row which helps to control water runoff. The tree rows are not the same distance apart throughout the orchard. Where the slope is steeper, the rows run closer together; where the slope is less, the rows are farther apart. In some places the rows may become so far apart as to allow a "spur" row to be inserted. On steeper parts the rows may come so close together that it may be necessary to discontinue a row.

The first contour line can be laid out about halfway up the slope. It is projected from a given point by using an engineer's level, and marking with stakes. At the steepest place on the slope, establish points at the minimum distance between rows above and below the first contour line. From these points project contour lines to the left and right. Similarly establish all the contour lines.

Where the tree rows come closer together the distance between the trees in the row can be increased; where the rows are farther apart the trees in the row can be closer together. In this way the number of trees per hectare (acre) is about the same as in the square system.

Terracing is used along with contour planting in areas where the rainfall is more than 84 cm (33 in.) a year and excess water must, at times, be diverted from the field. On land with a slope greater than 10%, bench terraces (which resemble large steps) may be used. However, they are expensive to construct and it may be wiser in such cases to use a permanent sod.

The channel floor of diversion terraces should have a grade of about 0.5–1%, depending on the slope, rainfall, soil type, and length of terrace. If the terrace has a water outlet at both ends, the highest point in the channel floor is about midway between the two ends. Keep waterways well sodded.

Terraces can be expensive to construct and demand careful maintenance so that they continue to operate properly. Use terracing equipment or a disk harrow set to throw the earth one way. Repair any breaks in the terrace immediately with soil 25% higher than the desired height of the mound to allow for settling.

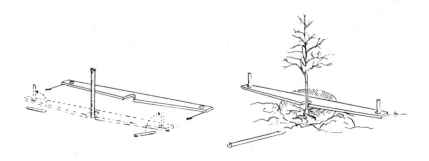

FIG. 1.4. THE USE OF A PLANTING BOARD ENSURES STRAIGHT ROWS OF ORCHARD TREES

Staking Out the Orchard

Determine the spacing of the rows and the location of the trees in the row by measuring and sighting, and set stakes at each tree location. If the field is too hilly or too large to sight from one side to the other, run a line across the center and stake the two parts separately.

After staking, a planting board may be used so that the stakes can be removed to set the trees.

Care of Stock

As soon as the trees are received, examine the roots. If somewhat dry, soak them at once and keep them moist until heeled-in or planted. If the stock is much dried out, try burying it completely in damp earth, leaving it there for 4−6 days. If stock is excessively dry when received, consider a prompt, proper protest.

If the trees cannot be planted immediately upon arrival, heel them in as soon as possible. Do this by digging a shallow trench in a shaded spot and covering the roots and a foot or more of the lower part of the stems or trunks with moist earth. Spread the trees in the trench and firm the soil around the roots.

Roots of young, dormant fruit trees cannot endure without injury the low temperatures to which tops may be exposed. They are also damaged by drying out in handling. The fact that they may be uncovered for a time in a moist packing shed or in the field during a moist still day should not encourage the belief that exposure to winds and dry air is not harmful.

Exposure of freshly dug trees for 15 min, if the day is dry and windy, and for more than 30 min on an average sunny day, may be injurious. While awaiting planting keep the roots protected.

When and How to Plant

Place orders early with the nursery firm to ensure prompt delivery of the trees at the time desired. Plant in spring as early as the land can be worked readily. Fall planting is more common in regions with comparatively mild winters than in regions with cold winters.

Make the holes for the trees somewhat larger than the spread of the roots after the broken roots and unnecessarily long portions have been removed. Plant the trees 2−5 cm (1−2 in.) deeper than they stood in the nursery. Dig the holes a little deeper than required; lay the topsoil to one side and scatter the subsoil away from the hole. In planting the trees, firm the earth around the roots by tramping, but take care to avoid injury to the roots. Use moist surface soil to fill in around the roots. Do not let the young trees dry out before or during planting.

SOIL MANAGEMENT

The most suitable system of soil management for any particular area depends largely on the type of soil, amount of precipitation, topography, and certain other factors.

The main objectives with any soil management system are: to increase, or at least maintain, the level of the humus content of the soil; to conserve moisture and prevent runoff and erosion; to maintain a high level of soil fertility; and, in short, to grow trees that are capable of producing a maximum crop of top quality fruit at . minimum cost.

Cultivation

The first few years in the life of an apple tree are very important. Make every effort to get the young tree established quickly and to have it make good growth. For this reason it is generally wise to follow a system of limited cultivation for the first 4−6 years after planting.

Shallow cultivate in spring during the growth period to conserve moisture by eliminating competition from weeds and to facilitate incorporation into the soil of organic matter and fertilizer. By early summer, stop cultivation and grow a cover crop such as oats, buckwheat, soybeans, or weeds. The cover crop aids in hardening the trees for winter by taking up moisture and N. It acts also as a protective cover for roots during winter, and when disked into the soil in spring it adds organic matter to the soil.

A 20-year comparison between normal (to mid-July) and minimum cultivation (to mid-May) in an orchard on a fine sandy loam in Ontario showed that, besides more economical yield of fruit, the soil with minimum cultivation had better structure, more replaceable K, and more organic matter than the soil under normal cultivation (Dickson and Upshall 1948).

Moisture content is higher in soil that is cultivated than in soil under sod, but is not as high as in soil under a deep mulch.

Sod System

After the trees have become well established and have . tained good growth, it is often wise to seed the orchard to a permanent grass cover. Under sod, a good soil structure is developed, soil erosion is prevented, penetration and percolation of moisture into and through the soil are facilitated, orchard operations when the soil is wet are made more easy, fruit color is generally improved, and the cost of maintenance may be less than if a limited cultivation system is followed. When trees in a sod orchard are mulched, tree growth and yield are increased.

Certain possible disadvantages attend the sod system of management. With sod, more attention must be given to the N needs of the trees. In nonirrigated orchards the moisture content of the soil during drought periods is less under sod than under cultivation. Mouse hazard is greater in a sod orchard than in a cultivated one. Scab control may be made more difficult in a sod orchard since infected leaves from the previous season

FIG. 1.5. THE FERTILIZED AND MOWED SOD (RIGHT) WILL
USE MORE MOISTURE BUT PRESERVE SOIL STRUCTURE

are not buried. Buffalo tree hopper and curculio may be more troublesome in a sod orchard, particularly alfalfa sod.

Various grass and legume mixtures may be suitable for permanent sod
in apple orchards, but the choice depends on the region, soil, and climate.
Grasses often used for this purpose are red fescue, Kentucky bluegrass,
Canada bluegrass, orchard grass, red top, brome grass, and timothy.
Legumes such as ladino and white dutch clovers are desirable in the
mixture from the standpoint of their beneficial effect on the soil.

Fertilize the sod cover annually so that it continues healthy and
vigorous. Keep it mowed to 7−10 cm (3−4 in.) high and leave the
cuttings in the orchard. In this way the grass remains green throughout
the growing season, orchard operations are facilitated, more organic
matter is returned to the soil, fire hazard is reduced, and the mowing
operation itself is made easier. The rotary type mower does a fast,
efficient job. The hammermill type of brush chopper has also been used
successfully in the orchard for cutting grass.

Chemical Control of Grass and Weeds

Grasses can be troublesome in orchards maintained in permanent sod.
Dense grass next to the trunk is difficult to mow and offers cover for
mice. Grass under trees may compete seriously for water and fertilizer,
particularly with young trees. Even with tillage, it is difficult to
manually control grass growth close to tree trunks.

Dalapon (2,2-dichloropropionic acid) can be used at the recommended
rate either to suppress or eradicate grass infestations. Spray the grass
when it is 10−20 cm (4−8 in.) high and growing well. The safety of
dalapon close to very young trees in nursery or orchard is questionable.
Established apple and pear trees are very tolerant to dalapon.

FIG. 1.6.   THE GRASS BENEATH THE TREE WAS SPRAYED
WITH DALAPON USING 4.5 KG TO 380 LITRES (10 LB TO 100
GAL.) ON MAY 15 AND RESULTED IN ALL GRASS KILLED
WITHOUT INJURY TO THE TREE

Aminotriazole is effective against grasses, poison ivy and other
broad-leafed weeds. Paraquat acts as a "chemical mower" and when used
with either diuron or simazine, which remains active in the surface soil,
the combination gives long-lasting weed control. Excellent results are
obtained from heavy mulching along with restricted use of chemical
herbicides.(Waywell *et al.* 1977). Simazine and amitrole alone and in
combination alter the protein and RNA content of apple leaf tissue
(Solecka *et al.* 1969).

Mulching

If a mulch of straw, hay, strawy manure, sawdust, or other organic
matter is spread over the entire orchard floor or just beneath the trees,
and maintained to a depth of at least 7.5 cm (3 in.) when settled, it
produces a beneficial effect on the trees in several ways: the humus
content of the soil increases the moisture-holding capacity so that
porosity and aeration are improved; erosion and water runoff are
eliminated, and there is less fluctuation in soil temperature; nutrient
elements are added to the soil (Table 1.2), and a high percentage of
windfalls is undamaged.

TABLE 1.2
NUTRIENT LEVELS UNDER ALFALFA MULCH AND IN THE OPEN

Depth (cm)	K (ppm)	P	N (ppm)	Moisture (%)
Under mulch				
0–15	50	High	313	10.9
15–30	20	High plus	137	8.8
30–45	20	Very high	95	4.9
Between trees				
0–15	15	High	55	5.7
15–30	5	High plus	28	6.8
30–45	5	Very high	22	4.5

Source: Archibald (1966).

Where mulch had been accumulating for 35 consecutive years in Ohio, a solid mat of mulch material 7–15 cm deep covered the soil. Throughout this mat were thousands of fibrous roots, which were so thick that a regular count could not be taken from the side of the trench. However, 2.5 cm from the trunk, 13 cm² of mulch was lifted, and the roots were separated from the decomposing mulch and weighed. The depth of the mat at this point was very little over 0.3 cm and the roots therein weighed 59 g. With few exceptions, these roots were all less than 1 mm in diameter; one was greater than 3 mm (Beckenbach and Gourley 1932).

Possible disadvantages of mulching are the slightly increased frost damage to blossoms, the cost and lack of availability of the mulch material, and the greater mouse hazard. If, however, the mulch is applied in late fall after mice have found their winter quarters, it will generally be wet and unattractive to mice the following spring and summer. Also, a regular, annual control program prevents a buildup of mice population with any system of soil management.

Mulch materials vary in dry matter content and in value as sources of plant nutrients (see Table 1.3 under Fertilizers). Hay is of greater value than straw in this respect; sawdust and wood shavings are inferior to hay or straw. A legume hay is high in N as is also pea silage. Barn manure may vary greatly in nutrient-element content depending on its source and care, and is unpredictable as to its N release. Manure has been associated sometimes with bitter pit of apples and with excessive, soft shoot growth. With sawdust and shavings more N must be applied to the trees. Even with straw or non-leguminous hay, symptoms of varying degrees of N starvation may become noticeable unless extra N fertilizer is applied, especially during the first 1–2 years after a previously cultivated orchard has been put under a sod and mulch system.

Waste tire fabric as a mulch added little to the soil in way of nutrients but compared favorably with hay and straw in conservation of soil moisture and control of soil temperature fluctuation (Teskey and Wilson 1975).

In New Hampshire, in the absence of fertilizer applications, hay mulch provided nearly ideal conditions for tree development and performance; seaweed ranked second. Sawdust mulch provided excellent conditions of soil moisture, but yield and tree growth were reduced and the foliage was pale in color. Sod culture without added mulch resulted in the lowest yield (Table 1.3) and in failure of the trees to make many fruit buds.

TABLE 1.3
YIELD PER TREE (KG) OF APPLES FROM YOUNG TREES
UNDER DIFFERENT SYSTEMS OF SOIL MANAGEMENT

Year	Sod	Sawdust	Seaweed	Hay
1943	4.90	10.70	22.14	21.05
1944	15.42	43.54	40.10	50.80
1945	2.54	4.17	4.17	4.17
1946	14.52	24.86	35.02	61.51
Total 1943–1946	37.38	83.28	101.42	137.53

Source: Latimer and Percival (1947).

Variation in weed control, moisture supply, organic matter, and quantity of available nutrients in the soil with the different treatments largely accounted for the different responses to the mulches and sod culture.

Little difference was noted in the effect of the different treatments on fruit color and size, except that seaweed mulch resulted in less red than the other treatments. Fruit on seaweed-mulched trees matured latest. Foliage color was better with hay and seaweed mulches than with sawdust mulch or sod culture. Shoot length was greatest on the hay-mulched trees. Hay mulch resulted in an accumulation of available N, P, Mg, and K in the soil. Such an accumulation was not found with sawdust and sod. Phosphorus accumulated under the seaweed mulch. The pH of the soil and amount of available Ca present did not differ among the soil management methods (Latimer and Percival 1947).

In the USSR and in China, sand as a mulch resulted in an increase in soil moisture and a lower water deficit in the leaves. There was less soil temperature fluctuation, greater shoot trunk and root growth, and

increased yields. Chlorophyll content was slightly higher and photosynthesis during July was double that of unmulched trees (Devyatov *et
al.* 1976).

Irrigation

In arid and semiarid areas where annual precipitation is less than 55
cm, irrigation is a necessity. If the annual precipitation appears adequate but is poorly distributed throughout the growing season it may be
profitable to supplement the rainfall. Adequate moisture is especially
critical during the early growing season and in the period of fruit sizing.
In many areas the economic feasibility of installing an irrigation system
is questionable. In some cases a good mulching program is a more
practical means of obtaining satisfactory growth, i.e., 25−35 cm of
terminal growth for trees 10 years old or less in a normal season and
15−25 cm for bearing trees.

The soil should be wetted to a depth of 1−2 m per application, the
lighter soils requiring more. This may require as much as 187,500 litres
of water per hectare. Climatic conditions favouring high rates of
evaporation cause a need for more irrigation water. Greatest yield and
fruit size have been obtained by irrigating to 60 cm depth when the soil
moisture had dropped to 40% available water during the period of most
intensive fruit growth (Levin *et al.* 1972).

The old furrow method of irrigation is still used effectively but it has
been replaced by the overhead sprinkler system in many areas. Drip or
trickle method of irrigating has the potential advantage of supplying
each tree with its exact water requirements. With this system sediment-
free water is essential. Also, it requires a very precise layout and
arrangement (MacLaren 1971).

PRUNING

Pruning of apple trees has been practiced since the earliest days of
apple culture, yet it is still an orchard operation that is often mis-
understood in its practical application. Orchardists are not entirely to
blame for this situation, however, since it is only within the last quarter
century that carefully planned experiments have been conducted to
study the effect of pruning upon growth and fruiting. Even today
leading authorities do not agree on some of the minor details.

Purposes of Pruning

The pruner should understand the reasons for pruning and have a

**FIG. 1.7. APPLE TREE TRAINED TO THE MODIFIED LEADER
SYSTEM**

The well spaced, wide angled laterals are spiralled around the
central leader which has been cut back to the desired height.

definite purpose in mind. Pruning the young non-bearing tree is primarily to train or shape it so that a strong framework will develop to support maximum crops of top quality fruit when the trees reach bearing age. The aim in a commercial orchard is to produce a tree that is profitable for an optimum period of time.

Light is vital for photosynthesis and fruit bud formation. One objective in pruning and training is to keep the tree reasonably open to admit sunlight and ensure good aeration. This, in turn, helps promote good quality and color in the fruit, aids in the control of disease, and facilitates thinning and picking. Another purpose is to remove from the tree weak-growing wood that never produces fruit of satisfactory size and quality. Pruning should establish a balance between vegetative growth and flower induction and, along with other orchard practices, maintain good conditions in the trees for growth and fruit production (Proctor *et al.* 1975; Lasko and Musselman 1976).

Principles of Pruning

Pruning Dwarfs Total Growth and Delays Fruiting.—In general, the early shaping of the tree should be accomplished with the minimum amount of cutting. Other things being equal, the total growth attained by an unpruned tree is always greater than that of a pruned tree, regardless of the type and amount of pruning. Pruning not only reduces

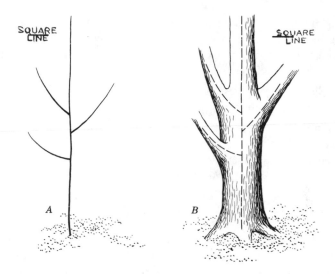

FIG. 1.8. SCAFFOLD LIMBS 5 TO 7.5 CM (2 OR 3 INCHES)
APART MAY SEEM TO BE WELL SPACED WHILE THE TREE IS
STILL YOUNG (A) BUT WITH ENLARGEMENT OF THE TRUNK
AND BRANCHES AS THE TREE GROWS (B) THIS SPACING
PROVES INSUFFICIENT AND CROWDING AND WEAK
CROTCHES RESULT

the total growth made by the above-ground portions but, as a consequence, total root growth as well. With young trees which have not yet
flowered, severe pruning delays flower-bud formation. A heading-back
cut seems to delay flowering more than a thinning-out cut. Once young
trees have begun to bear, pruning rarely inhibits fruit-bud formation;
once it is initiated, fruit-bud formation is not easily suppressed.
Nevertheless, some delay in production is usually necessary in order to
obtain a strong tree (Teskey *et al.* 1973).

Reduces Yield.—To prune bearing trees usually results in a reduction
in total yield. Exceptions occur with trees that are too dense, or
devitalized trees which set few fruits. The effect of pruning, however,
on the yield of marketable fruit is the important factor. Therefore,
growers are faced with the practical problem of determining the amount
of pruning which will produce favorable effects and yet not reduce the
marketable yield unnecessarily. Where the trees are so dense as to shade
the fruit considerably, a light to moderate pruning tends to increase the
yield of marketable fruit.

Stimulates New Growth in Older Trees.—Pruning, through its stimulation of more shoots and spurs, produces a most positive and beneficial effect on older trees which have borne heavily for a number of years. Cropping has an exhaustive effect and tends toward a decided reduction in growth of shoots and spurs. Pruning, by removing a number of flowers and stimulating a vigorous vegetative type of growth, results in new wood which in turn bears its quota of flower buds.

Increases Set of Fruit.—Pruning tends to increase the set of fruit. The elimination of certain growth points indirectly increases the supplies of water and nutrients available to the remainder. Although pruning does have this effect, fruit setting often is more satisfactorily maintained by application of fertilizer. With most cultivars, increase in set from pruning is usually less than from application of fertilizer.

Bulk pruning, which removes large limbs and branches, produces such a decided effect near the cuts that watersprouts (vigorous, succulent growths from above-ground parts of the tree) often become excessive. A less drastic method that removes smaller branches and twigs is to head back the large, undesirable limbs by degrees and not all at one time. Such a method dwarfs large branches gradually and usually results in fewer watersprouts. The thinning out of outer small branches is considered preferable, but the time and expense involved have prevented its widespread adoption in practice. During this operation certain dominant lateral branches may be cut back to advantage.

Effect Is Localized.—The stimulative and total dwarfing effect of pruning a branch or section of a tree is manifest in the branch or section pruned. Thus, by selective cutting, new growth can be developed where needed, or a branch can be made subordinate where desirable.

Cultivar Aspects.—Cultivar habit often decides the type of treatment to be given. For instance, with cultivars which are prone to produce long, rangy growth with few laterals, more heading in (cutting back the leader and branches) and careful handling during the training period is required than with cultivars which naturally produce many laterals. In the latter case the pruning treatment becomes more of a system of heading back certain tip growths.

Narrow-angled Crotches.—A scaffold branch that forms a narrow angle at the point of union with the trunk is structurally weak. This weakness is due to inclusion of bark in the narrow crotch between branch and trunk, and failure of the tissues of trunk and branch to unite and grow together. In a wide-angled crotch a woody structure forms in the crotch, uniting crotch tissues of the branch with adjacent tissues of the trunk. Such branches are strong and can support heavy crops of fruit without breaking at the crotch.

FIG. 1.9. A GOOD EXAMPLE OF CROWDED LIMBS, NARROW-ANGLED CROTCHES, AND A WEAK STRUCTURE THAT RE-SULTS FROM NO TRAINING DURING THE FORMATIVE YEARS

A further objectionable feature in the narrow-angled scaffold branch is that the tissues of such crotches are susceptible to winter injury.

The relative size of the two crotch arms is important in breaking strength, the weight required to break any given crotch increasing directly with the increase in size of the larger arm compared with the smaller. There is no evidence, however, that the crotch angle is of great importance if there are no bark inclusions. More than two side branches coming off the trunk at about the same level are likely to make a complicated and weak crotch structure. Better structure is secured when only one main side branch arises at any one level on the tree.

When all laterals are left on the young tree for the first 2–3 years they form larger angles with the trunk. The lower laterals, having the greatest number of laterals above them, form the largest angles. After the laterals develop, retain those most favorably placed to form a good framework and remove the others.

Some cultivars, e.g., Delicious and Stayman, are especially prone to develop narrow angles, whereas other cultivars, e.g., McIntosh, Jonathan, Rhode Island Greening, and York, normally form wide crotches. Wide crotch angles result from the action of a plant hormone formed in the growing points of the young tree and passed downward through the phloem to buds and developing shoots below where its action inclines the direction of growth of the shoots toward the horizontal.

Indolebutyric acid in lanolin paste applied to the upper surface of the basal internode of a young shoot when this internode still was elongating promoted a marked increase in the angle formed by trunk and shoot (Verner 1938). Sprays of 2,3,5-triiodobenzoic acid (TIBA) tend to promote flowering by causing a bending of the branches but the actual crotch angle is not altered (Edgerton *et al.* 1964).

Time of Pruning

Pruning in late winter or early spring gives the greatest response in vegetative growth, and relatively little dwarfing effect. This is also the time, or close to the time, when healing takes place most rapidly. Thus the more that new growth is desired, the more pruning should be done at this time of year; conversely, if little new growth is wanted, pruning at this time should be accordingly light.

Late spring pruning is the most dwarfing and results in little or no vegetative response. Intermediate times of pruning between early and late spring produce correspondingly intermediate results with regard to dwarfing and invigorating.

Summer pruning (July) is done to control excessive vegetative growth without too much dwarfing and with only moderate vegetative response the following year. At the same time summer pruning is often useful in opening up the tree to light and air.

FIG. 1.10. EARLY JULY PRUNING OF MCINTOSH/M. PREVENTS EXCESSIVE SHOOT GROWTH AND ALLOWS LIGHT INTO THE TREE

Fall pruning is not recommended because the wounds will not heal at this time. Also, the shorter the time between the pruning operation and occurrence of low temperatures, the greater the hazard of winter injury. Thus in districts where sub-zero temperatures may occur, late fall, early and midwinter pruning is very hazardous. Never prune when the wood is frozen.

Pruning at Planting Time

If trees are cut back severely when planted, the survival is greater. Trees heavily pruned at planting time receive less setback due to transpiration loss, and therefore may start into growth as much as three weeks before unpruned trees. Thus, the more severe the pruning the sooner the transplanted tree becomes established, and the greater the top growth produced by the end of the first and second years.

By the end of the third and fourth years the effect of heavy pruning at planting time seems to be negligible except with trees that have been cut to a whip. A positive correlation exists between weight of tree at planting time and weight 4 years later. Trees left unpruned or pruned to a whip require more pruning in the second, third, and fourth years to correct framework deficiencies than do trees pruned to two scaffold branches and a leader, either cut back or not cut back.

If trees are set as they come from the nursery and not pruned until the second, third, or fourth years, many of them will never have a good framework unless reformed by drastic pruning.

Pruning One-year Whips.—Head back 1-year-old whips at planting time to a good strong bud at a height of about 1 m (4 ft). This causes the bud below the cut to elongate and side shoots to form. The next spring, if a strong branch has developed 0.6−0.9 m (2−3 ft) from the ground, select it as the bottom scaffold limb. Select 2−3 additional branches spaced at least 15−20 cm (6−8 in.) apart along the trunk and extending in different directions from the trunk. If the top branch is weak remove it and develop the leader from a lower, stronger growing branch. It is not necessary to select all main scaffold limbs at the start of the second year. Additional strong laterals form in succeeding years to complete the framework. Leave 5−7 branches to form the head of the tree in the modified central leader system of training.

Disbudding.—This is the removal of all buds, in the year of planting, from the whip, except those where laterals are desired. This method produces an almost ideal framework from the standpoint of distribution and spacing of the scaffold limbs, but may result in many narrow crotches. The fewer buds allowed to develop the greater the growth rate and the narrower the crotch angle.

Pruning Two-year Trees.—The first step in pruning a 2-year-old tree at planting time is the selection of laterals for the scaffold limbs. Such trees have usually been headed back in the nursery and have produced a number of laterals closely spaced along the trunk. Select 2−3 of these laterals, well spaced. Retain the strongest upper lateral for the leader, head it back slightly, and remove the rest of the shoots. The laterals that remain are cut back to different lengths, depending on their location on the trunk and length of leader. Additional branches arise from the leader during the succeeding years and from these select the required number of scaffold limbs to complete the framework.

Pruning Young Nonbearing Trees

Prune lightly in the third and fourth years, chiefly by thinning out rather than heading back branches. Remove branches that form narrow angles with the trunk, and crossing and interfering branches. Some heading back must be done to maintain balance in the top.

Pruning Bearing Trees

Pruning young bearing trees involves very little removal of wood until the extent to which the tree is opened by bearing has become evident. The annual corrective type is preferable, and if the trees have been properly trained no large cuts should be necessary. Fruit on young bearing trees usually is of good size and color even when the tops are a bit dense. If, after 2−3 crops, the quality is not satisfactory, thin out the top.

Chief aims are to maintain production of a good yield and grade of fruit and to reasonably restrict tree height.

Apples mostly are borne terminally from fruit buds on spurs 2 years old or more; some may live 15−20 years. When a fruit sets, a new shoot arises from the side of the spur. A fruiting spur shows a zigzag growth; in a non-fruiting spur, growth is straight. An apple spur, unless the fruit is thinned, tends to bear every other year. A heavy crop exhausts the tree for the next year's crop. Since the spurs are long-lived and only about 10% of the blossoms need to set fruit in order to produce a commercial crop, apple does not need to make as much new growth as some other fruits. Young bearing trees should make 23−46 cm (9−18 in.) of terminal growth each year; older trees 15−25 cm (6−10 in.); light setters, 25−38 cm (10−15 in.).

So-called spur-type trees as occur in Delicious, Golden Delicious, and perhaps McIntosh and other cultivars, are thought to be mutations. With

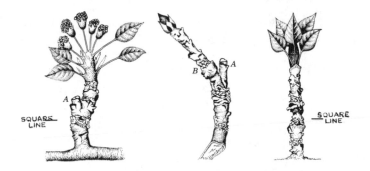

FIG. 1.11. HOW APPLE SPURS GROW AND DEVELOP

Apple spur (left) starting its fourth year's growth; fruit has
been borne at A. Center shows a 4-year-old spur; it fruited at
A and secondary growth from the cluster base extended to B,
terminated by a leaf bud. The next year a straight growth was
made which formed a terminal fruit bud. At right is a 6-
year-old apple spur which has never bloomed nor borne fruit.

such trees there is little branching, vegetative growth is limited, and
they are generally somewhat smaller than regular trees. Fruit is borne
mostly on spurs growing directly from the main limbs. Pruning is
reduced chiefly to a renewal system by removing a portion (about 10%)
of the oldest spurs each year.

One method of opening up the top to facilitate spraying and harvesting
and to improve fruit quality by admitting more light is by the "clo-
verleaf" system. Two 0.5 m (2 ft) channels are cut through the tree at
right angles to each other and diagonally to the direction of the tree row.
Each quarter of the tree consists of 1−2 large limbs. A modification of
this method is to divide the top into quarters by cutting wedges to the
center. Begin the operation early in the training period of the tree or,
with bearing trees, spread it over 3−4 years to avoid heavy pruning.

Light to moderate pruning is desirable in bearing trees. Moderate
pruning is followed by less fruit thinning, whereas light pruning tends
to emphasize a need for thinning. The amounts of pruning must be
regulated by the yield and quality of fruit produced. In trees that have
been bearing for some years, avoid the removal of large limbs as far as
possible; removal of small branches throughout the tree is better
practice.

Thinning Out.—This type of pruning removes entirely a shoot, spur,
branch, or other part. The cut is made where these grow out from the

parent stem structure or to a lateral. Thus, thinning out reduces the number of branches or other parts.

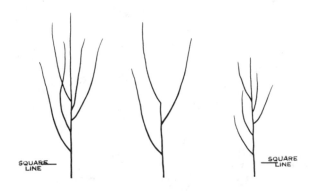

FIG. 1.12. THE SILHOUETTE OF APPLE TREES: (LEFT) UNPRUNED; (CENTER) PRUNED 50% IN EACH INSTANCE BY THINNING OUT; (RIGHT) HEADING BACK

Heading Back.—Remove only a portion of the branch, leaving another portion from which new growth can develop. This reduces the length of branches or other parts. Generally, the pruning of bearing trees consists of both thinning-out and heading-back cuts.

Thin Wood Pruning.—This is a light pruning system that involves the removal of wood which is small in diameter for its age and which develops poorly colored fruit of small size. This thin wood is the result of shading or competition and should be removed wherever it may occur. As a result of this operation the proportion of culls to crop is reduced and the harvested crop is restricted to the portions of the tree from which a high percentage of the fruits will be of good size and color.

Height of Head

Rising costs of spraying and picking necessitate establishing the height of tree head lower than formerly. At present, about 4.5 m (15 ft) may be the very maximum height for economical handling with a strong trend away from standard trees and toward dwarf and semidwarf trees (Chap. 3). Allow the central leader to continue upward growth until a well-formed lateral has developed at or near the desired height of the tree. Then remove the leader with a slanting cut immediately above the lateral. This cut establishes the height of head and opens up the top of the tree.

Courtesy of Ontario Dept. Agr. and Food

FIG. 1.13. AUTOMATIC EQUIPMENT IN THE HANDS OF A
SKILLFUL OPERATOR AND EXPERIENCED PRUNER CAN SAVE
UP TO 50% IN TIME AND LABOR

The tops of old trees can be lowered by cutting back 1 or 2 of the tallest limbs to a strong, well-placed lateral. Or, such limbs may be removed entirely. In the latter case, the severe pruning entailed and the opening of the top induces watersprouts. Sprouts can be controlled by the application of NAA (Raese 1976).

Small wounds usually heal rapidly without infection. Apply a wound dressing mostly to pruning cuts over 5 cm (2 in.) in diameter. Until such a wound is callused over it remains unprotected from weather effects and to attack by wood-rotting fungi. Certain asphalt compounds, applied cold, are widely used. Turpentine in a paint may kill exposed cells of the bark in fresh wounds.

Mold-and-hold

This is a new term applied to an old principle and practice. The head, framework, and height of the tree are determined by formative pruning (molding) during the developing years of the tree's life. It is then a relatively simple matter to hold the tree to this form by regular light, corrective pruning. Essentially, all shoots and suckers are headed back, while thinning out is reduced to a minimum. An attempt to apply this method to older, untrained trees may prove costly and frustrating because of the heavy pruning required and the difficulty or impossibility of correcting faults already developed.

FIG. 1.14. BRUSH DISPOSAL IS MADE SIMPLE BY THE USE OF
EQUIPMENT THAT LEAVES THE CHOPPED PRUNINGS IN THE
ORCHARD

MOUSE, RABBIT, AND DEER CONTROL

Much damage is caused every year to fruit trees in some areas by mice
and rabbits, and the loss to the grower sometimes is very serious. Most,
if not all, of this loss can often be avoided by taking proper preventive
measures. Mouse and rabbit populations tend to build up if unchecked,
reaching a peak about every 7–8 years. The most serious damage occurs
during these periods of peak populations, and it behooves the orchardist
to see that the numbers of these orchard pests are not allowed to increase
in his orchard.

Mice

Two kinds of mice attack fruit trees. The pine mouse has a small body,
short tail, brownish fur, and small, sunken eyes. The meadow mouse is
larger than the pine mouse, has a longer tail, gray fur, and prominent
eyes. Damage by the pine mouse is to the roots of trees since it burrows
underground. This type of injury is the more serious because it cannot
readily be detected. The meadow mouse, by far the more common of the
two, spends most of its time just beneath the surface litter or snow. Mice
chew the bark from roots or trunks of young fruit trees. Large wounds of
this nature weaken the tree, and heavy root damage or girdling of the
trunk can cause death of the tree.

For control, place a metal or plastic guard around the tree trunk at
planting time. Sink the base of the guard to root depth and fill in to a
little higher than ground level with crushed stone. The height of the
guard depends on the expected snow depth (meadow mice work just
beneath the surface of the snow in winter). Asphalted sheathing paper
also can be used to protect the trunks.

In early fall, clear all straw and grass away from the base of the trunk for at least 1 m (2 ft). A light disking at this time and again a month later helps control orchard mice by breaking up their runways, nests, and places of concealment. Mowing with a rotary mower is also helpful in mouse control.

Inspect the orchard in early fall for signs of mice such as fresh runways, grass clippings, droppings, and chewed apples. If the presence of mice is indicated, initiate a control program after consulting with local authorities. Orchards in sod particularly need special attention. Baits made from apple drops can be used successfully in early fall and until the temperatures go below freezing. In colder weather use grain baits.

Rabbits, Groundhogs

Rabbits chew the tender bark from stems and low branches of trees, often completely girdling the trunk. Buds and young shoots are often nipped off. The attacks usually occur in winter, but on the western Great Plains rabbits may cause much injury during summer droughts.

Groundhogs (woodchucks) (*Marmota monax*) tear the bark from trees as high up as they can reach by standing on their hind legs. Damage is most prevalent in the early part of the season and most severe on small trees.

Wire mesh or paper, plastic or metal guards as described for mouse control prevent rabbit injury to the trunks. However, in areas of very deep snowfall guards cannot protect the branches.

Repellent materials of several kinds brushed or sprayed on young trees give varying degrees of success. A mixture of rosin and ethyl alcohol applied to the dry trunks has given some good results. Other materials used are 40% nicotine sulfate, tetramethylthiuram disulfide, and tetraethylthiuram monosulfide. Water emulsions of asphalt, latex paint, and synthetic plastic emulsions are satisfactory carriers for the above materials, but a stabilizer such as a household detergent is necessary when asphalt emulsion is used as a carrier for nicotine sulfate. Apply the repellents whenever there is evidence of rabbits and before damage occurs.

Deer

Damage to orchards from deer occurs mainly in winter when deep snow and cold weather force the deer away from their normal winter ranges into agricultural areas. Controls include frightening devices, chemical repellents, mechanical controls, feeding to keep the deer from entering

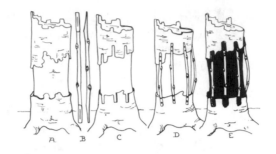

FIG. 1.15. BRIDGE GRAFT

(A) Wound trimmed; (B) scions for bridging; (C) channels cut to receive the scion bridges; (D) scions in place; (E) all wounded tissue protected with grafting compound.

possible damage areas, herding the deer from damage possibilities, and reducing the deer populations by authorized personnel.

The Utah Cooperative Wildlife Research Unit tried many different chemicals as repellents, but none effectively repelled starving deer. Some showed promise, however, of keeping deer away from orchard trees when other acceptable feeds were available.

In many cases, 2.5 m (8 ft) high fencing is the most economical solution to the problem from a long-range viewpoint. Commercial electric fences around orchards are effective.

Damage to small trees can be eliminated by wrapping the entire plant with burlap as high as the deer can reach. Besides protecting trees from deer and rabbit damage, the burlap protects them from frost and prevents them from being pushed out of shape by heavy snows.

Damage by big game frequently appears much more severe than it actually is. Damaged trees often outgrow the injury done them. Recovery can be speeded by trimming and pruning the trees.

Damage Repair

Carry out careful periodic inspections of the orchard for animal injury. As soon as injury is detected trim the ragged edges of the wound and apply a liberal coating of protective compound, such as beeswax or asphalt-water emulsion.

Girdled or semigirdled apple and pear trees can often be saved by repair grafting. Secure dormant scions long enough to bridge the wound, and store them in damp sand, sawdust, or equivalent material in a cool place. In late spring, as soon as the bark slips readily from the wood, graft the scions to the injured trunk so that they bridge the wound. Cut channels to fit the scions in the healthy bark below and above the wound. Insert the lower end of the scion, beveled on one side, into the channel below the wound and nail in place. Similarly graft the upper end of the scion above the wound. Place scions thus across the wound at 5 cm (2 in.) intervals. Cover all wounded tissue with protective tree dressing.

If the injury is to the roots, graft the lower ends of the bridging scions to the roots beyond the injury. Sometimes it is best to plant 3−5 seedling trees around the injured tree, and inarch their tops to the trunk or scaffold branches above the wound. Inarching is also useful where injury to the trunk has been caused by freezing, canker, or mechanical injury.

FIG. 1.16. INARCHING: (LEFT) YOUNG NURSERY TREES BEING PLANTED AROUND THE DAMAGED TRUNK; (MIDDLE) THE NURSERY TREES PLANTED AND THE TOPS "INARCHED" ABOVE THE WOUND; (RIGHT) THE OPERATION COMPLETE WITH ALL WOUNDED TISSUE TREATED WITH GRAFTING COMPOUND

Root- and stem-injured trees may have pale, underdeveloped foliage the following summer. Prune the injured trees heavily in the spring, and supply a little extra N to correct the balance between root and top and to promote vigor and growth.

LOW TEMPERATURE INJURY

Injury sustained during cold periods is a major hazard in commercial apple growing regions throughout the entire United States and Canada. In some areas, some minor injury may occur in any winter, especially in orchards where the trees go into winter in an unthrifty or immature condition or, as mentioned earlier, when the pruning has been untimely. A severe or "test" winter can be expected at irregular intervals.

Classes of Injury

Buds and Blossoms.—Apple fruit buds are especially hardy to winter cold. Only rarely do they suffer injury from low temperatures.

Apple blossoms may be severely damaged at temperatures below -2°C (28°F), but considerable recovery may occur after freezes as low as -3°C (26°F). When the young seeds are killed, however, the fruit abscisses. The more open the blossom, the greater is its susceptibility to cold injury.

With favorable conditions for pollination, a normal set of fruit may take place although there may be fewer than five functioning styles per flower. This explains the set of fruit sometimes obtained when frost damages the blossoms just as they are opening.

Terminal Shoots and Spurs.—"Dieback" of terminal shoots is a common form of winter injury in apples. The result is the same as if the shoots were pruned back and is not a permanent or serious injury. The cause is probably due to the fact that late growth often leaves the terminals in an immature condition in late fall. Fruit spurs also are occasionally injured in this way but not so frequently as pear fruit spurs. Do not confuse this form of winter injury with fire blight.

Large Limbs.—"Patch injury" is a type of winter injury which involves areas or patches on the larger limbs. These areas may be of any size or shape and occur on any part of the limb. The affected patch first appears as a sunken, water-soaked area. Later the bark dies, cracks, and peels. Small patch injuries sometimes may be repaired by cutting out the injured tissues, painting the wound, and covering it with a grafting compound.

There is a close correlation between this winter injury and the distance or obstruction separating the foliage from the lower parts of the tree. Injuries which partly cut off or distantly remove an area of the tree from the leaves may cause the region below the injury to be subject to winter-killing.

Crotches.—Winter injury to the main crotches of the framework is both common and serious. The cause is the same as for patch injury, i.e., remoteness of the crotch from the foliage. Since hardiness of the woody tissues is, in part, derived from the leaves, the crotch tissue, being the farthest point of the limb from the foliage, matures last.

Any crotch which is poorly supplied from the limb that forms it must receive most of its leaf-produced material from limbs attached higher on the trunk. Conductivity along the route from these upper limbs may be hindered by the structure or injury of the parts traversed. Any crotch which is closely below another crotch, following the grain of the wood, is more liable to winter injury than a similar crotch in other positions.

Two crotches occurring laterally close enough together, so that the two susceptible trunk regions depending on higher crotch conduction are side by side, create an extensive area subject to low temperature killing. In setting the tree, situations of this nature may chance to be placed at any point of the compass; consequently, trunk injuries which are due to this cause might occur on any side of the tree. Such a vulnerable spot on the southwest side of a tree trunk, because of two independent factors, poor conduction and winter sunscald, is very subject to winter injury.

The lower a crotch is located in a tree, other things being equal, the greater is the chance of winter injury, probably because the low crotch is so remote from the foliage. Trunks which divide at one point to form two or more large and nearly equal scaffold limbs with a common crotch between them make up a very susceptible area. A crotch of this nature constitutes an isolated region where translocation is sluggish.

Removal of large limbs often results in the death of rather extensive areas of the trunk below the limb scars. The trunk tissues dependent on the leaves of a limb are left subject to winter killing following the loss of the limb. Acuteness of the angle is closely correlated with injury to the crotch. Since wide crotches are most resistant to injury, encourage a young tree to form limbs that make a wide angle with the trunk.

In the more northerly areas of commercial apple growing, crotch injury has been successfully circumvented by topworking the cultivar onto a hardy framework such as Antonovka or *Malus robusta*.

Trunk.—Three common forms of frost injury to the trunks may occur in apple trees.

Frost Cracking.—This is a longitudinal crack in the trunk and sometimes extending into a main scaffold limb; it may involve only the bark or it may penetrate to the pith. Frost cracks usually occur after a sudden severe drop in temperature in late fall or early winter before the trees have attained maximum hardiness. A frost crack may close as soon as the temperature rises and thus be difficult to detect. If the bark is tacked down along the edges of the crack immediately after the injury occurs, the wound usually heals over. A dressing of grafting compound aids in

FIG. 1.17. SOUTHWEST INJURY ON APPLE CAUSED BY
EXTREME FLUCTUATIONS IN DIURNAL TEMPERATURES

rapid healing of the tissues. If the injury is left untended, the bark may
lift from the wood and so expose both tissues to desiccation. To repair
the resulting larger wound, cut away the dead bark, tack down the bark
along the edges, and apply a liberal coating of grafting compound.

Frost Rings.—In this form of winter injury to the trunks, the par-
enchymatous cells in the xylem appear abnormal when no external injury
is apparent. Frost rings tend to be more common on the north than on the
south side of the trunk and to be made more severe by heavy fall
applications of fertilizer.

Sunscald (Southwest Injury).—Sunscald occurs on the southwest side of
the trunk, especially in late winter during periods of still, sunny days
and cold, still nights. During such a period the difference in temperature
between the north and south sides of the tree trunks may be -1° to 4°C
(30° to 40°F). Thus, over a 24 hr period a very wide fluctuation takes
place in the temperature of the tissues on the sunny side of the trunk.
This temperature fluctuation causes death to the cambial tissues and
results in winter sunscald. Autumn applications of N fertilizer may
increase the possibility of winter sunscald. Trees growing on a southern
or southwestern slope may be more susceptible to the injury than those
on a northern or eastern exposure.

If the lowest limb is established on the southwest side of the tree, the resulting shade helps to prevent sunscald. In some areas, boards are leaned against the trunk on the southwest side. Painting the trunk and main limbs with a durable white latex compound has made a significant temperature difference (Martsolf *et al.* 1975). The addition of a repellent such as thiram gives double value to such a paint. Small sunscalded areas may be cut away and covered with wound dressing. Extensive areas of injury require bridge grafting or inarching.

Pith and Sapwood.—Very low temperatures may cause injury that results in the conducting vessels of the pith and sapwood becoming clogged with a dark brown gummy substance. The term "blackheart" is assigned to this form of winter injury. Affected limbs may continue growth and production if conditions are favorable, since the cambium remains alive and continues to lay down new layers of xylem and phloem. Such limbs, however, are likely to be weak and brittle. The dead cells are susceptible to wood-rotting fungi; and areas exposed by pruning should be dressed. The injury is especially serious in nursery stock.

Collar or Crown.—The transition area of the crown or collar is subject to low temperature injury because it is the most remote portion of the aboveground part of the tree from the foliage, and it is also adjacent to the root which may be the least hardy part of the tree. In this type of winter injury the bark tissue at the base of the trunk is killed, the dead bark forming a collar from 5 or more cm (2 to several in.) wide and sometimes extending completely around the trunk. The injury resembles collar rot, a disease caused by the fungus *Phytophthora*.

Where the injury girdles the trunk, the tree can often be saved only by bridge grafting or inarching. If injury extends for more than 1/4 of the trunk circumference, repair grafting might be in order.

Roots.—The roots that comprise the portion of the tree least exposed to low temperatures have acquired the least degree of hardiness. Even beneath the frozen surface soil, heat is always emanating from the soil below. Snow cover in winter acts as insulation to hold the heat in the soil. But in areas of low winter temperatures the roots may suffer from freezing if the ground is bare. Under such conditions a sod cover and/or mulching helps prevent root injury. Relatively heavy pruning of the top, along with N feeding, aids root-injured trees to recover, whereas trees suffering winter injury to the woody parts above ground should be lightly pruned. Different rootstocks vary in hardiness (Chandler 1954).

Factors Influencing Amount of Injury

Cultivar.—Apple cultivars differ in natural hardiness. Among the most hardy of the commercial cultivars are Duchess, Fameuse (Snow), Yellow Transparent, Wealthy, and Bancroft. Considerably less hardy are Stayman, Winesap, Rome, Jonathan, and York. Lying somewhere between these two groups are McIntosh, Cortland, Northern Spy, Yellow Newtown, Rhode Island Greening, Delicious, Spartan, Idared, and Mutsu.

The degree of hardiness in plants is not fixed. It is influenced by conditions which may offset the inherent hardiness of the species, variety, or cultivar.

Overbearing.—An excessively heavy crop, especially of late cultivars, delays maturity of the woody tissues and buds and may leave the tree low in carbohydrate reserves in late fall. A tree in such a devitalized condition is prone to winter injury. Chances of injury are greater if very low temperatures occur early in winter before the tree has attained sufficient hardiness.

A correlation exists between amino acid content and winter hardiness. Apple seedlings treated with abscissic acid had a higher soluble protein accumulation and were hardier than untreated trees (Holubowics and Boe 1970).

Late Growth.—Any orchard practice that favors growth of the trees in late summer or fall increases the hazard of winter injury. Therefore, avoid excessive pruning, late cultivation, and fall application of N in regions where low winter temperatures may occur. Trees growing in a heavy clay soil mature later than those in light soil.

Weak Foliage.—Hardiness in the tree is related to the amount of carbohydrate reserves built up by the leaves. Thus, any factor which adversely affects the photosynthetic efficiency of the leaves in summer results in decreased winter hardiness of the tree. Foliage should therefore be kept free of pests such as scab and mites. Spray injury can seriously reduce efficient leaf surface, as can also nutrient element deficiencies. For example, foliage that is poorly colored as a result of N starvation cannot function to full efficiency in carbohydrate elaboration.

Trees that suffer from drought are more susceptible to winter injury than trees with an ample moisture supply. Imperfect drainage and poor aeration are causes of unthriftiness in fruit trees, and such trees are lacking in winter hardiness.

Climatic Conditions.—Tissues may be killed by freezing temperatures. Rate of temperature drop and duration of the freeze, however, may be more serious than the actual degree of cold reached. McIntosh, for example, may withstand $-7°C$ (20°F) without injury if that temperature occurs after a gradual drop over a long period. Even a lower temperature may result in little injury if the period of duration is but a few hours. During periods of very low temperatures, especially on windy sites, the tree tissues may be injured as much by desiccation as by freezing. Sustained temperatures below 0°C increase cold resistance more than temperatures fluctuating from above freezing to very low (Ketchie and Beeman 1973).

After the rest period has been broken (see also Injury to Buds, Prolonged Dormancy, Chap. 4), the buds and other tissues of the tree are ready to become active as soon as the temperatures rise high enough to favor cell activity. Pruning lowers the resistance of the tree to low temperatures. Much of the loss in hardiness as a result of pruning may be regained in a few weeks, but if low temperatures follow the pruning operation considerable injury may result. Hot, droughty soil conditions in summer weaken the tree, rendering it more prone to winter injury.

FERTILIZERS

Apple trees require all the mineral elements needed by other fruit trees, but usually only N, P, and K must be constantly replaced in normal orchard soils (Table 1.4). If the nutrition of the apple tree could be considered separately from the various soil and environmental conditions, a definite formula could be given for its feeding, but trees are grown in soils of widely variable physical and chemical composition, each different soil condition having an influence on availability of nutrients to the tree. Availability of nutrients in a given soil is itself affected by variations of climate and weather, especially rainfall (Archibald 1966).

TABLE 1.4
NUTRIENTS CONTAINED IN FRUIT CROPS

Crop	Yield (kg)	Part of Crops	N (kg)	P (kg)	K (kg)
Apple	7200	Fruit,	9.0	3.2	13.6
		leaves, wood	4.5	1.4	2.3
Peach	11250	Fruit,	13.6	6.8	24.9
		leaves, wood	24.9	4.5	20.4
Pear	5400	Fruit	3.4	1.4	6.8
Plum	4500	Fruit	6.9	2.5	9.7
Cherry	3600	Fruit	7.3	1.8	9.1

With all the essential elements available in sufficient but not excessive amounts, production of maximum crops will not be difficult. With any element missing or available in insufficient quantity, serious nutritional difficulties may arise. Excess amounts of any mineral element cannot be stated definitely since relative amounts are often more important than the actual amount available. Whether the correct relation or balance between the different elements exists in a given soil depends on several factors, such as the original soil content, the cultural treatment the soil has received, and the fertilizer program that has been followed.

Nutrient Balance

Deficiency symptoms of one element are sometimes affected by the application of another. The terms "antagonism," "nutrient balance," "ionic ratio," and others have been used in various ways to explain changes in tissue content of one ion as a result of the application of another to the nutrient medium. Of the many reciprocal relationships or "ion balances" reported, those caused by N and K fertilizers have probably received the most attention. N fertilization generally reduces the foliage content of P and K and frequently increases that of Ca and Mg of fruit trees. Likewise, K fertilizer generally reduces the Mg content of the foliage, and there is frequently a reciprocal but not a directly proportional relation between these two elements in the foliage of plants (Boynton 1954).

Many of the changes in the percentage of one nutrient ion in the plant tissue, as a result of the addition of another to the nutrient substrate, can possibly be explained either by growth dilution or changes in distribution as a result of stimulated metabolic activity and differential rates of translocation.

Certain nutrient deficiencies are reflected in the composition of the foliage and its analysis as a means of detecting many deficiencies. But leaves are not the only organ of the plant in which an essential nutrient may be utilized or stored, nor does every change occurring therein reflect directly or consistently similar changes in nutrient absorption by the roots.

Nitrogen (N)

Of all the elements, N is the most effective for growth. At the same time, in most soils it is present in only small amounts. It is built up very slowly in any soil, even under the best soil management system, yet it is readily lost through leaching, erosion, and crop removal.

All the mineral elements must be in balance with N, and when one element becomes excessively low in relation to N, further applications of N to "improve" growth simply make matters worse.

In a mature orchard of standard trees planted 12 × 12 m (40 × 40 ft), the roots of one row may meet those of the next and occupy all the orchard soil space. Therefore, a fertilizer containing N should be applied regularly to the entire orchard floor. The grass cover can often be used as one indicator of the N requirements of the orchard.

Nitrogen Mineral Balance.—Constant use of N alone may upset the nutrient balance in both soil and tree with respect to the needs for K, P, and possibly one or more of the minor elements. Many instances have been reported where deficiencies of other nutrients have occurred. Leaf tests of trees fertilized with high amounts of N have indicated that K was deficient in the greatest number of cases (Table 1.5). Sometimes Mg, B, Zn, Fe, Cu, and Mn have been deficient.

TABLE 1.5
AVERAGE POTASSIUM AND MAGNESIUM LEVEL IN THE
FOLIAGE OF MCINTOSH LEAVES FROM TREES
FERTILIZED WITH MODERATE AND HIGH
AMOUNTS OF NITROGEN

	No. of Trees	N	K	Mg
			(% of Dry Weight)	
Medium N	6	2.8	1.06	0.28
High N	6	2.40	0.83	0.32

Source: Weeks and Southwick (1957).

The tree itself is the best index of whether N is needed and, if so, how much to apply. Growers should learn to observe and recognize certain signs and symptoms which permit them to adjust their N fertilizer program accordingly. If mature trees average 15–25 cm (6–10 in.) of shoot growth each year and if the leaves by mid-July are a true green in color, it is almost certain that they are receiving adequate N. If shoot growth is less than 15 cm (6 in.) long and the leaves become yellow-green, the N supply is too low. Excessive growth, large green leaves, and production of very large, poorly colored fruit indicate that the N supply is too high. By application of these principles the keen, observing orchardist can largely solve his own N fertilizer problems.

Effect on Vegetative Growth.—N can influence the entire process of metabolism of the tree and in this way can control both growth and yield.

Growth of apple trees is directly proportional to the nitrates in the soil up to the point where N in the soil becomes excessive. N deficiency causes restriction in growth as expressed by length and weight of terminal shoots and trunk and branch diameter. All other factors being equal, girth increase will be less on trees bearing a heavy crop of fruit than on trees with a light crop. Since N fertilization affects both growth and yield, to some extent, the one may nullify the effect on the other. Size of top in a N-deficient tree is curtailed by less elongation of branches and by the fact that fewer buds are formed and develop into branches (Boynton *et al.* 1950).

In general, apple trees less than 10 years old should produce 25–35 cm (10–14 in.) of terminal growth in a normal season. The terminal growth of bearing trees over 10 years of age should be 15–25 cm (6–10 in.). If growth is shorter than normal or is weak and spindly, N is commonly deficient. Excessive N produces succulent terminal growth of greater than normal length.

Evident responses to N are larger leaves and a deep glossy green foliage. N deficiency symptoms in apple trees quickly become apparent as small, pale green foliage.

Effect on Firmness and Color.—In general, applications of N produce larger, less highly colored apples, with a higher water content, a greater amount and N percentage, and higher catalase activity. The relationship between N and color is very evident in red cultivars but, though less dramatic, exists also in the ground color and in green and yellow cultivars (Raese and Williams 1974). Fruit from trees treated with up to three times normal amounts of N have not shown any greater tendency to internal breakdown than fruit from trees receiving no N.

N fertilizers may delay maturity, but at the rates generally applied, fruit quality and storage quality at maturity are not seriously affected.

Adverse effect of N on color is due partly to delayed maturity, partly to shading as a result of excessive vegetative growth, and partly to a higher chlorophyll content in the fruit. An increase of 0.1% N in dry weight of leaf sample was associated with a decrease of 5% in the surface of the fruit colored (Magness 1939). There appears to be no material difference in color development of fruits from trees fertilized with N in spring, summer, or fall.

There is not always a direct correlation between amount of N applied and effect on firmness or ground color, but there is almost always an inverse correlation between leaf N and fruit firmness and ground color.

Effect on Preharvest Drop.—Preharvest drop is increased by conditions causing high N in relation to carbohydrates during the latter part of the fruit maturation period. Fertilization with N, late cultivation, and prolonged mulching with hay increase the nitrate N in the soil at the end of the summer and also increase the amount of preharvest drop (Hoffman 1939).

Any factor such as diseases or insects that cause a reduction in efficient leaf surface also favors preharvest drop. Excessive shading and dull weather toward harvest time produce a similar effect.

Effect on Fruit Set.—N applications often promote an increase in the percentage of fruit set. But such increase can take place only where N is the limiting factor. Thus, in orchards deficient in N, early spring applications of N fertilizer could increase fruit set. If the N level is adequate but deficiency of some other element has become dominant in the complex of factors affecting fruit set, further applications of N fertilizer would tend to worsen the situation rather than help it.

Where N is to be applied to promote increased fruit set, the kind of N fertilizer used seems to make little difference. Foliar applications of urea, however, produce faster response than soil applications. Since the first wave of fruit abscission takes place shortly after bloom, any treatment to be effective in increasing fruit set must exert its influence early. Urea applications to the leaves several days after bloom may be as effective in increasing set as soil applications of N made several weeks before bloom (Overholser and Overley 1940).

TABLE 1.6

COMPARATIVE LEVELS OF LEAF NITROGEN (N) AND POTASSIUM (K) FROM MATURE MCINTOSH APPLE TREES THAT RECEIVED EITHER DRIPLINE, BROADCAST, OR INSIDE (WITHIN 1 M OF TRUNK) APPLICATIONS OF FERTILIZER

	Treatment	1964	1965	1966	1967	1968
				% N		
Dripline	0.9 kg NH_4NO_3 + 1.4 kg KCl	1.91 b[1]	1.80	2.06 b	2.17	2.05
	1.8 kg $NaNO_3$ + 1.4 kg KCl	1.91 b	1.78	2.05 b	2.12	2.06
Inside	1.8 kg $NaNO_3$ + 1.4 kg KCl	1.94 a	1.75	2.29 a	2.08	2.04
	0.9 kg $NaNO_3$ + 0.7 kg KCl	1.95 a	1.74	2.11 b	2.16	1.95
Broadcast	0.9 kg NH_4NO_3 + 1.4 kg KCl	1.90 b	1.78	2.08 b	2.12	2.08
				% K		
Dripline	0.9 kg NH_4NO_3 + 1.4 kg KCl	1.40	1.34	1.43	1.39	1.32
	1.8 kg $NaNO_3$ + 1.4 kg KCl	1.47	1.35	1.47	1.46	1.37
Inside	1.8 kg $NaNO_3$ + 1.4 kg KCl	1.41	1.28	1.30	1.32	1.34
	0.9 kg $NaNO_3$ + 0.7 kg KCl	1.45	1.30	1.42	1.36	1.32
Broadcast	0.9 kg NH_4NO_3 + 1.4 kg KCl	1.53	1.35	1.48	1.38	1.39

Source: Michelson *et al.* (1969).
[1]Means of the same column for each element followed by different letters are significantly different at the 5% level. Each value is the mean of 10 trees.

Effect on Fruit-bud Formation.—The N level in the tree has a direct influence on the initiation of fruit-bud primordia. N stimulates the formation of fruit buds; and with low-N trees, if applied early enough, it might increase the number of flower buds initiated. Trees with a high N level would not show this response to N applications.

N applications early in the "off" year have a slight tendency to promote annual bearing in alternate-bearing trees having a low N level. This reaction is due to a possible increase in fruit set and to the indirect effect that N has on size and efficiency of the leaf surface (Boynton *et al.* 1950).

A high N level in young nonbearing trees may delay fruiting. This delay is generally of little consequence with most cultivars, but with those like Northern Spy that are slow to come into bearing, the delay may be serious. For this reason it is sometimes wise to reduce the amount of N to Northern Spy trees after they have attained good size and are approaching bearing age. This can be done by reducing or eliminating N applications or by putting the orchard in sod.

Effect on Fruit Size and Yield.—Size of fruit at harvest and total yield depend directly on the amount and efficiency of the leaf surface during the growing season. N is only one of the many variables that contribute to the development and maintenance of an efficient leaf area; but, as such, it can be the limiting factor in the production of a heavy yield of good-sized apples. At the same time, N, by increasing fruit set, may give increased total yield but a decrease in fruit size and color.

Midsummer sprays of urea tend to increase the size of fruit, but reduce fruit color. The effects of sprays of urea N on yield and color of McIntosh apples depend on the timing of the sprays and on the dosages. Timing of these sprays may permit better control of fruit set and fruit color, but the commercial value of such practices is doubtful. Response in yield and fruit size to N applications may vary greatly from year to year and between orchards in the same year. However, unless there is sufficient available N present, a profitable crop of good-quality apples cannot be expected, regardless of other factors.

Sources of Nitrogen.—There are a number of fertilizers available that are good sources of N. These differ in cost, analysis, and the carrier employed, and any one may be best suited for a particular set of conditions.

Calcium Cyanamide.—When used as a source of N, calcium cyanamide (20% N, 70% Ca) usually is applied to apple trees in the fall as a broadcast application. This is an advantage since the performance of an apple tree in any given year depends primarily on N reserves already stored in the tree, rather than on the N applied the same spring. However, because of its high cost and possible toxicity, calcium cyanamide has been replaced largely by ammonium nitrate for orchard use.

Calcium Nitrate.—This source of N (15.5%) is comparatively costly and not readily obtainable in many areas. For N in the soil to be converted by nitrifying microorganisms to the nitrate form for tree use, the soil should not be strongly acid. Since Ca plays an important role in regulating soil acidity, Ca level in the soil is very closely related to N release. Therefore, although Ca content is incidental to the main reason for its use, calcium nitrate could be a particularly useful source of N on soils that tend to be too acid.

Ammonium Nitrate.—This is a readily available form of nitrogen (33.5% N). Applied in early spring just as vigorous growth is starting, it produces a quick response with apple trees. Apply about 112 g (1/4 lb) per year of the tree's age, to a maximum of 1.8–2.2 kg (4–5 lb) per tree.

If applied in late fall, most or all of the ammonium nitrate is leached away by spring, although some may be absorbed by the tree roots from soil with temperatures as low as 0°C (32°F). Being highly soluble, it may be broadcast on the soil, applied in a narrow band around the outer periphery of the tree, or along the row at the outer edge of the trees. If the orchard is in sod, the band method allows the fertilizer to penetrate more rapidly to the feeding roots.

Ammonium Fertilizers.—Those applied in solution are: aqua ammonium (20% N), anhydrous ammonia (82% N), ammonium nitrate ammonia (38 or 41% N), and ammonium nitrate urea (28 or 32% N).

Ammo-phos.—This is a good source of N when P also is needed; it is available in several analyses such as 11-48-0, 13-15-0, 16-20-0, and 18-46-0.

Farm Manure.—Manure is a valuable source of N and other nutrient elements as well as organic matter. The response of trees to manure is much greater than would be caused by the actual N it contains. The nutrient value of manure depends to a considerable extent on its source and the care it has received.

Legume hay and straw are higher in N and other mineral elements than are animal manures (Table 1.7). At the same time, residues other than manure are higher in organic matter and are usually more easily obtained by orchardists. Use manure with caution on apple trees because the N contained is slowly available and may cause late maturation of trees and fruit and adversely affect tree hardiness and fruit color.

TABLE 1.7
MANURIAL VALUES OF FARM RESIDUES

Crop	N	P	K	Dry Matter
		Approximate kg per Tonne		
Alfalfa hay	24.5	5.0	21.0	916
Soybean hay	23.0	7.0	11.0	914
Clover hay	21.0	5.0	20.0	871
Bean straw	19.0	3.9	12.8	895
Pea straw (threshed)	19.0	3.9	12.8	895
Poultry manure	16.3	15.4	8.5	440
Timothy hay	12.5	5.5	10.0	908
Pea straw (from viner)	10.4	3.5	9.9	430
Marsh grass	8.6	5.3	27.0	860
Sheep manure	8.3	2.3	6.7	360
Leaves, mixed	8.0	3.0	3.0	916
Peat moss	8.0	1.0	1.5	[1]
Corn stover	7.5	4.0	9.0	800
Buckwheat	6.5	3.6	12.1	901
Oat straw	6.0	2.0	13.0	908
Horse manure	5.8	2.8	5.3	300
Wheat straw	5.5	2.0	10.0	904
Mixed stable manure	5.0	2.6	6.3	241
Hog manure	4.5	1.9	6.0	270
Cow manure	3.4	1.6	4.0	230
Sawdust shavings	2.0	1.0	2.0	[1]

Source: Archibald (1966).
[1]Moisture content variable.

Ammonium Sulfate.—Because ammonium sulfate (20% N) tends to lower soil pH it may be preferable on comparatively high pH soils. Apply it at about the same rate as sodium nitrate in either late fall or spring.

Sodium Nitrate.—This is a readily available source of nitrogen (16% N). Apply it to apple trees in the spring, at the basic rate of 0.1−0.2 kg (1/4−1/2 lb) per year of tree age. Trees growing in sod require a higher rate.

Urea.—Either as a soil or foliar application, urea (38.5% N) usually produces a favorable response in apple trees.

Fertilization of apple trees by spraying with urea solutions is practiced in some areas. The lower surface of the apple leaf absorbs urea more rapidly than the upper surface, and young expanding leaves more efficiently than mature leaves. The presence of a wetting agent in the urea solution may enhance absorption. Weather conditions causing a low vapor pressure deficit in general favor absorption; 1/2−3/4 of the absorbed N may be translocated out of the absorbing leaves.

Favorable responses from urea foliar sprays have been observed in orchards situated on unfavorable soil sites where past soil fertilizer applications were relatively ineffective. Application of 0.59 kg (5 lb) of urea per 100 litres (100 gal.) of water causes no injury to the foliage; 1.18 kg (10 lb) causes injury. Midsummer sprays of urea tend to increase fruit size and reduce fruit color at harvest. Prebloom sprays increase fruit set as well as fruit color. Spray treatment results in N and chlorophyll levels that are greater early in the season for a given amount of foliar nitrogen than for soil application of N.

Urea foliar sprays on 20-year-old McIntosh trees in Vermont not only increased yields but enhanced leaf size, tree growth, and fruit-bud development. Some of the yield increase was due to more spurs maturing 2−3 apples as compared with 1 apple borne by most McIntosh spurs. The quick response on mature trees seemed to be due, in part, to N assimilation by the leaves, thus minimizing their dependence upon translocation of assimilated N from the roots (Shim *et al.* 1973).

Phosphorus (P)

The P requirements of apple trees are small, but this element is an essential participant in the metabolic processes of the plant. In growing plants, P is the most abundant in meristematic tissue as a component of nucleoproteins and other P-containing compounds. P is interrelated with N in plant metabolism. Excessive amounts of available phosphates in the soil may interfere with absorption of N (Meyer *et al.* 1960).

Apple trees seem able to absorb adequate P from soils with strong P-fixation tendencies, soil in which many other plants could not be grown without liming or heavy P fertilization, or both. P deficiency symptoms are not common in the orchard. In apple trees grown in sand culture, symptoms of severe P deficiencies are weak growth, small leaves, fewer leaf buds, delayed bud break in the spring, and dull, dark green foliage that becomes bronzed except in the midrib region.

Application of Phosphorus.—There is little experimental evidence that P fertilization produces economic response in apple trees. However, sod or cover crops may benefit from P fertilizers. In soils very low in available P and with high fixation powers, heavy applications of P mixed with the soil in the bottom of the tree hole before planting may promote better growth.

Manure applications, especially poultry manure, can supply some P to orchard soils (Table 1.7). Mulches of hay or straw release P to the soil as a product of decomposition.

In many areas, P in the form of superphosphate, ammonium phosphate, or rock phosphate broadcast on the soil or into the grass cover becomes fixed in the top 5−7.5 cm (2−3 in.) of soil and does not become available at the depth of the main root area.

P seldom moves very far from the point of application because of its rapid fixation with the hydrated oxides of aluminum and iron to form relatively insoluble compounds in acid soils, or as calcium phosphate in highly alkaline soils. A soil may have an abundance of P and yet have little available to the trees.

This difficulty may be overcome by deep placement of superphosphate (20% P_2O_5) by means of a subsoiling or chiseling machine with a fertilizer attachment. Somewhat the same result is obtained by sowing the fertilizer in the bottom of a plow furrow and replacing the furrow slice. The deep placement machine or plow is run just outside the outer branches of the trees and the operation is done in late fall to minimize root damage. Root damage often may be greater than any accrued advantage. Breaking up of the heavy subsoil, where this exists, is in itself beneficial in improving drainage and aeration, and in allowing better root penetration into subsoil which may be rich in mineral nutrients.

A mulch of wheat straw had a slight effect in increasing the depth of penetration of readily available P into the soil when applied as 9 kg (20 lb) of superphosphate per tree per year for 3 years (Table 1.8). The P content of Stayman leaves was increased more through the use of superphosphate on sod than on mulch although all phosphate treatments increased leaf P (Wander 1943).

TABLE 1.8
PHOSPHORUS IN STAYMAN LEAVES

Treatment	P Oct. 16, 1944 (%)	P Oct. 10, 1946 (%)
Control, no fertilizer	0.095	0.104
Sod plus 9 kg 20% superphosphate each year since May 1944	0.124	0.256
Wheat straw mulch established May 1944 plus 9 kg 20% super phosphate each year since May 1944	0.144	0.163
Wheat straw mulch established May 1942 plus 9 kg 20% superphosphate each year since May 1944	0.103	0.147

Source: Wander (1947).

Foliar sprays of P may give better results than ground applications on soils of high fixing power. Water-soluble P fertilizers applied to foliage and small branches of apple trees were absorbed and translocated to other parts of the trees. However, foliarly applied P did not adequately supply all the P needs of the trees. It is possible, though, that trees receiving suboptimum supplies of P from the root media may be benefited by foliar applications of this nutrient. Rates of absorption of P by apple leaves vary with age of the leaves and source of the element. Young leaves absorb P faster than old leaves. The use of glycerine in the spray solution increases P absorption.

Potassium (K)

An apple tree producing 363 kg (20 bu) would utilize about 0.8 kg (1-3/4 lb) of muriate of potash in the fruit alone. Added to this amount would be the requirements for roots, branches, and leaves (Table 1.4).

K is essential in the manufacture and translocation of sugars and starches, and is a factor in the uptake and loss of water by the tree. Moderate K deficiency results in poor root development and in fruit of poor color and quality. More serious deficiency is characterized by olive brown leaf color, rolling of the leaf margins upward toward the midrib, and finally a marginal chlorosis or burning of the leaf. These symptoms appear first on the basal leaves of the current growth and progress upward.

Potassium Balance with Other Elements.—The balance between K and N is very important. The greater the supply of available N up to a point, the greater is the rate of tree growth and the greater is the demand on all other essential elements, especially K. Thus, N applications alone can readily lead to K deficiency symptoms. The amount of K required is governed more or less by the amount of available N. Excess K in the soil does not generally do any harm. But in soils low in Mg, excessive K may aggravate the Mg deficiency. K deficiency sometimes occurs as a result of an excess of P. The uptake of K and P increases in the leaf tissue of apple with increasing moisture and with increasing K and P levels. There is a direct interrelation between K and P intake.

Too much lime in the soil may interfere with the uptake of K by apple trees so that continued use of lime without K is not wise (Forshey 1963).

Potassium and Color.—Poor fruit color is often associated with K deficiency. Color of apples can be improved by K applications where the cause of poor color is lack of K. But the K supply is not the only factor that influences red color development in apples and when one of these other factors is involved, K applications do not give a favorable response.

Effect on Firmness and Storage Life.—Applications of K may influence the firmness and storage quality of apples only indirectly. Where excessive available N is present, a commensurate supply of available K can result in more rapid growth and late maturity of the fruit. However, even in excessive amounts, K seems to have no direct effect on the fruit. Under various soil and climatic conditions the use of K fertilizers alone, or in combination with N or N and P, does not seem to affect the firmness or keeping quality of apples.

Sources of Potassium.—Supplemental K can be supplied to apple trees in a complete fertilizer, in combination with P, or as potassium sulfate or muriate of potash. Organic matter may also be a good source of potash.

Farm Manures.—Relatively small amounts of K are found in manures (Table 1.7), but if available in sufficient quantities manure probably supplies the K needs of most apple orchards. Hay and straw are higher than manure in K and when used as a mulch are an effective means of getting available K down to rooting depths.

In Ohio, K, P, Ca, Mg, and B had been increased in the soil beneath a heavy mulch as compared with cultivated land. K was considerably increased under mulch at all depths sampled. Mulching increased the K and P contents of the leaves. Ca was reduced in the fruit but Mg was increased. B was slightly increased in leaves of mulched trees but remained the same in the fruit from both treatments (Wander 1943).

Potassium Chloride or Muriate of Potash.—This contains 50−60% K_2O and can be applied to the soil at the rate of 1−2 kg (3−4 lb) per 20-year-old tree. Applied to the entire orchard area, the rate would be 220−550 kg (200−500 lb) per 0.4 ha (acre) of mature trees, depending on severity of the deficiency. Muriate of potash may also be applied as a spray to apple trees. The chloride fertilizers are not favored so much as formerly.

In Indiana, trees in cultivation to which KCl was applied alone or in combination with superphosphate or sulfate of ammonia did not show as much K in their terminal leaves as trees mulched with straw or strawy manure. Application of KCl to the soil did not increase the K content of the leaves significantly (Baker 1949).

Potassium Sulfate.—Potassium sulfate (48% K_2O) in a spray of 118 g (1 lb) per 100 litres (100 gal.) of water applied to the foliage can correct K deficiency in the season in which it occurs. K deficiency symptoms in the trees have disappeared after spring applications of potassium sulfate have been made to the soil. A basic rate of application is about 1−2 kg (3−4 lb) per 20-year-old tree.

Poor soil aeration hinders the uptake of K by apple roots, whereas undisturbed soil with good structure, such as that found beneath a mulch, favors root activity and K absorption. Thus, trees growing in poorly drained soil may show K deficiency symptoms, whereas trees growing in well aerated soils will not show such symptoms even though the latter soil may be lower in replaceable K.

Other Supplements

Boron (B).—Boron deficiency symptoms in apples vary with the cultivar, climatic conditions, stage of development at which they appear, and vigor of tree. The symptoms may appear on isolated trees in an orchard or on only a single branch of a tree.

A deficiency of B causes internal cork (dark brown corky lesions in the flesh that may not be seen from the outside) in the fruit of some cultivars. External cork or "drought spot," which is also caused by B deficiency, is often associated with dry weather. Twigs suffering from B deficiency are shortened and have a leaf rosette or bunching of the leaves. Dieback of twigs is another form of the disorder (Burrell 1940).

Internal bark necrosis (measles) is a disorder in which high concentrations of Mn, deficiency of B, and toxicity of Fe and Al have all been implicated as causal factors. Among the evidence, some conflicting, a Mn-Ca relationship is involved but not pH of the rooting medium (Shannon 1954; Cooper and Thompson 1972).

B deficiency may become apparent in a dry season, yet the same trees appear normal in seasons of adequate moisture. Poor drainage also aggravates a low B condition. A high level of available N is often associated with B deficiency symptoms, probably as a result of an increased demand for all nutrients. There may be sufficient B in the soil for normal tree growth but not enough for increased growth caused by N feeding.

Disorders caused by B deficiency may be corrected by a soil application of 0.1–0.5 kg (1/4–1 lb) of borax[1] per tree. The treatment is usually effective for at least 3 years. Since borax is highly soluble and there is little fixation of boron in the soil, broadcast applications of borax can be made. Sprays of 236 g (2 lb) of borax per 100 litres (100 gal.) of water per 0.4 ha (acre) at the calyx and first 2 cover sprays are effective in some areas (Bramlage and Thompson 1962). Some growers prefer annual spray applications of highly soluble sodium pentaborate combined with 1 or 2 pesticide sprays after petal fall (Oberley and Forshey 1969).

---

[1]Borax contains 11.34% boron or 17.9% boric acid; hence 4 kg of borax, or 3 kg boric acid provide 300 g of boron.

Magnesium (Mg).—Magnesium is present in apple leaves in only small concentrations and the amount remains constant during the growing season. Mg deficiency symptoms resemble those of K deficiency, appearing on the basal leaves of the current growth rather late in the year and gradually progressing upward. At first there is a yellowing between the leaf veins; later, patches of dead tissue develop. Early defoliation takes place, beginning with the basal leaves.

On acid soils, a long-range liming program favors an increase in Mg availability; dolomitic limestone is more beneficial than calcitic limestone. "Sul-pomag"—$MgSO_4$ (MgO 18%) plus KSO_4 (K_2O 22%)—is a water-soluble source of both K and Mg. Temporary correction of Mg deficiency can be obtained by spraying the foliage with 2.36 kg (20 lb) of magnesium sulfate (Epsom salts) per 100 litres (100 gal.) of water.

Manganese (Mn) and Iron (Fe).—Foliage symptoms of both manganese and iron deficiency are rather similar, i.e., leaves turn mottled green, sometimes becoming pale yellow or almost white, with the veins remaining green. These symptoms are called "lime-induced chlorosis" and are generally associated with soils high in lime; however, the condition may occur in trees on acid soil. Mn and Fe have been implicated in the disease known as internal bark necrosis (see Boron) (Fisher *et al.* 1976).

Mn deficiency may be controlled by soil applications, but except in acid soils with low fixing power, foliar sprays of 1% manganese sulfate are more effective.

Iron is plentiful in most soils and its apparent deficiency is usually the result of other factors rendering it unavailable to the trees (see Chap. 4). With the apple, trunk injections of Fe sulfate may be more reliable than soil applications or foliage sprays of Fe.

Calcium (Ca).—Calcium accumulates in apple leaves in about the same concentrations as K, but unlike K, it then becomes relatively immobile. Ca deficiency in the apple tree rarely occurs because most soils have lime in their parent material. Too high Ca, however, may promote lime-induced Mg chlorosis.

Ca deficiency in the leaves and fruit has been correlated with the incidence of cork spot. Ca concentrations in apple fruits is now believed by some to be the most important factor influencing the incidence of storage disorders, e.g., bitter pit and breakdown (Shear 1972).

Proof is lacking of apple trees responding directly to lime applications. In humid regions considerable lime is applied in sprays, and in drier regions the soil is usually well supplied with Ca. In low pH soils the sod cover may benefit directly from liming and the trees benefit indirectly.

Copper (Cu).—Copper deficiency in apple trees in North America has been reported only from coastal California. The symptom is annual dieback of the terminal growth which in time results in a short, bushy tree. Control has been effected by foliar spray containing copper.

Zinc (Zn).—A characteristic symptom of Zn deficiency is a rosetting of the leaves. Zinc deficiency in apple orchards occurs mostly in semiarid regions under irrigation, although it may occur in any area. Since all mineral soils include some Zn, and since trees rarely respond to soil applications of Zn, it would seem that lack of Zn in the tree tissue may be the result of some soil condition that renders this element unavailable. However, recent evidence adds weight to the probability that a mycorrhizal fungus may be involved in Zn uptake (Benson and Covey 1976).

TABLE 1.9

EFFECT OF VARIOUS ZINC TREATMENTS ON ZINC CONTENT AND SYMPTOM
EXPRESSION OF 8-YEAR-OLD NORTHERN SPY AND DELICIOUS
TREES IN ONTARIO

Treatment	1952 Symp. Index	1953 Zinc (ppm)	1953 Symp. Index	1954 Zinc (ppm)	1954 Symp. Index	1955 Zinc (ppm)	1955 Symp. Index	1956[1] Zinc (ppm)	1956[1] Symp. Index
Dormant spray, $ZnSO_4$, 4% mid-April	2.2	11.6	3.9	12.8	4.4	14.9	5.0	15.8	4.9
Zinc oxide, 0.2%, May 30	1.2	12.2	5.0	16.0	4.9	16.4	5.0	16.0	4.8
$ZnSO_4$, 0.15%, May 30	1.9	14.2	4.9	16.3	4.8	17.2	4.7	18.0	4.7
Control, no Zn	1.4	9.5	2.4	9.8	1.3	9.6	1.0	14.9	4.5
LSD[2] (P 0.05)	NS[3]	2.3	1.9	2.9	2.1	2.8	1.8	2.1	NS

Source: Heeney et al. (1964).
[1]All treatments, including the check treatment, were sprayed with 0.15% zinc sulphate on May 28. Other zinc treatments were not applied.
[2]Least significant difference.
[3]Not significant.

Foliar Sprays

Foliar sprays are useful in prompt treatment of deficiencies such as Fe, Mg, Mn, N, K, and Zn. Also, they may be superior to ground applications on calcareous soils or soils of high fixing power (Tukey 1952).

In wet, cool soils, nitrification may proceed too slowly to provide the N essential to satisfactory fruit set. Foliar sprays offer the possibility of supplying supplemental N at this critical time. But, soil of high N level in a season of adequate moisture and conditions favorable to vigorous growth may result in fruit of poor color. With foliar applications of N available to meet N deficiency symptoms, an orchard may be safely maintained at a sufficiently low N level to favor the development of good fruit color.

With an exact and prompt control of nutrients, climatic variations may be easily adjusted from year to year and annual bearing may be more nearly realized than is usually the case. Also, the possibility is offered of supplying specific nutrients at the time most needed. Possibly, applications of P might be desirable when this element is being used heavily in fruit and seed development.

Spraying dormant trees with certain materials may aid quick recovery from winter injury and other adverse conditions. Also, various pesticides need to be evaluated in terms of nutrition. "Foliage feeding" may involve other portions of the plant, such as trunk, branches, and shoots, as well as foliage. To illustrate, among the first applications of commercial N as fertilizers to fruit trees were sprays to dormant trees, and old horticultural practices include coating trunks and branches of fruit trees and vines with various manurial and mineral substances.

Radioactive isotope studies have shown that materials like orthophosphoric acid, potassium carbonate, and urea may move readily into the above-ground parts of plants, not only into leaves but also into branches, even in midwinter. P and N tend, to a limited degree, to move toward regions of greatest metabolic activity in the plant, whereas K is more generally distributed.

SOIL AND TISSUE TESTS

Tissue analysis is widely accepted as a diagnostic tool in nutrition work. With fruit trees, total amounts of nutrients in the leaves are determined and standards on critical and desired levels have been established. Another method is to use the soluble nutrient content as an index of nutrition.

In studies on the degree of correlation between total quantities of N, P, K, Ca, and Mg in apple foliage (ash) and amounts of these elements in extracts of leaves, a positive correlation has been found between the total and extract fractions, except for N. They provide a relatively cheap, rapid, and accurate diagnostic method for use alone or in conjunction with soil tests in determination of nutrient levels in apple trees and in making fertilizer recommendations.

Spectrographic analysis involves ashing of the plant material and the analysis of radiation in a spectrum. This method is fast and has become of value as a service to growers for diagnosing the nutritional status of fruit trees.

The relation between chemical nature of the soil and nutritional status of fruit trees is complex, since the nutritional status of fruit trees is influenced by factors such as rootstock, pruning, injury, and cultivar which have no bearing on the availability of soil nutrients. Changes in the cation content of apple leaves take place with the advancement of the growing season, size of fruit crop, and with differential applications of N fertilizer. The concentration of a given nutrient in the leaves of fruit trees is often influenced by the concentration of one or more other nutrients in the same tissue, even though the concentration of the element in question has not changed at the absorbing surface of the roots. However, the occurrences of certain element deficiency symptoms of apple trees have been correlated with the supply of these elements in orchard soils.

Chemical analyses of the soil and of leaves of apple trees used together give more reliable information than either analysis by itself. Soil and tissue analyses, when supplemented by careful observation and intelligent symptomatic diagnosis, provide the best basis for fertilizer recommendations. Soil samples for analysis must be representative of the orchard. The position and depth of soil sampling with respect to the tree deserves careful consideration.

FRUIT-BUD FORMATION

Flower induction in the apple bud takes place about the time when the leaf that subtends it becomes nearly full-grown. This is almost a year before the flower opens. The exact time of flower initiation varies with the cultivar, location, season, and to some extent with soil and other factors.

Development of the floral parts is in the order of calyx, anthers, pistil, ovary cavities, and petals. In Virginia, primordia of Duchess flower buds appeared June 15–30, petals not until November 15; tetrad formation did not take place till March; differentiation was complete by April 1. However, it is possible under magnification to recognize the initials of floral parts in a flower bud a few weeks after the subtending leaf reaches full size.

Some time quite early in the season the change from the vegetative to the reproductive state occurs. This transformation is the result of some physiological change not yet understood. However, the treatment to which a tree is subjected has an influence on whether a bud becomes a shoot bud or flower bud (Salsbury and Ross 1969).

Carbohydrate Accumulation

Carbohydrate accumulation has long been associated with fruit-bud formation. But this is only one factor in a complex, and carbohydrate analysis alone is not sufficient basis for conclusions.

Of the various carbohydrate and N fractions determined in the new secondary and vegetative spur growths, and in the bud meristems, higher starch percentages alone were correlated with the initiation of bud primordia. Although it was not concluded that starch, as such, was the cause of floral induction, the evidence indicated that the factor, or group of factors, responsible for this change in the bud meristem was intimately associated with synthesis and deposition of starch (Harley 1941).

A factor in fruit-bud initiation is the foliage/fruit ratio. An adequate leaf area must be maintained so that a carbohydrate reserve, sufficient for fruit-bud formation over and above the demands made by the developing fruit, can be built up in the tissues. Shading, a heavy crop, poor leaf surface, and other influences that reduce the tendency to form fruit buds also reduce the accumulation of starch. However, a great accumulation of starch in the tissues does not necessarily result in abundant flower-bud formation. Other factors are at work.

Nitrogen Supply

N can be a limiting factor in fruit-bud formation. If the N supply to the tree is very low the rate of metabolism is reduced, little carbohydrate is synthesized, and fruit buds do not form. With normal light, adequate moisture, and healthy foliage, the number of fruit buds formed increases with the N supply almost to the point of maximum vegetative growth and darkest green leaf color.

However, a N supply great enough for maximum growth and darkest green leaf color inhibits fruit-bud formation, especially on young trees having all other conditions for rapid growth. With dull weather, or a leaf surface reduced by pests or spray injury, a N supply much less than that required for maximum growth may be sufficient to reduce fruit-bud initiation.

Carbohydrate-nitrogen Relationships

The balance between carbohydrates (starches and sugars) and N in plants has been referred to as the carbohydrate-nitrogen (C/N) relation (Kraus and Kraybill 1918). Although a definite relation between vegetation and reproduction is evident, the actual amounts of carbohydrates and N are much more important for fruit-bud formation than is the ratio of carbohydrates to N.

With regard to carbohydrates and N, the best condition for flower induction is adequate N for growth and production and relatively high amounts of carbohydrates in the fruiting wood. In this case growth is good and there is a surplus of carbohydrates for flower-bud production.

If N is high but the supply of carbohydrates is restricted, growth is weak and fruit-bud initiation is inhibited. This condition is unusual but might occur if foliage has been weakened by pests.

When carbohydrates are adequate only for growth and N is abundant, vegetation is vigorous but few fruit buds are formed because all or nearly all the carbohydrates are used in growth. A too-heavy dormant pruning could cause the condition by removing excessive amounts of stored carbohydrates and leaf buds.

If N is deficient, growth is weak, the foliage is yellowish, and flower production is poor even though there may be an excess of carbohydrates. Such a condition is found in "starved" orchards that need fertilization.

Water Supply

When the amount of vegetative growth of the tree is reduced as a result of decreased soil moisture a greater proportion of buds develop flower

Courtesy of Wenatchee Chamber of Commerce

FIG. 1.18. AN APPLE ORCHARD IN FULL BLOOM ALONG AN
IRRIGATION DITCH SHOWING THE IRRIGATION GATE

primordia. Thus, a moderate but continuous reduction in water supply from the normal amount stimulates fruit-bud formation. Young apple trees, deprived of normal water supply, sometimes differentiate flowers several years before they would ordinarily. However, a severe deficiency of water adversely affects flower formation. The condition is aggravated by high temperatures and sunny weather since such circumstances result in excessive depletion of carbohydrates. If, at the same time, the tree is bearing or has borne a heavy crop there is still further depletion of food materials and fruit-bud production suffers accordingly. During a dry season, irrigation does not directly increase fruit-bud formation on trees bearing a heavy crop.

Defoliation

Tissues of defoliated shoots and spurs show a marked reduction in total carbohydrates. When a leaf subtending a bud is removed, differentiation of flowers in that bud is prevented even though adjacent leaves are present. A reduced leaf surface caused by insects, disease, or spray injury can reduce or even prevent fruit-bud formation.

Biennial Bearing

Bearing a heavy crop is exhaustive to the tree. In the year of a heavy crop the demand on the carbohydrate supply is such that few fruit buds are formed. During the following year, when little or no crop is being borne, the carbohydrate reserve in the tree builds up and an excessive number of fruit buds tends to form. Thus, the tree acquires a biennial bearing habit which, unless broken by a program of fruit thinning, or in other ways, may persist for many years.

Cultivars differ in tendency to biennial bearing. Wealthy, Yellow Transparent, and Duchess are very strongly biennial in habit; others, e.g., McIntosh, Jonathan, and Rhode Island Greening, incline toward annual bearing. Nevertheless, even "annual" bearers tend to differentiate fewer fruit buds in the year of a heavy fruit set. Also, any cultivar can become biennial bearing in habit if some external factor such as frost seriously reduces the crop in any year.

Pruning

Excessive pruning acts somewhat like excessive N applications in inhibiting fruit-bud formation. Pruning stimulates vegetative growth by reducing the number of growing points and readjusting the relation between N and carbohydrates. In old trees low in N, a heavy pruning tends to act like a N application and increase fruit set; but in vigorous trees very heavy pruning produces vegetative growth at the expense of fruiting.

However, moderate pruning to allow more light to reach the potential bearing points is conducive to fruit-bud formation. The deleterious effect of shade on fruitfulness, as well as on growth and color, emphasizes the need to regulate pruning and planting practices with light as one consideration.

Ringing

A tree can be made to flower earlier in its life and more prolifically than normal by slowing the rate of flow of carbohydrates from top to roots. Ringing, an ancient practice, involves girdling the trunk or branch before blossom time by a cut to cambium depth. The cut is generally about the width of a saw blade or less. Ringing can be hazardous if not done with care. If fire blight is prevalent, treat the wound with a disinfectant. Scoring involves only the cutting of the phloem with a knife blade, removing no bark.

POLLINATION AND FRUIT SET

Apple cultivars are commercially self-unfruitful. That is, they require another apple cultivar for cross-pollination in order to set a commercial crop of fruit. Cultivars vary considerably in their inherent tendency toward self-fruitfulness. Jonathan, Rome, Duchess, Wealthy, Golden Delicious, Yellow Newtown, Grimes, and Yellow Transparent tend to be partly self-fruitful.

Most fruits require the presence of an embryo and endosperm for their setting and growth. Without pollination, fertilization, and formation of the zygote, the ovule, ovary, and associated tissues will not develop further and the flower usually abscisses.

The polyembryonic apple fruit usually does not "set" unless fertilization has occurred in several of the ten ovules. Under self-pollination, where self-fertility exists, the average number of seeds per fruit may be 3−5 or fewer, and with good cross-pollination may be 5−8, or up to 10. When the crop is fairly heavy, fruit with fewer than three seeds usually abscisses.

Embryo abortion, although visibly observable in naturally- or artificially induced drops of young apples, is probably caused by disturbances in, and metabolic inefficiency of, the endosperm. This temporary but important 3n tissue produces at least one powerful indigenous hormone (ethyl ester of indoleacetic acid). This, or a derivative of it, seems to be an indispensable catalyst for metabolism of the embryo and tissues making up the fruit. Induction of embryo abscission, through disturbances in development and function of the endosperm, is brought about when α-naphthaleneacetic acid is sprayed on trees to reduce an excessive fruit set. The endosperm is especially sensitive to abnormal disturbances during the free nuclear stage of development (Murneek 1947).

Factors Affecting Fruitfulness

Causes of a poor set of fruit may be related to inadequate pollination or conditions existing after pollination has taken place.

Defective Pollen.—Pollen produced by certain cultivars may be sterile, weak, or abortive. Some cultivars, such as Rhode Island Greening, produce scanty amounts of effective pollen.

Incompatibility.—A certain combination of cultivars may be unsatisfactory in cross-pollination because of an inherent incompatibility factor. However, both cultivars may possess normal reproductive parts and may effect cross-pollination when combined with other cultivars. The pollen tube will not grow in the stylar tissue of a plant carrying the same incompatibility factor. For this reason, bud sports are usually not good pollinators of the cultivar from which they arose. Incompatibility should not be confused with sterility, where fertilization fails to take place because one or more of the reproductive organs is nonfunctional. Varying degrees of incompatibility may be produced by different cultivar combinations (Table 1.10).

Irregular Chromosome Behavior.—Apple cultivars have developed through the centuries from ancestors that have intercrossed. This development has given rise to a complex hybridity and an intricate polyploid constitution. With Stayman, for example, abnormally small and large pollen grains arise as a result of irregular meiotic divisions (Shoemaker 1926). Among the triploids (51 chromosomes) which are generally poor pollinators are Baldwin, Gravenstein, Rhode Island Greening, Stark, Stayman, and King. The diploid condition (34 chromosomes) is more common and cultivars of this type are generally good pollinators.

Climatic Conditions.—Optimum temperatures for pollination and fertilization are 21°–27°C (70°–80°F) with moderate relative humidity (Table 1.11). Temperatures below 4°C (40°F) inhibit both bee activity and pollen germination. Some pollen germination takes place at 4°–10°C (40°–50°F). When temperatures are constantly below 10°C (50°F) at pollination time, the set of fruit may be decreased because of slow pollen-tube growth, the embryo aborting before the pollen tube reaches it.

TABLE 1.10
COMPARATIVE EFFECTIVENESS OF POLLEN CULTIVARS ON
MCINTOSH AND SOME OF ITS SEEDLINGS

Pollen Cultivar	Cultivar Pollinated					
	McIntosh	Melba	Milton	Cortland	Macoun	Early McIntosh
	% of Blossoming Spurs Setting and Maturing Fruit					
McIntosh	2	51	76	59	88	72
Melba	64	22	65	50	82	100
Milton	69	70	0.5	62	84	74
Cortland	66	71	66	3	75	3
Macoun	—	81	67	60	1	74
Early McIntosh	60	70	60	5	71	1
Starking	76	82	75	66	72	—
King (triploid)	3	0	6	6	0	0

Source: Weeks and Latimer (1938).

Temperatures of 15°–21°C (60°–70°F) give satisfactory results. Pollen germination is again inhibited by temperatures above 27°C (80°F).

The more open the blossom, the more susceptible it is to frost injury; the longer the cold period, the greater the injury. Blossoms are killed at -4° to -2°C (25°–28°F). The pistil is the first part of the blossom to be killed by low temperatures. A frost-injured pistil appears brown when the blossom or bud is cut. This discoloration may be used as a test in estimating the amount of blossom injury. Sometimes a good set of fruit occurs after earlier examination of the blossoms seemed to indicate a crop failure. Only about 5% of the blossoms on an apple tree in heavy bloom are required to set fruit in order to produce a commercial crop.

Pollen can remain viable below 0°C (32°F), as in cold storage, for use in a succeeding year in breeding work. In fact, pollen collected for artificial pollination has been stored successfully at 2°–8°C (35°–46°F) and 50% moisture for as long as 4 years (Nebel 1940). Even greater longevity can be obtained by keeping pollen under vacuum.

Very high relative humidity prevents proper release of pollen; low humidity reduces pollen germination and may cause desiccation of stigmas. Rain during bloom may retard bee activity and prevent proper pollen release. Strong winds reduce bee activity, especially at temperatures below normal, and may cause drying out of stigmas and mechanical injury to the blossoms.

Vigor in Fruiting Wood.—The function of the leaves in close proximity to the delicate fruit embryo plays an important part in maintaining an uninterrupted supply of food materials that, in turn, enable the spur to hold its fruit. Fruit setting thus depends on a high level of food supply.

Any treatment, such as the use of fertilizer, irrigation, pruning, and

TABLE 1.11
EFFECT OF TEMPERATURE ON PERCENTAGE OF POLLEN GERMINATION

	Degrees C							
Cultivar	8	12	16	20	24	28	32	36
McIntosh	8.0	15.6	32.2	40.4	36.1	32.1	19.3	Tubes burst
Wealthy	0.0	8.3	32.7	41.2	54.7	50.0	31.1	Tubes burst
Winesap	0.0	6.7	11.8	25.4	19.2	16.9	—	Tubes burst

Source: Roberts and Struckmeyer (1948).

thinning, that contributes to normal vigor of bearing trees increases the set of fruit, and any factor which inhibits vigor reduces the set. It is possible to overstimulate bearing trees, particularly young trees, by heavy fertilization and pruning so that fruiting is reduced. Such a condition is amenable to control by cultural methods.

Agents of Pollination.—Pollen of apple and other fruit trees is heavy so that wind plays a negligible role in pollination. Cross-pollination is achieved by insects, especially honeybees, and a lack of insect pollinators results in reduced pollination and fruit set. Wild bees sometimes are important in pollination, but cannot be relied on by the commercial grower because they often are lacking sufficient numbers at blossom time. In most areas, lack of natural shelter and extensive use of insecticides has so reduced wild bee populations that their contribution to pollination is small. Bumblebees work during weather too cold for the honeybee and are thus important in some years, but their numbers are never large.

Honeybees are the chief agents of pollination. They are specific in their habits (visiting only one species of flower on each field trip), carry through the winter in large colonies, and their numbers in the orchard at any time can be controlled by man. Pistils are not injured by repeated visits of bees.

Bees in the Orchard

Many orchardists make it a regular practice to rent colonies of bees for pollination purposes.

Number of Hives.—In general, one strong, healthy colony will pollinate each 0.4 ha (acre) of fruit bloom. The actual number required varies somewhat with the age of trees, amount of bloom, planting distances, and number and placement of pollinizer cultivars.

Placement of Hives.—The efficiency of bees in an orchard increases, to a point, with their concentration. Therefore, place colonies in small groups (up to ten) in a sunny location. Bee activity is reduced or prevented by cold winds. Bees tend to work into the prevailing winds.

Move colonies into or near the orchard when the blossoms are beginning to open. Remove colonies as the petals begin to drop and before poison spray applications begin. Bees often prefer other plants to fruit trees. If they appear to be working some other crop such as dandelions, alfalfa, or maples, replace them with colonies from another location.

Cost of Hives.—Honeybees are chiefly after pollen when working fruit trees. In carrying out their vital service of cross-pollination in orchards, bees seldom gather more than enough honey to meet their own daily requirements. Only apple occasionally provides a slight surplus. To make bee colonies available for pollination requires extra time and labor by the beekeeper. Colonies must be examined, prepared for moving, loaded, often transported many miles, unloaded, and sometimes visited and given special attention in the temporary location. Unless the beekeeper receives a satisfactory rental fee for his colonies he will not find it attractive to supply bees for pollination.

Pollinizer Cultivars

Since apple cultivars are generally commercially self-unfruitful, it is necessary to provide suitable pollinizer cultivars in the orchard. This should be done when the orchard is planted.

TABLE 1.12
AVERAGE BLOOM PERIODS OF APPLES AT COLUMBIA, MO.

	April									May				
	11	13	15	17	19	21	23	25	27	29	1	3	5	7	9
Duchess														
Gano														
Jonathan														
Arkansas														
Delicious														
Grimes														
Winesap														
Winter Banana														
Yellow Transparent														
Wealthy														
Stayman														
York														
Rome														

Source: Murneek (1947).

Choice of Cultivars.—A pollinizer cultivar should produce plenty of good pollen (diploid) and be compatible with the cultivar to be pollinated. It should have commercial value and the two cultivars should come into bearing at the same time. The bloom periods of commercial cultivars usually overlap sufficiently for effective cross-pollination.

Planting Plan.—Set out the orchard so that no tree is more than 24 m (80 ft) from a pollinizer. If the pollinizer cultivar is of lesser commercial value, it may be every third tree in every third row, i.e., 1 pollinizer to 8 of the main cultivar. However, there are advantages from the standpoint of spraying and harvesting in having trees of one cultivar together in solid rows or blocks rather than scattered. Thus, every fifth row might be of the pollinizer cultivar. Where 2 or more cultivars of equal value comprise the orchard, plant them in alternate blocks of 3 rows (Table 1.13).

TABLE 1.13
PLANTING PLANS TO PROVIDE FOR CROSS-POLLINATION

Every Third Tree in Every Third Row a Pollinizer	Every Fifth Row a Pollinizer Cultivar	Cultivars in Alternate Blocks of Three Rows (D—Delicious; S—Spy, M—McIntosh)
X X X P X X X	P X X X X P X	D D D S S S M M M
X X X X X X X	P X X X X P X	D D D S S S M M M
P X X X X X P	P X X X X P X	D D D S S S M M M
X X X P X X X	P X X X X P X	D D D S S S M M M
X X X X X X X	P X X X X P X	D D D S S S M M M
P X X X X X P	P X X X X P X	D D D S S S M M M

Using Bouquets.—If adequate provision for cross-pollination has not been made at planting time, some temporary measure must be taken until grafted or newly planted trees come into bloom. Place large bouquets of suitable cultivars through the orchard. The bouquets should consist of large branches and have some dehiscing anthers as well as unopened blossoms.

Put the bouquets in containers with water and hang the containers in the trees; the required maximum is one bouquet per standard tree. Provide a large concentration of bees in the orchard while the trees are in bloom and the bouquets are in place.

Topworking.—Where cross-pollination is a problem in an established orchard because of a lack of pollinizers, topwork some of the trees to a suitable cultivar. Single branches or parts of trees can be grafted throughout the orchard. A better method, perhaps, is to change over entire trees to another cultivar. Quickest results can be obtained by a frameworking method (Teskey 1976).

Artificial Pollination

Artificial methods of cross-pollination have been used for a number of years in apple orchards.

Hand Pollination.—Pollen may be bought from a commercial firm or obtained from any apple cultivar known to be a good pollinizer for the cultivar in question. Gather the blossoms when many of them are in the "balloon" stage and remove the anthers by rubbing the flowers over a fine screen. Cure the pollen by holding the anthers in shallow trays at room temperature for about two days. Avoid exposure to direct sunlight. If the pollen cannot be used immediately after it is cured, place it in bottles and hold in a dry, cool place.

Bee-collected pollen can be readily obtained in large quantities and may serve as well as hand-collected pollen in hand pollination. A pollen trap yields as much as 1.1 kg (2.5 lb) of pollen per day per colony of bees. The trap consists of a perforated screen grid devised to scrape pollen from the legs of bees as they go through it while entering the hive. Although the viability of this pollen is rapidly lost at room temperature, it can be stored for several days at 0°C, and for a much longer period at extremely low temperatures.

About 212 g of pollen are required per hectare (3 oz per acre) of mature orchard. This is diluted in lycopodium powder (28 g pollen, 57 g lycopodium) and applied to the blossoms by means of a small, stiff brush. Flowers that have been open only a few hours are most receptive. In apple orchards with a heavy bloom, one flower in every fifth cluster is sufficient to give a heavy set. Pay particular attention to the shady and windy side and to the top of the tree since insect pollinators tend to work best in the warmest and most protected portion of the tree.

A skilled operator can pollinate a 20-year-old apple tree in 1/2–1 hr; 40–50 manhours are required to pollinate 0.4 ha (an acre) of mature orchard. Considerably more time is required to hand-pollinate a cherry tree where a much higher percentage of blossoms must set fruit for a commercial crop. At best, hand pollination is time consuming and costly.

Spraying and Dusting.—Many labor saving methods for applying pollen have been developed and tested. Bombs and shotgun shells containing pollen have been employed. Dust and liquid mixtures of pollen have been applied by airplanes and with conventional spray and dusting equipment. Although pollen can be transferred from a liquid medium to the flowers, apple pollen is seriously macerated when subjected to the pumps and fans of modern sprayers.

Pollen Dispenser.—A semiartificial method of cross-pollination involves pollen dispensers or inserts designed to force honeybees to pass through prepared pollen as they leave the hive, the idea being that the bees will pick up the pollen and spread it through the orchard (Townsend *et al.* 1958).

TABLE 1.14
AVERAGE BLOOMING PERIOD OF APPLES AT THORNBURY, ONT.

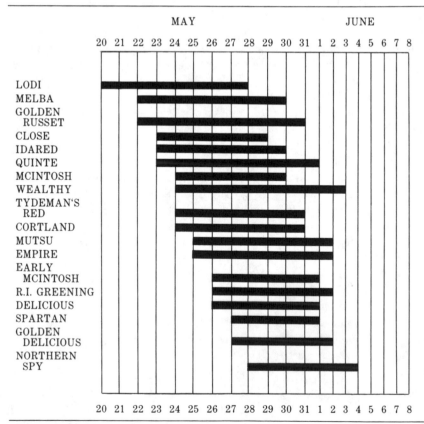

Source: Teskey (1976).

TOPWORKING

Topworking is the practice of changing all or part of a tree's top to a different cultivar by means of grafting. Fruit trees are topworked to a more preferable cultivar and sometimes to provide better conditions for cross-pollination.

In grafting, the scion should be dormant and both scion and stock should be healthy and vigorous. The cambial layers of scion and stock must be held in close contact until the union is made. If done too early, the scions may dry out before uniting with the stock. All wounded tissue must be protected against desiccation.

Wood Grafting

Insert a dormant scion with 3−6 buds into a cleft or cut in the wood of the stock so that the cambium of the scion is in contact with that of the stock. Graft in early spring, generally in March, April, or May in northern regions. There are three principal methods.

Cleft Method.—The method usually known as the "cleft graft" has been widely used in topworking apple and pear trees. Cut off the end of a branch leaving a stub 2−5 cm (1−2 in.) in diameter. Cleft the stub about 7 cm (3 in.) and hold the cleft open while two wedged scions of equal diameter (one on each side of the cleft) are inserted (Fig. 1.19). Tension of the stock holds the scions firmly in place. Cover all wounded tissue thoroughly with grafting compound. If the two scions "take" in the graft, leave both until well established and then remove one of them.

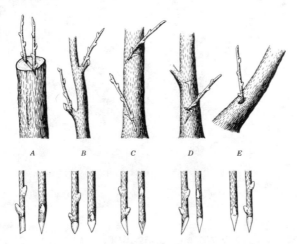

FIG. 1.19. SEVERAL METHODS OF GRAFTING: (A) CLEFT; (B) STUB; (C) SIDE; (D) INVERTED "L"; (E) AWL

A cleft graft is relatively quick and simple to make, but much of the framework of the tree is destroyed, and considerable time is required before a new top is grown and the tree is back again in commercial production. For this reason the practical use of the cleft graft is limited.

Stub Method.—Stub grafts may be used in frameworking apple and pear trees. First prepare the tree by removing all branches smaller than 1 cm (1/2 in.) diameter, as in the case of the inverted "L" and awl graft described later. Select branches up to 2.5 cm (1 in.) diameter for the grafting.

Make an oblique cut halfway through on the upper side of the stock, about 2.5 cm (1 in.) from the main branch. Open the cleft by bearing down on the branch being grafted, and insert a beveled scion. Tension holds the scion in place so that no taping is required. Cut off the branch just beyond the graft and protect the wounds well with grafting compound.

To framework a fully grown apple tree may take up to 200 or more scions and require several hours of work. However, the framework of the tree is retained, the tree suffers no serious setback, and commercial production of the new cultivar may be realized in 3 years.

FIG. 1.20. YOUNG APPLE TREE (LEFT) BEING FRAMEWORKED IN APRIL AND THE SAME TREE (RIGHT) IN JUNE WITH THE NEW SCION CULTIVAR FORMING A FULL LEAF SURFACE

Side Method.—This method is useful where there are no suitable laterals for stub grafting. This type of graft may be used in conjunction with stub grafts in frameworking apple and pear trees. Make a cleft in the side of the limb to be grafted. Insert a scion with a long, thin wedge into the cleft for maximum cambial contact. No nailing or tying is necessary, only a coating of grafting compound.

Bark Grafting

There are several types or methods of bark grafts. In one type, cambial contact is made by slipping a wedged scion beneath the bark of the stock. The wood of the stock is not cut in making a bark graft. Do the work in late spring when the bark slips readily from the wood.

Veneer Method.—Stubs too large for cleft grafting may be grafted by this method. However, the practicality of such a method is doubtful.

The dormant scion with 5−8 good buds is beveled on both sides. Prepare the stock by cutting off the branch to leave a stub 15−20 cm (6−8 in.) long. Make a longitudinal cut in the bark of the stub and insert the beveled end of the scion beneath the raised bark. The scions may be tacked in place or held by grafting tape. Apply grafting compound liberally to all wounded tissue.

Invert "L" Method.—In preparing a large tree for frameworking, remove all branches smaller than 1 cm diameter. Scions may be placed anywhere on the larger limbs.

Make an inverted L-cut with an obtuse angle in the bark of the stock. Lift the bark from the wood and insert the beveled end of a 6- or 8- bud scion beneath the bark. Tack the scion in place and cover all wounded tissue with grafting compound.

Awl Method.—This is more quickly and simply done than the inverted "L" method. Use a tool such as a bent screwdriver to make a slit in the bark. Insert the beveled end of the scion into the slit and apply grafting compound. No taping or tacking is necessary.

Repair or Bridge Method.—Damage to apple and pear trees that involves total or partial girdling of the trunk can be repaired by bridging the wound with scions or inarching suckers or rooted seedlings above the wound.

FRUIT THINNING

Good orchard planning and practices are directed toward providing for a heavy fruit set every year. The crop is then regulated when necessary by thinning the blossoms or young fruits.

Reasons for Thinning Fruit

Thinning is done to increase the percentage of top-grade fruit, increase size, and improve color (Table 1.15), induce annual cropping, promote tree vigor, reduce winter injury, avoid limb breakage, and minimize the handling and storage of low-grade fruit.

TABLE 1.15

EFFECT OF THINNING ON SIZE, GRADE, AND VALUE OF APPLES
(12-YEAR-OLD WINESAP TREES) IN OREGON

Spacing of Fruit in Thinning Treatment (cm)	Grade	Yield: No. Boxes per Tree	per Tree ($)	per Acre ($)
None	Extra fancy	1-1/2	13.25	343.22
	Culls	10-1/2		
7–10	Extra fancy	3	25.28	1364.58
	Culls	5-1/4		
15–18	Extra fancy	4-1/2	35.10	1825.69
	Culls	1-3/4		
23–25	Extra fancy	4	33.98	1764.12
	Culls	2		

Source: From Brown (1926) with values updated to 1977.

A certain amount of thinning takes place naturally during the season from calyx stage until harvest time. This abscission occurs in waves, the first wave being the unfertilized flowers at or shortly after the calyx stage, and the second (June drop) about six weeks later.

Effect on Fruit Size.—The most pronounced effect of thinning is on fruit size. The larger the area of foliage per fruit, up to a point, the greater is the size of the apple, the wood growth, and the carbohydrate stored in the fruit. With most cultivars, the optimum number of leaves per fruit is 30–40, which is the equivalent of 15–20 cm (6–8 in.) spacing of the fruit on the tree. Delicious requires 40–50 leaves per fruit for optimum size. The earlier the thinning the greater is the effect on size (Table 1.16), but late thinning, even after the June drop, gives some response. Spot picking of early cultivars like Yellow Transparent results in a gain in size of the remaining fruits.

Effect on Color.—Fruit color is generally poorer with trees bearing an excessively heavy crop than with trees bearing a normal load. Even with red sports, the inherent red is duller and of poorer quality on overloaded trees. Red skin color in apples is directly related to the amount of carbohydrates (sugars) manufactured in the leaves (see Color). Therefore, more leaves per fruit, up to a point where excessive shading occurs, results in better color. Excessive thinning, however, results in too great a decrease in total yield.

Effect on Quality.—Thinning affects quality of the fruit in the same way that it affects color since there is a definite relation between red color and quality. With 30–50 leaves per fruit more carbohydrates are stored in the fruit than when the tree is overloaded. A high percentage of sugar is one attribute of quality in fruit. Apples receiving an adequate supply of sugars from the leaves mature at the proper time, have firm flesh and good color, and tend to keep well in storage.

Effect on Tree Vigor.—Maturing a heavy crop is an exhaustive process for the tree and may leave the tree in a weakened condition. Such trees are slow in maturing their wood in the fall and may not acquire sufficient hardiness before killing temperatures occur. Subsequent winter injury may leave the trees low in vigor, resulting in poor yields and possibly shorter commercial life.

Reduction of the crop by thinning lessens the drain on the tree's carbohydrate reserve. Normally, a tree can bear a moderate crop of fruit to maturity and itself be well hardened before winter. Fruit thinning to regulate the size of the crop is one means of promoting and maintaining tree vigor.

Effect on Annual Bearing.—Fruit buds of apple are initiated shortly after bloom, almost a full year before they develop into blossoms. Actual date of bud initiation and differentiation varies with the cultivar, season, and region.

Carbohydrates are very important in the production of fruit buds. A leading factor in fruit-bud initiation is the ratio of foliage to fruit.

In order to achieve annual bearing, the stored energy from the previous season must be conserved. Fruit buds are not formed only by the growth activity of the current season. The outflow of energy (carbohydrates) during the critical period of fruit-bud formation must be so regulated and controlled that sufficient energy is present to differentiate fruit buds. Early thinning enables the tree to convert the energy present into fruit buds which would otherwise be wasted in the growth of surplus fruit.

Size of crop is a major factor influencing annual bearing. Either a severe reduction in crop because of unfavorable weather conditions or an excessive set may promote biennial bearing.

Attempts to break the biennial bearing habit by N fertilization have generally not been successful. In Massachusetts, McIntosh trees during a 6-year period on different levels of N were not able to maintain uniform annual production any better than trees at low N levels (Weeks and Southwick 1957).

In thinning a 19-year-old Wealthy tree, it was necessary to remove 4775 fruits, which filled 8 bushel baskets (282 litres) and weighed 159 kg (350 lb). Unthinned Yellow Newtown trees in the 6-week period from bloom to thinning time carried a load 300% greater than necessary to produce a full crop. This fact illustrates the tremendous waste of carbohydrates which may occur in the period following bloom, where the set of the fruit is excessive and allowed to develop to the end of June.

The earlier thinning is done, the more effective it is in promoting annual bearing (Table 1.16). Late thinning has some effect on size, color,

TABLE 1.16

EFFECT OF THINNING MCINTOSH (1 FRUIT TO 50 SPUR LEAVES) ON FRUIT–BUD FORMATION AND YIELD THE SUCCEEDING SEASON

Date of Thinning, 1935	Days After Bloom	Spurs Blossoming, 1936 (%)	Yield per Tree, 1936 (kg)	Yield per Tree, 1935 and 1936 (kg)
June 20	21	24.6	170.55	193.23
July 15	36	11.8	76.20	130.64
August 10	62	4.0	17.69	88.00
Check		5.7	28.12	170.12

and quality of apples, but unless thinning is done before fruit-bud initiation is completed, it will not produce any effect on the bearing habits of the tree.

Thinning to increase the leaf area per fruit results in increased fruit-bud formation where vigorous trees are thinned heavily before mid-June, but not where they are thinned July 1. With less vigorous trees, thinning about June 10 has less effect on fruit-bud formation than with more vigorous trees. Thinning the less vigorous trees after June 20 has no apparent effect on fruit-bud formation. In vigorous trees, thinning within six weeks of bloom is effective in increasing fruit-bud formation for the following year.

Hand Thinning

Hand thinning of the young fruits offers the ideal method of spacing apples on the tree and, at the same time, of eliminating potential culls. However, the labor and time involved, especially for early thinning, make the operation too costly.

Chemical Thinning

The idea of chemical thinning goes back several decades when it was found that there was a "critical period" in blossom-bud formation at the time of fruit setting, and that the time of hand thinning is too late to affect fruit-bud formation.

Caustic Sprays.—Dinitro (DN) compounds have given fairly consistent results when the application has been timed to coincide with the development of full bloom on such heavy setting cultivars as Wealthy (Table 1.17), Yellow Transparent, Baldwin, Duchess, Golden Delicious,

TABLE 1.17
EFFECT OF ELGETOL APPLIED DURING BLOOM ON FRUIT SET OF WEALTHY

Treatment	No. of Blossoming Spurs	Blossoming Spurs with Fruit (%)	Fruits per 100 Blossoming Spurs	Injury
0.1%, May 27	188	20.7	22	Slight
0.2%, May 27	83	0	0	Medium
0.1%, May 27 and 29	116	6.0	6	Slight
Control	149	45.0	58	—

Source: MacDaniels and Hoffman (1941).

Melba, Rome, and Yellow Newtown. However, with other cultivars, notably McIntosh, Rhode Island Greening, and Delicious, the results have been rather erratic. This may be due partly to improper timing of the spray with respect to the processes of pollination, but the dinitros seem to be more toxic to the flower parts of some cultivars than to others.

Dinitros are applied at full bloom, 120−240 ml (1−2 pt) per 100 litres (100 gal.) of water. The actual rate varies with the cultivar, area in which it is grown, and the season. DN thins by destroying the functional parts of the side flowers in the cluster after the central flower which opens first has been pollinated. If sprayed 48−60 hr (depending on the temperature) after pollination, destroying the stigmas does not prevent fertilization and fruit set.

Cutting off stigmas next to the ovary at various intervals up to 24 hr after pollination prevented setting of fruit. Cutting them off after 48 hr had little effect when daily maximum temperatures were above 33°C. With daily maximum temperatures of about 21°C, this treatment prevented set until after an interval of 60 hr.

Dinitros as thinning agents permit early application which is important in regulating annual bearing. They may be more effective than other materials in thinning heavy setting cultivars.

Disadvantages of DN materials are: careful timing of application is required; it is often difficult to recognize the full-bloom stage because of the uneven opening of blossoms; at bloom there is no way of knowing for sure what the actual set will be since cold, wet weather following pollination can seriously reduce fruit set; some injury to the young foliage in the form of yellowing and epinasty may follow dinitro sprays, but usually the injury is slight and temporary.

Auxin Sprays.—Two distinct mechanisms are involved in the use of auxins as thinning agents for apples: (a) the effect of the spray during bloom is to prevent fertilization of freshly opened flowers by inducing an incompatible condition between the pollen tubes and the stylar tissue; (b) postbloom applications thin by increasing the drop of young fruits after fertilization has taken place as a direct result of seed abortion induced by the treatment.

Possibly another factor, that of nutrition or spur vigor, is also involved. Developing seeds in fruits on weak spurs with a limited food supply may be more susceptible to the influence of thinning agents than those with an abundant food supply. This may account for the heavier thinning of fruit from weak wood in shaded positions of the tree. The elimination of clusters may also be explained on the basis of competition for food.

The sodium salt of naphthaleneacetic acid (NAA) causes shedding of flowers during bloom and also induces abscission of young fruits at the calyx stage and later, well past the pollination period. The drop of young fruits is influenced by many factors, such as thoroughness of pollination, the supply of water, N, carbohydrates, and possibly other materials not yet identified. These factors vary with the season and orchard conditions and thus may greatly influence the amount of thinning caused by a given treatment of a material like NAA.

In thinning apples with NAA, the objective is to apply the spray while the fruits are still susceptible to the thinning effect, but after the leaves have passed that stage in their development where injury occurs. With summer cultivars this may be a very short period. NAA (20 ppm) is generally applied 7−10 days after the calyx stage and at the rate of 0.5−0.9 g per litre (7−12 oz per 100 gal.) of water.

Using Tween 20 at 0.125% as an adjuvant, the rate of leaf absorption of NAA was 6-fold. The degree of fruit thinning was also in the same relative proportion, thus indicating a more rapid and more complete utilization of NAA with Tween 20 as the bridging solvent. The under surface of leaves was 60% more efficient in absorbing NAA than the upper surface (Harley et al. 1957).

When cool, cloudy weather predominates during the early developmental period, NAA may result in serious overthinning of some cultivars even though the set on untreated trees is excessive. Early thinning is most effective, but NAA can cause dwarfing of foliage and young shoots when applied at the calyx stage. Late thinning (2 or 3 weeks after calyx) may cause premature ripening of some early cultivars.

NAA is generally preferred to a dinitro for thinning most cultivars since exact timing is less important. Also, a better chance is provided to estimate the crop from observations of bee activity during bloom and weather conditions following pollination.

TABLE 1.18
RESPONSE OF YELLOW TRANSPARENT TO CHEMICAL THINNING

Treatment	ppm Conc.	Date Applied	Fruits/100 Blossoming Clusters July 10	Thinning Degree	Foliage Condition
Control	—	—	57	—	—
NAD	20	May 27 PF[1]	25	Satisfactory	Good
NAD	40	May 27 PF[1]	18	Very good	Good
NAA	15	May 27 PF[1]	3	Excessive	Severe epinasty
NAA	15	June 13 late	51	None	Slight flagging

Source: Hoffman *et al.* (1955).
[1]Petal fall, June 13, 17 days after petal fall.

Naphthaleneacetamide (NAD) can be as effective as NAA for thinning some apple cultivars under certain conditions (Table 1.18). It can be used at the calyx stage with no apparent injury to foliage or young growth, and so affords the advantage of early thinning.

NAD is milder and, therefore, safer for thinning than NAA. When the set is excessive there is a relatively wide range of concentration over which the NAD does not seriously overthin or defruit the trees. Under rapid drying conditions, NAD may not adequately thin heavy setting cultivars like Wealthy. But if the weather is calm, cloudy, warm, and humid, NAD may be the best material to use.

In blocks of three acres or more of Wealthy, with scanty provision for cross-pollination, good thinning and annual bearing were obtained with a single amide spray during full bloom. With heavy setting cultivars like Wealthy and Golden Delicious subjected to thorough cross-pollination, adequate thinning with these compounds was difficult (Hoffman *et al.* 1955).

1-Napthyl-N-methylcarbamate (carbaryl), first used in orchards as an insecticide, if applied from petal fall to three weeks and after is an effective and safe thinning agent which has much less dependence on weather conditions than the hormone sprays. Carbaryl appears to thin by interfering with the movement of vital chemicals in the vascular tissue (Teskey and Kung 1967).

Timing of the thinning spray according to size of young fruit rather than to the number of days after petal fall has considerable merit (Table 1.19). Sensitivity to thinning agents depends on fruit development, which, in turn, is affected by environmental conditions. Average fruit diameter reflects these yearly variations in growing conditions. During a backward spring, fruits may require 12 to 13 days after petal fall to reach the NAA-sensitive stage, whereas during a warm spring this stage may be reached within 6 days. Thus, fruit size may provide a more accurate timing index. However, controversial evidence exists regarding the effectiveness of thinning sprays as related to fruit size at time of application (Batjer *et al.* 1968; Donoho 1968).

Several cultivars including Spartan, McIntosh, Cortland, Mutsu and Spy can be thinned by 2-chloroethyl phosphonic acid (ethephon) at rates from 100 to 1000 ppm applied 20 to 40 days after full bloom, the higher rates causing almost complete abscission in some cases. When succinic acid 2,2-dimethylhydrazide (daminozide [or SADH]) is combined with ethephon, particularly at the lower rates of both, the effect is moderated (Table 1.20). Ethephon is physiologically a very potent agent and although not truly a plant hormone is a powerful growth regulant. Because of the great variability of results depending on age and vigor of tree, weather and cultivar, recommendations are not given for its use as a thinner (Lord *et al.* 1975).

TABLE 1.19

FRUIT SIZES WHICH RESULTED IN IMPROVED TIMING INDICES FOR CHEMICAL THINNING THREE APPLE CULTIVARS IN EASTERN ONTARIO

Cultivar	Av. Diameter	
	(mm)	(in.)
McIntosh	8.0−9.5 mm	5/16−3/8 in.
Delicious	6.5−8.0 mm	1/4−5/16 in.
Northern Spy	10.0−11.0 mm	25/64−7/16 in.

Source: Leuty (1971).

The bearing of fruit has a dwarfing effect on young growing trees. The precocity of apple on certain rootstocks is such that if early fruiting is not controlled, the tree may never attain the desired size and form. It is especially important to prevent the central leader from "fruiting out," i.e., losing its dominance by cropping too soon. Fruit set can be prevented or greatly reduced by spraying the tree, particularly the leader, with a defruiting agent such as NAA plus carbaryl or NAA plus oil according to local recommendations.

Factors Influencing Effectiveness of Thinning Agents

Results of chemical thinning are affected by climate, weather, soil, tree age and vigor, orchard management practices, and other factors. The effectiveness of a hormone for thinning apples is influenced by its rate of absorption by the leaves. Cultivars differ in their ability to absorb these materials and thus vary in response to chemical thinning sprays. The amount of absorption by the leaves is influenced also by weather. If the spray is applied on a warm, bright, windy day, the rate of drying may be so rapid that there is time for very little absorption. Ideal conditions for spraying are high humidity, no wind, overcast, and temperatures in the 30s (°C). Any recommendations on rates, therefore, can be only general. Insufficient thinning one year does not necessarily warrant increasing the concentration the following year. Each grower should proceed on a trial basis until he has found from experience the best methods for his particular conditions.

Chemical thinning can be done with any good sprayer. Thinners can be applied successfully in low concentrate form, if special care is taken. Thinning materials are compatible with most insecticides and fungicides used in orchard work but it is generally best to apply the thinner by itself.

TABLE 1.20
EFFECTS OF SADH AND ETHEPHON 26 DAYS AFTER FULL BLOOM ON
FRUIT SET OF "CORTLAND" APPLE

Treatment and Rate ppm	Fruit Set, 1971 %	Blossom Clusters, 1972 per cm Limb Circ.
Control	61	3.4
SADH, 1000	55	4.3
Ethephon, 1000	2	9.1
SADH, 1000 + ethephon, 1000	2	13.1
SADH, 500 + ethephon, 500	16	10.4

Source: Lord *et al.* (1975).

COLOR IN APPLES

With red-skinned cultivars such as McIntosh, Stayman, Delicious, and Rome, red color is very important in determining the grade of the fruit. The consumer associates red color with good quality. Red apples are attractive and the fresh fruit market preference is given to the better colored fruit. A relation often exists between red skin color and sugar content. Red color generally indicates maturity and full flavor. However, even with red sports, it costs more to produce apples with a high percentage of good red color than greener ones, and although sales may be increased 72% on the average by increasing the area of solid red color from 15 to 20%, the cost also increases and total dollar returns may decrease. Therefore, in establishing price premiums for redness the effect of price on sales volume must be considered. Lack of color does not adversely affect apples for processing (Dayton 1963).

Color Sports

Probably the greatest progress in color improvement of apples has been by selection. Certain bud sports or mutants such as Starking Delicious produce fruit that is more highly colored than that of the cultivars that gave rise to the sports. The fruit of such a sport is redder than that of its parent even though the optimum picking date is the same for both. Vegetative propagation of bud sports has resulted in redder Stayman, McIntosh, Rome, Jonathan, Winesap, Delicious, and other cultivars. Red sports tend to color even in the shaded parts of the tree. The fruit of some sports is such a dark, dull red color as to be undesirable.

Atomic radiation involving thermal neutrons has been more effective in producing gene mutations than have X-rays. Such radiation treatments of dormant scions with 2−8 hr exposure has caused both sectorial and complete color sports. The color sports, as in those of Cortland, are often very dark red and readily distinguished from the normal color.

Color Development

Chemical composition of the fruit and exposure to light are two outstanding factors that seem to influence color development in apples. Although light exposure is essential to the development of anthocyanin pigments in apples, it is by no means the only factor involved.

Anthocyanin pigments arise from chromogen which, in turn, is associated with the amount of carbohydrates (sugar) that is synthesized in the leaves. Hence, the problem of color is related directly to the nutritional condition of the tree and is affected by cultural factors such as fertilization, cultivation, pruning, thinning, and spraying.

Apples will not color without light and it is the ultraviolet rays of sunlight that are essential (Table 1.21). These rays are absorbed by smoke, dust, and water particles. Hence, apples grown at higher elevations and dry conditions develop more red color and fruit tends to color

TABLE 1.21
COLOR DEVELOPMENT OF DETACHED APPLES UNDER GLASS AND
IN DIRECT SUNLIGHT

| | Percentage of Solid Red Color | | | | | |
| | When Started | | After 5 Days | | After 12 Days | |
Cultivar	Under Glass %	Full Sunlight %	Under Glass %	Full Sunlight %	Under Glass %	Full Sunlight %
Jonathon	6.8	8.6	10.5	37.9	44.0	96.0
Delicious	4.5	5.4	5.6	24.2	17.3	74.0
Rome	2.5	2.0	4.0	8.0	9.3	36.3

Source: Magness (1939)

more rapidly after a rainy period. When green apples within the canopy of the tree were exposed to supplementary light, they synthesized anthocyanin in proportion to the added incident energy. The minimum energy required varied with cultivar and weather. Supplementary light to mature detached apples, however, does not appear to increase anthocyanin synthesis unless induction has already occurred (Proctor 1974; Bishop and Klein 1975).

Night temperature is closely correlated with red color development in apples, probably because carbohydrates (sugars) are elaborated only in the presence of light while the breaking-down process of respiration is continuous. On cool nights the rate of respiration slows down, making possible a buildup of sugars. Thus, many cultivars, e.g., McIntosh, are better colored when grown in Northern regions and/or at higher altitudes. However, some cultivars attain fairly good color under relatively high temperature conditions.

Leaf Area and Color

Since sugars are produced in the green parts, a tree with insufficient leaf area to properly mature its crop produces fruit of relatively poor color. An increase in color is developed as the number of leaves per apple increases (Table 1.22). Therefore, up to a point, the more leaves the higher the color, provided they do not excessively shade the fruit. Excessive crops are commonly poorly colored.

The amount of light as a factor in red color development of apples emphasizes the need of pruning and fruit thinning as a practical means of admitting more light to the fruit. When bearing trees are invigorated by N fertilizer, with the consequences of more fruit on the inside of the tree and greater foliage on the tree, the lessened amount of light to such inside fruit is a limiting factor on color and size.

TABLE 1.22
RELATION OF LEAF AREA TO SUGAR CONTENT AND COLOR OF
DELICIOUS APPLES, WENATCHEE, WASHINGTON

No. of Leaves	Sugar Content Reducing (%)	Nonreducing (%)	Total (%)	Solid Red Color of Surface (%)
10	9.19	0.45	9.64	23
30	9.96	1.51	11.47	42
75	10.84	3.29	14.13	58

Fruit on trees that suffer from drought in summer may not have as bright red color as fruit on trees with a continuous moderate water supply. Irrigation sometimes may brighten the color of apples.

Fertility in Relation to Color

An abundant N supply may reduce the amount of red color by increasing the chlorophyll in the skin, and by causing a thick growth of leafy shoots that shade the fruit. Also it may delay maturity and color development so that the fruit is picked before it is well colored. Fruit grown on clay soils may have poor color because of delayed ripening. As long as the nitrates entering the tree combine with practically all the available sugars and form organic N, there is no accumulation of carbohydrates from which chromogen is formed.

The reduction in color often associated with liberal use of N in orchards has been attributed also to a reduction in carbohydrate supply to the fruit because of greater usage in growth when abundant N is present. Fruit maturity is delayed on trees high in N.

TABLE 1.23
PERCENTAGE OF RED SKIN COLOR, FLAVOR AFTER STORAGE, AND SIZE OF
APPLES GROWN UNDER 4 LEVELS OF N (1957–1961)

	N per Tree (kg)				
	0.23	0.45	0.91	1.81	F Test
	Red Skin Color (%)				
Mean	76	72	70	67	5.37
	Flavor[1] After Storage				
Mean	6.0	5.7	5.9	6.0	10.3
	Fruit Size (g per Fruit)				
Mean	139	138	139	137	0.10

Source: Mason (1964).
[1]Flavor index: Scale 1 to 10; 1—very poor, 10—excellent.

Increasing the N fertilization increased leaf N and decreased leaf K and was accompanied by less color and poorer keeping quality of fruit. When 2.27–4.54 kg (5–10 lb) of K_2O per tree were included with the N application, leaf N remained the same, but leaf K increased and also fruit color and keeping quality (Weeks *et al.* 1958).

Color development may be slightly better on fruit from trees sprayed in the early season with urea than on fruit from trees receiving soil application of ammonium nitrate.

Soil applications of sugar "cerelose" increase the percentage of fruits having a high percentage of color. Fruit from trees so treated matures about two weeks earlier than that from trees not receiving sugar and shows an increase in reducing sugar content (Raese and Williams 1974).

Effect of Preharvest Sprays

Auxins, such as NAA and 2,4,5-trichlorophenoxypropionic acid (2,4,5-TP), used as preharvest sprays to control fruit drop, tend to increase the rate of maturation of apples. At normal harvest time the fruit is both redder and softer than unsprayed fruit because of its more advanced stage of ripening. Besides this indirect effect of color, hormone sprays may exert a direct influence on anthocyanin formation in the fruit. Of the two materials mentioned, 2,4,5-TP seems to have the greater effect on fruit color development. Alar appears to have little effect on color of apples.

If preharvest sprays are applied as generally recommended and the fruit is picked at the normal time, increase of red color is slight and the difference between sprayed and unsprayed fruit is negligible in softness of flesh. However, apples showing a marked increase in red color as a result of a preharvest spray should not be stored for a prolonged period (Teskey and Francis 1954).

Ethephon applied in August significantly increases anthocyanin development and shortens maturation time by as much as 2 weeks. Low rates of ethephon with daminozide are more effective in producing desired fruit color and firmness (Pollard 1974).

The most effective way to obtain the desired red color in apples in a given cultivar is to follow practices in pruning, thinning, and fertilization that promote its development and not depend on growth regulating sprays.

PREHARVEST DROP

Apple trees drop some fruits throughout the entire season from the calyx stage to harvest time. Fruit drop occurs more or less in waves with certain peak periods, especially at the calyx stage, about six weeks after bloom, and, with some cultivars, just before maturity.

Factors Affecting Drop

The extent of fruit drop depends on several factors, such as the cultivar, number of fertile seeds per apple, weather conditions especially following bloom and toward harvest time, and amount of damage to seeds by insects and frost.

The number of seeds is positively correlated with the date of drop of apples. Drop is delayed by the presence of many seeds. Conversely, any factor that reduces the number of seeds tends to increase the amount of preharvest drop.

Applications of N fertilizer and conditions which permit the trees to obtain excessive amounts of nitrates late in the growing season increase preharvest drop. When cultivation is carried on into midsummer or later, more drop results. The relation between cultivation and fruit drop is probably associated with the amount of available soil nitrates (Dickson 1939).

The fact that preharvest drop is usually greater from the more shaded parts of the tree may mean that a high portion of N compared with carbohydrates in the woody tissue tends to hasten fruit abscission.

McIntosh is especially prone to preharvest drop. Some others that may drop their fruit badly are Wealthy, Winesap, Duchess, Delicious, Stayman, Yellow Newtown and Rome.

Loss from Drop

When sound apples drop from the tree before they can be picked financial returns are reduced. Loss from preharvest drop can be very serious, often amounting to 50% or more of the crop and occasionally as high as 95%. In an orchard where the drop was 30% of the yield, the loss was 2.14 kl (243 bu) per 0.4 ha (acre).

Physiology of Drop

The mechanism of leaf and fruit abscission is not clearly understood. It is generally assumed, however, that the abscission layer at the base of the leaf petiole or fruit pedicel plays a role in leaf fall or fruit drop.

Abscission of flowers and immature fruits is preceded by the formation of an abscission layer by secondary cell division. This layer is formed either within the limits of or distal to the abscission zone at the base of the pedicel. Cell separation results from the breakdown of pectic compounds in the middle lamella and primary wall. Although an abscission zone persists in the base of the mature pedicel, this zone does not predetermine the path of abscission. Abscission occurs either in the abscission zone or in the pedicel distal to this zone.

Control of Drop

Success with auxins such as naphthaleneacetic acid and naphthaleneacetamide for preventing preharvest drop of apples was first reported in 1939.

2,4-Dichlorophenoxyacetic Acid (2,4-D).—Now a well-known herbicide, 2,4-D gave good control of preharvest drop of certain cultivars, e.g., Winesap and Stayman, but tended to cause damage to both tree and fruit, and created dangerous drift hazard. The effectiveness of 2,4-D and of NAA as preharvest sprays is compared in Table 1.24.

TABLE 1.24
RELATIVE EFFECTIVENESS OF NAPHTHALENEACETIC ACID AND
2,4–DICHLOROPHENOXYACETIC ACID IN RETARDING FRUIT
DROP OF WINESAP APPLES

Material (10 ppm)	Date Sprayed	Average Injury Rating	Average Accumulated Drop on October:			
			14th %	21st %	25th %	26th %
2,4-D	Aug. 9	2.0	0.1	0.2	0.8	1.7
2,4-D plus lead-cryolite	Aug. 9	3.0	0.1	0.2	0.2	0.4
NAA	Sept. 25	0.0	0.3	0.7	3.4	8.8
Unsprayed		0.0	1.2	10.8	27.3	32.3

Source: Batjer and Thompson (1947).

Naphthaleneacetamide (NAD).—One of the first materials tested for control of fruit abscission, NAD gives good control of dropping, but is generally not so satisfactory as NAA for this purpose (Gardner *et al.* 1939).

2,4,5-Trichlorophenoxyacetic Acid (2,4,5-T).—Compared with other growth regulants, 2,4,5-T is quite effective in controlling preharvest drop of apples. In some areas, 2,4,5-T tends to result in too much injury.

Naphthaleneacetic Acid (NAA).—At 20 ppm, NAA gives good control of drop when applied at the first sign of sound, fully seeded fruit dropping from the tree, or about ten days before picking time. If longer control is wanted, spray 14–20 days before picking and repeat the application 7–10 days later. NAA does not injure the tree at this concentration. NAA gives good drop control up to ten days, but it is not so effective as some other materials in either degree or duration.

2,4,5-Trichlorophenoxypropionic Acid (2,4,5-TP).—This chemical has a longer lasting effect than NAA. When applied at 20 ppm 14 days before NAA, or before sound fruit drop, it controls drop for as long a period as NAA. However 2,4,5-TP tends to hasten maturation and fruits sprayed 2 to 5 weeks before picking may be redder but past mature if left on the trees until normal picking date. 2,4,5-TP gives good control of apple drop provided application and harvesting are properly timed.

Succinic Acid 2,2-Dimethylhydrazide (Daminozide).—The growth regulant (SADH) acts as an abscission control agent and can effectively prevent preharvest drop of fruit when applied 60–70 days before harvest. Daminozide causes a decrease in the metabolic rate; thus, fruit from sprayed trees can be firmer than controls at normal harvest time. Because of its growth retarding properties, daminozide tends to result in smaller fruit; the closer to full bloom that it is applied, the greater is this effect on fruit size. When applied as recommended for preharvest drop control, the effect of daminozide on fruit size is slight (Forshey 1970).

2-Chloroethyl Phosphonic Acid (Ethephon).—Opposite to daminozide in action, ethephon stimulates fruit abscission and tends to counteract the stop-drop effect of hormone sprays previously applied. This characteristic may prove useful in preparing the crop for mechanical harvesting (Edgerton and Blanplied 1968).

Effects of Preharvest Sprays

Auxins used to control preharvest drop may have other effects on the fruit and tree.

Maturation.—Sprays of NAA or 2,4,5-TP may directly stimulate the respiration and ripening rates of apples. Fruits so sprayed should not be left on the tree past maturity if they are to be stored for any length of time. With most late cultivars, if the material is applied within a week of sound fruit drop, and the fruit is harvested at the normal picking date, the effect on maturation is slight and probably not of commercial importance.

However, when 2,4,5-TP is used as a preharvest spray, watch the rate of maturation of summer and fall apples closely. In many instances, it may be of economic value to hasten maturity and color development of apples for an early market. Spot picking of the outside fruit would then seem wise. High temperatures seem to increase the effect on rate of maturation. 2,4,5-TP stimulates the rate of maturation more than NAA if applied a week or more before NAA, but if applied at the same time there is little differnce in results. In Ontario, McIntosh apples treated with 2,4,5-TP were slightly softer at harvest time than untreated fruit, but there was no difference between the sprayed and unsprayed in rate of softening during and after storage (Teskey and Francis 1954).

In Illinois, 2,4,5-TP induced abnormal maturation in Transparent, Lodi, Red Duchess, and Wealthy. Red color developed earlier and in greater amount than in check fruits. But the advantage of greater color was more than offset by the abnormal softness that developed at the same time (Lott and Rice 1955).

Daminozide, a growth retardant, results in fruit that is firmer at harvest time than fruit from unsprayed trees. In this respect it has an advantage over other stop-drop materials for apples designated for long-term storage. Ethephon sprays 6 days before harvest do not appear to affect fruit firmness or other storing properties of apple (Teskey et al. 1972).

Color.—The development of red skin color is sometimes affected by preharvest sprays of 2,4,5-T and 2,4,5-TP. The increased amount of red color is probably associated with more advanced maturity as a result of the auxin sprays, but there may also be a direct stimulation of red pigment development.

2,4,5-TP and 2,4,5-T, singly or in combinations, increase the amount of red color in apples at harvest time in direct proportion to the concentrations employed and to the length of time the fruit is left on the tree. Storage life of fruit so treated is shortened proportionally.

Summer application of daminozide tends to counteract the fruit softening effect of ethephon applied one month before harvest, and results in highly colored, firm fruits.

The effect of preharvest sprays on the development of red color in apples is slight and of doubtful commercial importance if the fruit is harvested at maturity. Stop-drop materials should not normally be used for improving fruit color, especially with apples that are to be stored for any length of time. Choice of red sports along with good cultural practices is the surest and safest method of obtaining good fruit color.

Storage Behavior.—A preharvest spray of NAA or 2,4,5-TP does not adversely affect the storage quality of late fall and winter cultivars of apples if the fruit is harvested at optimum maturity. McIntosh sprayed with 2,4,5-TP 10−14 days before normal picking date may be slightly softer than unsprayed fruit picked at the same time, but there is little difference in shelf-life between sprayed and unsprayed fruit (Franklin *et al.* 1970).

2,4,5-TP resulted in inferior keepability, decreased firmness, stem-end cracking, and deficient flavor of Golden Delicious. These effects were less likely to occur when it was used on trees with fruit which at maturity were normally hard, crisp, and juicy. Drop of Golden Delicious was controlled well with NAA (Lott and Rice 1955).

Fruit from ethephon-treated trees tend to be less firm at regular harvest time but the differences are lost in sorage. Where trees had been previously sprayed with daminozide and the fruit harvested about 20 days after ethephon application, the fruit was well colored, firm and stored well for as long as 4 months (Miller 1975; Hammett 1976).

Preharvest sprays of daminozide result in fruit that is more firm than unsprayed fruit at the normal harvest date. This advantage of firmness persists for some time in storage. There is some evidence that daminozide reduces certain physiological diseases. In British Columbia, apple trees of 5 cultivars were sprayed with 500-2000 ppm daminozide. There was a striking effect on drop prevention with treated fruit still firmer than check fruit harvested at the normal time. Although treated fruits were firmer at harvest date, they were as mature as checks (Fisher and Looney 1967).

In Massachusetts, daminozide applied at 2000 and 5000 ppm to Delicious apple trees in mid-July and mid-August reduced fruit flesh softening prior to harvest, and a firmness difference generally persisted through the storage season. Daminozide delayed the development of water core and reduced the occurrence of internal breakdown but had no effect on storage scald (Lord *et al.* 1967).

Daminozide increased firmness; delayed best harvest date for three cultivars in Ireland, but not for McIntosh in New York; delayed onset of respiratory climacteric; had no effect on post-climacteric respiration rates; delayed ethylene peak; frequently decreased scald; increased core browning in regular but not in CA storage (Blanplied *et al.* 1967).

Flavor.—Any change in concentration of sugars or acids affects flavor. Flavor is also affected by any factor that affects the characteristic aroma. NAA, 2,4,5-TP, daminozide, and ethephon appear to have no effect on flavor.

General Considerations

Apples from trees sprayed with auxins for preharvest drop control should be picked when mature if they are to be stored for any length of time. Since the spray compounds are absorbed by the leaves, the foliage must be healthy if optimum results are to be obtained. Trees suffering from insects, diseases, or spray injury and those low in vigor as a result of malnutrition, drought, or poor aeration do not respond well to preharvest sprays. Warm, bright days favor good results. Cool, dull weather, or too high temperatures between application of the spray and harvest reduce the activity of the leaves and lessen the effectiveness of the spray.

If applied too early the effect does not last till harvest; if abscission of sound fruit has started, preharvest sprays are too late to prevent drop. Too high concentrations make picking difficult, and with some chemicals may have too pronounced an effect on rate of maturation and ripening. Thorough and even coverage is necessary for best results.

HARVESTING AND PACKING

Careless harvesting, handling, packing, and storage practices may decrease the sales value of fruit that otherwise would be of choice quality. The general appearance and condition of apples in retail markets influence the demand. Cuts, bruises, decay, and other blemishes detract from the sales appeal and lessen the chances of apples competing successfully with other fruits on the market.

FIG. 1.21. THIS MECHANICAL HARVESTER FOR PROCESSING APPLES HAS A SHAKER ARM AND A CATCHING FRAME EQUIPPED WITH A BAFFLE SYSTEM TO REDUCE BRUISING

Much of the responsibility for fruit reaching the consumer in good condition lies with the wholesaler and retailer. Nevertheless, it is the producer who decides when the fruit is to be picked, who supervises the harvesting, and much, if not all, of the grading and packing (Franklin *et al.* 1970).

Equipment

All equipment that will be needed should be checked and made ready well in advance of the harvest season.

Ladders.—For semidwarf and standard trees have ready sufficient ladders that are light, easy to handle, strong, and well balanced. Aluminum ladders, because of their strength and lighter weight, are preferred to wood. For the lower trees 2.5 m (8 ft) stepladders are sufficient, but for higher trees some rung ladders, flared at the base and tapered to a point at the top, are also needed. The cost of picking rises steeply as tree height increases beyond 3 m (10 ft). Trees of 6 m (20 ft) are probably uneconomical for handpicking. Therefore, direct pruning toward keeping the tops down to 4.5 m (15 ft) or less.

Picking Containers.—Several types of picking bags, baskets, and pails are used in orchards. The ideal container protects the fruit from bruising and is easy to pick into and to empty. A container with rigid sides gives the most protection to the fruit whereas the canvas-bag type results in the most injury. Draw the canvas tightly across the bottom of canvas drop-bottom containers; if let out to increase its volume, apples at the bottom below the rigid metal sides are bruised. Fruit from dwarf trees can be picked directly into field containers.

Apples bruise when squeezed, bumped, or dropped 2.5 cm. Such bruises may not be immediately apparent but show up later to lower the quality of the fruit.

Field Containers.—Crates with close fitting boards with ends projecting 8 cm above the sides give good protection, stack well, and are the most durable. Crates with slats about 5 cm wide and placed 2.5 cm apart on sides and ends with the top edges flush cause some bruising, especially to tender cultivars like McIntosh. Northeastern apple boxes are similar to the field crate but of lighter construction. Bushel hampers with liners, pads, and lids take up more space than crates or boxes and lack strength for stacking.

Bulk handling of apples is a successful attempt to reduce costs and has pre-empted other methods. Pallet boxes or crates of 3.5, 6.5, or 7 hl (20 bu) capacity, handled by a tractor lift, reduce handling time and labor, reduce bruising, and speed the picking and transfer of the fruit from orchard to cold room. Unloading is accomplished by mechanical dumpers. Pallet boxes outlast bushel (litre) crates, increase storage space, and can

Courtesy of Ontario Dept. Agr. and Food

FIG. 1.22. APPLES IN BULK BINS CAN READILY BE MOVED
AND STACKED USING FORKLIFT EQUIPMENT

be disassembled or nested for off-season storage.

Unless containers are moistened and additional moisture added to the storage room, losses may result from shrinkage and shriveling of the fruit. Moisten the field containers before using them in the orchard.

Picking

Hand Labor.—Provide for adequate help before harvest begins. Year-round employees usually make up the nucleus of the harvest labor force. Added to this is the transient or "floating" labor which is generally available in established orchard areas. The labor needed for harvesting varies with factors such as age and height of trees, amount and quality of crop, and topography of the orchard floor. Under average conditions and standard-size trees, an experienced worker can pick 53 hl (150 bu) a day.

Paid by piecework, a worker harvests 1/3−1/4 more fruit per day, but he tends to treat both trees and fruit more roughly. On a flat rate of so much per hour pickers work more slowly but with less damage to trees and fruit. Under competent management and supervision both systems may work well. In a large orchard a piecework method of payment may be necessary. A "pick-your-own" business alleviates most of the problems of labor shortage but, of course, some other problems are introduced.

Mechanical Harvesting.—Harvesting with hand labor being one of the most costly items in orchard management, considerable interest and research is being directed toward mechanization. Picking costs rise in direct proportion to the size of the tree.

Mechanical aids of several types and designs have been tested. These devices are variations of the principle of moving the pickers along the row and positioning them within reach of the fruit. They may take the form of mobile, multi-level platforms, hydraulically-operated chairs on accordion-type arms; or platforms on hydraulic or pneumatic hoists. Chief drawbacks to the mechanical aids to date are: (a) they are too cumbersome, (b) trees are too wide and too uneven, (c) trees are too far apart, (d) the speed of the picking crew must be timed to the slowest picker.

Harvesting machines that shake the fruit off can be successfully used for apples going for juice or cider, and for other processing apples, providing the fruit is processed almost immediately. Within a few hours, bruises and cuts turn brown and the value of the fruit depreciates.

With dwarf and semidwarf trees growing in hedgerows, a feasible mechanical harvester may very well be developed. Such an overhead harvester might operate on the system of baffles and vibrators (like the grape harvester) or on an entirely different principle such as suction. The New Zealand Lincoln canopy system offers such a new concept. The dwarfed trees are trained to form a canopy. The handling system consists of two low trailers towed in parallel on either side of a row of canopy trees and carrying collecting heads and retractable cascades. Light tractors with offset drawbars are used so that their collecting heads are beneath the canopy (Dunn *et al.* 1976).

FIG. 1.23. APPLES BEING FLOATED INTO THE GRADER FROM THE AUTOMATED DUMPER; WAXING MAY BE DONE IN THE SAME PROCESS

Packinghouse

In determining suitable size and type of packinghouse the operator must estimate how much space he will need for unpacked fruit, packed fruit, empty containers, and the grader. The total floor space required averages about 6 m² per hl (25 ft² per bu) packed per hour. For each 68 kg (150 lb) of fruit marketed provide 0.1 m² (11 ft²) of floor space. Therefore, for marketing 112 tonnes (125 tons) of fruit annually provide about 153 m² (1700 ft²) of packinghouse floor space, as well as some place other than the main floor for storing containers.

The floor area of packinghouses should be readily accessible and of convenient shape. At least two sides of the packing area should be designed to permit easy access for trucks and tractors. Receiving and delivery can be done most rapidly when packinghouse floors and loading docks are at truck and trailer level.

The packinghouse should have adequate lighting and enough equipment to handle the fruit quickly and efficiently. To save time and labor, roller conveyors, hand trucks, or pallet and pallet trucks can be used for transportation within the plant. The pallet and pallet truck method is the most efficient. There should be a good water supply and provision made for disposal of waste fruit and other refuse. Some system of heating is essential if the building is used for late fall or winter packing.

Indices of Maturity

It is important to pick fruit at the proper stage of maturation. Apples picked while immature lack flavor and quality, tend to shrink and shrivel in storage, are subject to disorders such as storage scald, bitter pit, and brown core, and may lack attractiveness. Using a combination of maturity indices may be better than any one alone.

Seed Color.—Most apple cultivars have brown seeds when ready for harvest. This is an unreliable index of maturity, however, because seeds may become brown several weeks before proper picking maturity.

Size of Fruit.—Several factors such as nutrition, thinning, and climate influence the ultimate size of fruit. Therefore, fruit size is of little value in best determining when to pick.

Skin Color.—Surface red color has limited value because its development is associated not only with ripeness but with leaf area per fruit, exposure to light, temperature, and nutritional conditions in the tree. Also, the entire surface of red sports may redden when the fruit is still very immature.

Ground color changes, however, from green to yellow are valuable with some cultivars. In New York, McIntosh apples were at best maturity for storage when the ground color was 2.5–3.0 (U.S. Dept. of Agr. color chart for western apples and pears (Southwick and Hurd 1948).

Firmness of Flesh.—The range in pressure from year to year at the optimum harvest date varies considerably, and apples may remain within this range 10–21 days or more. The variability in pressure-test readings from year to year indicates that the growing season and nutritional conditions influence firmness of flesh as well as maturity of the fruit.

Days from Full Bloom.—In some areas the number of days from full bloom is a fairly reliable index of maturity.

A study of indices of early picking maturity for several cultivars has shown the pressure test, ground color, seed color and starch test to be unreliable. Regardless of locality and seasonal variations, the number of days from bloom to maturity is rather constant from season to season and is a more reliable index of maturity than the other factors.

However, in areas with marked differences in growing seasons from year to year this index may be of little value. Climatic conditions immediately after petal fall are most important in determining the length of the growing season. Based on weather data, computer programs can be developed for predicting with reasonable accuracy correct harvest time (Dilley 1976).

Separation from Spur.—As apples reach maturity they tend to hold less tightly to the tree. If, in picking, many spurs are removed, and stems pull out, the fruit has not reached proper maturity for harvest. However, differences occur among cultivars in this respect. McIntosh, for example, tends to drop before other indices indicate maturity; Cortland often hangs on the spur till the fruit is overripe. Also the N level, moisture supply, and temperature may influence the rate of abscission. Thus, ease of separation of the fruit from the spur is a valuable but not wholly reliable index of maturity.

Calendar Date.—The calendar dates for harvesting any cultivar in a particular area at the optimum stage of maturity vary only 5–10 days over a period of years. Approximate maturity dates may therefore serve as a general guide to the correct times to harvest cultivars, especially for persons lacking experience in this field. A chart indicating the approximate picking dates of the cultivars grown in a particular area and their interrelationship enables the grower to plan future plantings with greater confidence.

Index Number.—The index number represents the quantity of dissociated hydrogen per unit of displaceable hydrogen. The degree of dissociation of hydrogen as represented by the index number rises to a maximum about two weeks before picking maturity and then decreases rapidly. Several factors concerned are not fully understood. However, the time, kind, and method of N fertilization affect the index number.

Starch Test.—The starch content of apples changes to sugars as the fruit approaches maturity. The change begins in the core and progresses toward the skin. A measure of the starch at any given time indicates the degree of maturation of the fruit. It is a reasonably reliable test particularly when used in conjunction with other tests such as the days from full bloom.

The cut apple is placed in an iodine solution made by dissolving 10 g potassium iodide in 75 ml of water and adding 2.5 g of iodine crystals. Dilute the solution to one liter with water. The blue-black color in contrast with the normal white reveals the starch areas (Poapst *et al.* 1959).

Respiration Rate.—The respiration rate of apples varies with the physiological age, and if the relationship can be determined this could be a useful index of maturity.

The respiratory rate as measured by CO_2 and ethylene evolution is directly related to storage life, but the optimum harvest dates have not been found to be associated with any definite position on the climacteric curve. Furthermore, measurement of the climacteric is often difficult with some cultivars in some seasons. Major ethylene production, however, coincides with the beginning of ripening and is thus an indication of maturity (Blanplied 1969; Smith *et al.* 1969).

Soluble Solids Content.—Percentage of soluble solids as an indication of sugar content is an unreliable index of maturity when used alone. However, a range of soluble solids versus firmness is somewhat more useful (Mattus 1966).

Taste.—Tasting a few representative apples may provide information for a decision on picking. This method depends on personal judgement and is not infallible. A cultivar may taste sweet not because the sugar content is high but because the acid content is low.

Experience.—There is no single index that can be used to accurately indicate optimum picking date. The grower decides when to pick by using most of the indices mentioned above based on experience.

Terminology of Fruit Maturation and Ripening

Mature and Its Derivatives.—These terms pertain to the fruit while it is still attached to the plant in accordance with the following definitions (Lott 1945).

Mature.—It is that stage of fruit development which ensures attainment of maximum edible quality at completion of the ripening process. Only fruit which is mature at harvest can attain maximum edible quality (become ripe). It is incorrect to refer to one lot of fruit as being more mature than another; it can only be more nearly mature.

Maturity.—This is the condition of being mature. There is only one stage of maturity, that at which the fruit is mature. Therefore, stages or degrees of maturation should be referred to, not stages or degrees of maturity.

Maturation.—This is the developmental process by which fruit attains maturity. The term should be restricted to the last third or fourth of the interval from bloom to harvest.

Optimum Maturation.—Since the method of handling fruit after harvest affects the rate and degree of the physiological changes, it is desirable to employ this term with a qualifying statement concerning the method of handling and proposed use. For example, there is an optimum stage in the maturation process to pick Jonathan apples if they are to be stored until late December or early January. But if they are to be used soon after harvest for fresh consumption or for processing, quality of the produce can be improved by allowing maturation to proceed to maturity.

Postmature.—This refers to any particular stage in the changes occurring in fruit after it has reached maturity, but is still attached to the tree.

Ripe and Its Derivatives.—These terms pertain to physiological changes and conditions which occur in fruit following harvest in accordance with the following definitions.

Ripe.—This is the condition of maximum edible quality attained by fruit following harvest. Only fruit which becomes mature before harvest can become ripe. Fruit picked when still immature cannot become ripe, but can proceed through the ripening process only to some stage of ripening which will be determined by the degree of maturation at harvest.

Ripening.—This refers to the postharvest physiological process by which the fruit attains ripeness. If the fruit is immature when harvested, ripening can not proceed to ripeness but only to some stage of ripening.

Postripe.—This refers to any particular stage of the change occurring in fruit after it has reached ripeness. It should be used with qualifying terms to indicate the degree of postripeness. It expresses conditions more accurately than overripe.

Packing and Grading

It is of great advantage to apple growers and to the fruit industry as a whole to put up a pack that has both quality and attractiveness. Government legislation protects the good grower and packer as well as the public by setting minimum standards for fruit offered for sale.

Time to Pack.—Whether to pack immediately following harvest or out of storage prior to sale depends largely upon the storage facilities available, time and method of marketing, and labor supply.

Packed fruit can be stored in less space than "orchard run" fruit; poor grades and culls can be disposed of; apples are firmer at harvest time and may suffer less from handling. On the other hand, unpacked apples can be moved into storage more quickly and therefore may keep longer; a smaller packinghouse and smaller grading equipment are required when packing is spread over a longer period; the packer has the opportunity to cull out any deteriorated apples just prior to sale.

Grading Equipment.—Several types of mechanical graders are available. The size and type of equipment most suitable depends largely on the amount, kinds, and cultivars of fruit to be handled.

A grader is an assembly line where each crew member has his own particular job. It mechanically conveys the fruit from the feed belt to the sorters and through sizing devices to the packers.

Most graders size the fruit by one of the following methods: (1) An endless belt with openings of gradually increasing size through which the apple falls when its diameter is less than that of the opening. (2) The fruit is conveyed in individual cups which release it into the packing bins when the weight of the apple exceeds that of a set of counter balances. (3) The apples are spun off an endless conveyer belt into packing bins when their size is great enough that they contact one of a series of revolving wheels or spools adjusted at desired heights from the belt. (4) Sizing takes place between a pair of diverging belts which act as conveyors and sizers. As the belts move forward, their distance apart increases and eventually the fruit supported by them falls through into receiving trays. The latter is one of the best and most widely used types of graders, being simple, low priced, compact, and efficient. It lacks accuracy when fed indiscriminately with fruit which is not spherical.

Packing.—Standards for federal, state, and provincial grades consider condition, freedom from blemishes, color, and size. Containers carrying packed fruit must be clearly marked as to cultivar, grade, and name of packer.

Containers used for apples include boxes and crates holding approximately 0.4 hl (1 bu). The Northeastern apple crate has a capacity of about 0.5 hl, the Northwestern apple box, 0.4 hl, and they are designed for high quality, hand-wrapped, and hand-packed apples. Bushel (litre) baskets, mostly so-called tub baskets with stiff sides, are used in some areas. Their chief disadvantages are that, when stacked, more bruising to the fruit results, and they occupy more space than boxes or crates.

Any container is potentially a consumer package but generally only those of less than 35 litre (1 bu) capacity are used for this purpose. Paper cartons used for apples may be uncompartmented, have one or more horizontal partitions, or have individual compartments for each fruit. Their use as master cartons for small packages may reduce handling costs and bruising.

Consumer package baskets are usually 0.09−0.18 hl (1/4−1/2 bu) in size or woven or veneer baskets with handles and holding 3.8−15 litres (4−16 qt). Polyethylene bags of 1.4, 2.3, and 4.5 kg (3, 5, and 10 lb) capacities are widely used. They are well adapted to prepacking, have sales appeal, and tend to keep low grade fruit off the market, but bruising increases.

STORAGE OF APPLES

Apples remain alive even after they are picked from the tree. The length of postharvest life of an apples varies with the stage of maturation at picking time, length of time the fruit remains at ordinary temperatures, nutritional level of its tree, season, and the cultivar.

The greater the rate of respiration, the shorter is the storage life. Refrigeration slows down the rate of respiration and thus prolongs the life of the apple. This is accomplished either by holding fruit at a low temperature, or by low temperature combined with decreased oxygen. Low temperatures also slow enzyme activity and inhibit the growth of rot organisms.

Removing Field Heat

Remove the field heat from apples to be stored and lower their temperature soon after harvesting. The greater the time lag between picking and placing in temperature-controlled storage the shorter the potential life.

Even when apples are placed immediately in cold storage after picking it may be some time before the temperature of the fruit is reduced to that of the storage atmosphere. Unless proper provision is made for complete air circulation in the storage, fruit in the center of the stack may not reach storage temperature for days or weeks.

Precooling.—It is important to quickly reduce the fruit to, or close to, storage temperature before storage or transport. Precooling slows down the life processes quickly, thereby prolonging storage life. Precooling of apples often is accomplished in a separate room or chamber.

If a precooler (based on the principle of heat extraction by rapid movement of cold air over the produce) is used, the rate of cooling of the apples largely depends on the type of package and method of stacking. For maximum efficiency both factors should permit heat removal from the apples at the greatest possible rate. Containers should let a maximum of air come in contact with the fruit, and they should be stacked so that air can circulate freely among them. They should not be too close to the walls or to each other.

The rapid air circulation method of precooling is not suitable for holding fruit in storage because of the drying effect of the moving air. But, if the operation is carefully controlled, the field heat can be removed from a load by precooling in a few hours with negligible damage from drying of the product.

The disadvantage of precooling in a separate chamber is the double handling and higher costs involved. A common compromise is to use the storage room itself for precooling. This requires a larger cooling capacity. Rapid cooling is usually accomplished either by increasing air velocity or by reducing the temperature of the air from the cooler.

Temperature and Humidity

The main consideration in storage is temperature. Low temperature slows the rate of respiration as well as enzymatic action and activity of organisms. Transpiration is inversely proportional to the percentage of relative humidity of the storage atmosphere (Table 1.25). With most apple cultivars the longest storage life is obtained at -1°C to 0°C (30° to 32°F) and RH of 85—90%.

In commercial storages, 10—25% extension in marketing period may result by storing at 2°C below the standard 0°C. Differences of 2.6°C in storage temperatures markedly slow down the rate of quality loss. But -1°C is relatively close to the average freezing point of apples (-2°C) and accidental freezing might occur. The effect on quality of lower temperatures may be nullified by the onset of core flush.

TABLE 1.25
TRANSPIRATION OF GOLDEN DELICIOUS APPLES SUBJECTED
TO VARYING HUMIDITY

RH%	Nov.	Dec.	Jan.	Feb.	March	Total
			Transpiration Loss (%)			
95	0.59	0.55	0.51	0.50	0.48	2.63
90	1.06	0.99	0.87	0.80	0.77	4.49
85	1.41	1.55	1.07	1.00	0.91	5.54
75	2.18	1.59	1.33	1.30	1.20	7.60
70	2.60	1.87	1.43	1.40	1.29	8.59

Source: Phillips *et al.* (1955).

Inspect the apples frequently so that they may be removed and sold while still in good condition and with sufficient shelf-life to permit marketing. Ripening and deterioration are greatly accelerated by the higher temperatures met with after removal from storage. Ripening at 5°C is about twice as fast as at 0°C; at 10°C, about twice as fast as at 5°C; and at 21.1°C, about twice as fast as at 10°C.

Types of Storage

There are three main types of storages used commercially for long-term storage of apples. The objectives in each case are the same: to slow down the life processes of the product; prevent shriveling and shrinkage of the product; inhibit development of rot organisms.

TABLE 1.26
EFFECT OF POSTHARVEST TEMPERATURE ON
POTENTIAL LIFE OF APPLES

Days from Picking to Placing in Storage with Field Temperature at 21°C	Storage Temperature (°C)	Potential Life (Days)
0	0	150
5		75
10		0
0	4.4	75
5		37
10		0

Source: Phillips *et al.* (1955).

Common (Nonrefrigerated).—There is no refrigeration unit in the common storage. It makes use of the differences between night and day temperatures. The insulated building is equipped with a differential thermostat that registers inside and outside temperatures. When the temperature outside is a degree or two below that of the storage room, a fan near the ceiling is activated, forcing warm air out through a louvered opening. Simultaneously cooler air is sucked in through a similar opening near the floor. The action continues until the inside and outside temperatures are the same.

If the temperature outside lowers to freezing, an automatic safety cut-off switch inactivates the system. A heating system prevents the inside temperature from going below freezing and a humidifier is usually required to maintain the necessary high humidity. The common storage is relatively cheap and functions efficiently. In a warm fall, however, the common storage may not reach a satisfactory low temperature in time for the first of the apple crop.

Refrigerated.—This type of storage is just a very large-sized refrigerator. The refrigerant is usually ammonia or one of the freons. The evaporator may take many forms but is commonly placed near the ceiling in the completely insulated room.

In the case of jacketed storage, the evaporator is outside of the room and the cooled air is circulated around the outside of the storage room in the space between the walls of the storage and the insulated jacket. The advantage claimed for this system is that condensation in the storage room and within the insulation in the walls and ceiling is eliminated.

The storage may have a humidifier and a filter to remove volatiles given off by the respiring fruit.

Controlled-atmosphere (CA).—This is a refrigerated, gastight room. After the load of apples is put in and the room sealed, the oxygen level is slowly reduced and the percentage of CO_2 increases because of respiration. If the room were left sealed and no fresh air added, the percentage of oxygen would theoretically reach zero with subsequent damage to fruit from anaerobic respiration and from the high CO_2 levels, but the percentages of these gases are controlled by ventilation aided by a scrubber. The scrubber may be calcium or sodium hydroxide, water, or activated carbon and molecular sieves.

Yellow Newtown apples had good texture and were free of internal browning and scald after storage for 25 weeks at 5.5°C in 10% oxygen and 10% carbon dioxide; those stored at 5.6° to 7.2°C in normal air for the same period were overripe and severely scalded. McIntosh apples have been kept until July 1 with complete freedom of brown core by the use of controlled-atmosphere (Lott 1945).

Several types of atmospheric generators have been introduced which generate directly the desired atmosphere and deliver it to the room under slight pressure (Smock and Blanplied 1965).

TABLE 1.27
NORMAL AND MAXIMUM COLD STORAGE PERIOD FOR CERTAIN
IMPORTANT APPLE CULTIVARS AND THEIR SUSCEPTIBILITY
TO STORAGE DISORDERS

Variety	Storage Period Normal Months	Maximum Months	Tendency to Storage Scald	Other Disorders Likely to Occur in Storage
Gravenstein	0–1	3	Slight	Bitter pit, Jonathan spot
Wealthy	0–1	3	Slight	Soft scald, Jonathan spot
Grimes	2–3	4	Severe	Soggy breakdown, bitter pit, shriveling
Jonathan	2–3	4	Slight	Jonathan spot, water core, soft scald, internal breakdown
McIntosh	2–4	4–5	Slight	Brown core, soft scald
Cortland	3–4	5	Medium	Breakdown
R.I. Greening	3–4	6	Severe	Bitter pit, internal breakdown
Golden Delicious	3–4	6	Very slight	Shriveling, soggy breakdown
Delicious	3–4	6	Slight to medium	Bitter pit, water core, internal breakdown
Stayman	4–5	5–6	Severe	Internal breakdown, water core
York	4–5	5–6	Severe	Bitter pit
Arkansas	4–5	6	Severe	Bitter pit, water core
Northern Spy	4–5	6	Slight	Bitter pit
Baldwin	4–5	6–7	Medium to severe	Bitter pit
Rome	4–5	6–7	Medium to severe	Bitter pit, brown core, soft scald, internal breakdown, Jonathan spot
Ben Davis	4–5	8	Medium	
Yellow Newtown	5–6	8	Slight	Bitter pit, internal browning (in California)
Winesap	5–7	8	Medium	Water core, shriveling

Source: Wright *et al.* (1954).

Although CA storage has proved successful in many instances, it has certain disadvantages, including the expense of making the storage airtight and the difficulty of maintaining proper concentrations in the atmosphere. Also, such a storage must be kept closed until all fruit is removed and can not be entered from time to time during the storage period. The exception is when the room must be entered for repairs in which case the CA is re-created afterwards with an attending time penalty added to the required storage period. Cultivars respond differently to CA storage and the same storage conditions do not suit all cultivars (Dewey *et al.* 1969; Liu 1977).

Sealed Package.—Atmosphere modifications within the package have been accomplished by using lining material of limited permeability (sealed liners) such as polyethylene, in which CO_2 increases and O_2

TABLE 1.28
EFFECTS OF PREHARVEST ALAR AND ETHEPHON TREATMENTS ON C₂H₄
PRODUCTION (μLITRES/KG·HR) OF MCINTOSH APPLES AT 20°C AFTER
STORAGE FOR 152 DAYS AT 0.5°C

Treatment	Days at 20°C				
	1	4	6	10	11
E800−6	100.5 ab[1]	144.6	136.0 a	114.8 abc	105.5 ab
E400−6	109.1 a	146.9	137.8 a	118.3 ab	111.1 a
AE800−6	66.6 c	85.6	81.1 b	82.1 b	78.6 c
AE400−6	57.6 c	83.8	84.8 b	91.6 cd	86.1 c
AE800−3	79.0 bc	96.8	91.9 b	92.7 cd	85.3 c
AE400−3	69.9 c	77.2	78.5 b	87.2 d	83.8 c
Alar	75.6 bc	103.3	91.9 b	95.8 abcd	91.0 bc
Control	100.9 ab	135.9	129.6 a	120.5 a	110.1 a

Source: Teskey *et al.* (1972).
[1]*abcd* Significance calculated within days. Means followed by the same letter are not significantly different ($P < 0.05$).

decreases as a result of respiration of the fruit.

Successful results have been obtained with several apple and pear cultivars stored in heavy gauge polyethylene liners at 0°C and 4.4°C. Yellow Newtown, Rome, Grimes, and Jonathan stored for six months remained greener, firmer, and developed less scald than those stored at the same temperature without liners. The atmosphere in the sealed film liner boxes on removal from cold storage was 8−12% O_2 and 3−6% CO_2. No impairment of flavor or increase in decay was encountered (Hardenburg and Siegelman 1957).

Factors Affecting Storing Quality

Nutritional Effect.—Certain disorders of stored fruit are caused by nutritional disturbances. Fertilizer treatments, particularly N, have an influence on physiological disorders. In general, apples from trees of low vigor tend to store better than those from more vigorous trees, but it would be unwise to lower the thriftiness of the trees and reduce yield and fruit size in order to gain better storage behavior. When the N level in McIntosh and Northern Spy leaves exceeds 2% (dry weight) the storage quality of the fruit is impaired.

Increases in N fertilizer result in decreased firmness of fruit at harvest time and in retardation of normal development of yellow ground color. There is not always a direct correlation between amount of N applied and effect on firmness or ground color, but there is almost always an inverse correlation between leaf N and fruit firmness and ground color. Soluble solids in the fruit are reduced by the higher N levels (Smock and Boynton 1944).

P has little effect on storage behavior. The effect of K seems to be related to the ratio between K and N. The greater the N to P ratio, the poorer is the quality and storage behavior of the apples. N to K ratio over 1.25 reduces quality.

Weather.—Storage quality within a cultivar differs greatly, depending on such factors as the locality where grown, seasonal conditions in the locality, cultural and spray practices, maturity when picked, care in handling, time between picking and storage, and storage practices. As an example of how cultivars vary when grown in different localities, McIntosh apples grown in the Middle Atlantic States mature as early fall apples and are suitable for only 2–3 weeks of cold storage; when grown in New England and Canada, they mature late and can be stored as long as 4–5 months. In the same locality, storage disorders vary with growing conditions during different years. The amount of storage scald varies from year to year. Severe scald has been associated with ample soil moisture during the growing season. When picked in an early stage of maturation, apples are more subject to scald, whereas those picked late are more likely to develop breakdown in storage. High moisture and low temperature during the maturation period of McIntosh may be conducive to development of brown core.

Apples vary from year to year in both eating and keeping quality, sometimes due to weather. High susceptibility in brown core in McIntosh is associated with low solar radiation and low mean temperatures during the last six weeks of the growing season. Scald susceptibility in McIntosh and Rhode Island Greening is associated with high mean temperatures during the last six weeks of the growing season (Smock 1953).

The influence of climatic conditions on storage behavior of apples is complex but extremes in any direction may be harmful. When weather during the growing season, and particularly toward harvest time, has deviated from normal, maintain a careful watch on apples in storage.

Preharvest Sprays.—McIntosh sprayed with 20 ppm 2,4,5-TP or NAA are redder and softer than the controls at harvest time but there is no difference between the 2 groups in rate of change during or after storage. The difference between sprayed and unsprayed fruit in this respect is slight if both are picked at the proper time. Apples left on the tree after they have reached maturity should not be stored as long as those picked at or slightly before maturity. If the delay in harvest afforded by these sprays extends beyond maturity, storage quality may be adversely affected.

Storage life of fruit is extended commercially through control of environmental factors, most of which directly influence the rate of respiration. Maleic hydrazide appears to have the ability to decrease respiration. However, preharvest sprays of 150, 300, and 450 ppm on Delicious trees did not extend the storage life of the fruit. Daminozide-treated fruit is firmer at normal harvest date than unsprayed fruit and may tend to remain somewhat more free of storage disorders (Batjer 1964).

Storage Volatiles.—Interest in the chemical composition of the odoriferous substances in apple arises chiefly from two sources: (a) their usefulness as flavoring agents, and ((b) the detrimental physiological effects of their accumulation in fruit storages. Flavors and volatile substances have been removed from apple tissue and a large number of compounds have been identified: CO_2; amyl esters of formic, acetic, caproic, and caprylic acids; propionic and butyric acids; also acetaldehyde, acetone, methanol, ethanol, butanol, caproaldehyde, and 2-hexanol. Ethylene evolution is correlated with respiration rate, and so the storage life of apples can be analyzed by gas chromatography (Lougheed *et al.* 1969). The composition of storage volatiles differs from that occurring in apple juice (Table 1.29).

TABLE 1.29
COMPARISON OF THE ESTIMATED COMPOSITION
OF APPLE STORAGE VOLATILES WITH THE
COMPOSITION OF APPLE JUICE VOLATILES

Compound Class	Apple Storage Volatiles %	Volatiles from Apple Juice %
Acids	1.4	—
Alcohols	0.05	92
Esters	68	2
Carbonyls	17	6
Others	13	—

Volatiles produced by one lot of apples can increase the ripening rate of another lot (Kidd and West 1932). Storage scald is caused by an accumulation of the volatiles emanating from the fruit. Fruit warehousemen have realized the need for a practical means of removing volatile materials from storage room atmospheres.

Storage Disorders

Scald.—This is probably the most serious disorder of apples in storage. In mild cases of scald small areas of the surface on the unblushed side are lightly tinted with brown and the skin remains firm; in more severe cases larger areas of the surface become dark brown and slightly depressed. Sometimes the browning extends into the flesh a short distance. Scald differs from rot in that it affects a considerable area of the apple to a rather uniform but shallow depth. Scalded tissue is usually not so soft and watery as decayed tissue. Delayed harvest has reduced the danger of storage scald on several cultivars.

Scald is more prevalent on large apples than small ones of the same cultivar; immature apples are more susceptible than those picked later; the green or unblushed side of the fruit is mainly affected. Any orchard practice or condition which favors late maturity of the fruit tends to increase the danger of scald with susceptible cultivars. But, there is some indication that preharvest hormone sprays such as Alar and certain fungicides may reduce scald during storage (Porritt and Meheriuk 1968).

In storage, scald is caused by the action of volatile substances given off by the respiring apples. The disorder is thus aggravated by poor air circulation and lack of ventilation in the storage. CA storage permits accumulation of the volatiles causing scald.

Cool scald-susceptible cultivars quickly and hold at the lowest practical temperature. Apples may develop a potentiality for scald before browning actually appears but scald develops rapidly on such apples after removal from cold storage. Test the fruit from time to time for scald development after storage. Apples liable to scald should be held at 8°–15°C during their shelf life.

Several materials have been used as postharvest dips and sometimes as preharvest sprays in attempts to control apple scald. Diphenylamine is the most effective but also most likely to cause injury. Ethoxyquin is recommended for sensitive cultivars like G. Delicious (Hardenburg 1963).

When the room air is circulated through activated coconut-shell carbon, some volatile emanations are removed from the fruit. Other ethylene absorbents, e.g., "Purafil," may be more effective. This air purification treatment may reduce scald and retard ripening.

Ozone introduced in the storage room atmosphere at the rate of 1–2 ppm is effective as a deodorizing agent and in controlling growth of molds on packages and on walls, but it does not control decay and scald. Concentrations as high as 3.25 ppm increase decay and impair flavor.

Core Flush (Brown Core, Stem Cavity Browning, Core Browning).— This originates in the core area. It is materially lessened in McIntosh apples by storing at 4°–5°C. However, the storage life of the fruit is shorter than when stored at 0°C, unless controlled atmospheres are employed.

Best eating quality (flavor and texture) of McIntosh apples was reached at about 3 months' storage at 2°C. Reducing temperatures to as low as 0°C had little influence. This phase was characterized by maximum chlorophyll content, slowing down of soluble pectin accumulations, maximum ethylene content, possible peak of CO_2 fixation rates, leveling of pressure test trends, and stabilization of extractable juice values. It was followed by quality loss (Phillips *et al.* 1955).

Internal Browning, Soft Scald, and Soggy Breakdown.—These are among other disorders induced by low storage temperatures. Yellow Newtown, Rhode Island Greening, Delicious, McIntosh, and Jonathan sometimes are stored at 2°–4°C to avoid the internal browning that may occur when stored at below 1°C. Storage for 8 weeks at 1°C before lowering the temperature to -1°C has controlled soft scald of Delicious. Soggy breakdown and soft scald are more likely to occur when storage is delayed. When these injuries occur the higher storage temperature and early disposal of the fruit are advised.

"Spot" Disease.—This physiological disorder, peculiar to Northern Spy held in cold storage, is probably the same as Jonathan Spot. It is checked by CA storage at 5°C. The symptoms are circular-to-ovate, brown, small, sunken lesions usually located on the stem end and red or blushed portion of the fruit.

Bitter Pit.—Common to Northern Spy and Delicious, bitter pit sometimes occurs in Gravenstein, Grimes, Rhode Island Greening, York, Rome, and Yellow Newtown. Large fruits from lightly laden trees and

FIG. 1.24. BITTER PIT ON NORTHERN SPY

fruits picked at an immature stage are more susceptible than are medium-sized fruits from trees with a good crop picked at the proper time. Bitter pit may not develop until after fruit is stored. It develops more rapidly in common storage than in cold storage and less rapidly at 0°C than at 5°C.

Bitter pit susceptibility is increased by any orchard treatment that increases leaf osmotic values at the expense of fruit osmotic values. Heavy N applications, heavy thinning, heavy shading, and delayed storage increase susceptibility. CA storage, waxing of fruit before storage, and high relative humidity in storage delays or checks bitter pit on susceptible apples.

Sprays of $CaCl_2$ at 590 g (5 lb) to 100 litres (100 gal.) of water about mid-July have given positive results in some cases. Daminozide also shows some promise for bitter pit control (Drake *et al.* 1966; Faust and Shear 1968).

Watercore.—The cause of watercore is in the orchard, when abnormally high growing temperatures are accompanied by erratic moisture conditions. The disorder is one that occurs internally without external symptoms. Affected fruits have a higher specific gravity than normal fruits and can be detected by flotation method. Measuring of light transmittance through intact fruits has provided a means of detecting watercore and other internal disorders. Watercore usually decreases as the storage period lengthens, and if not severe it may entirely disappear from firm cultivars, such as Winesap and Yellow Newtown. Watercore in cultivars such as Jonathan, Delicious, Stayman, and Rome often develops into breakdown, and, if this condition is found, market the fruit early (Batjer and Williams 1966; Francis *et al.* 1965).

Fumigating the Storage

Methyl bromide has been used to control rodents and insects such as the Japanese beetle in fruit storages. Fumigation with methyl bromide may cause injury to some cultivars, e.g., Williams, Wealthy, Delicious.

Fumigation is best delayed until apples have been stored a few weeks after harvesting since the storage life may be shortened if the fumigant is used before the climacteric rise in respiration has begun.

INSECTS AND DISEASES

Sanitation is a general control practice for pests in the orchard. Destroy hawthorns and other wild fruit trees on the property, since these are hosts for many orchard pests. Prune sufficiently to facilitate proper spraying. Remove dead or dying trees and limbs. Store firewood and containers in a mothproof place. If props must be left in the orchard spray them with insecticide. Remove culls from the orchard or bury culls and pulp. Refer to the local spray calender for specific controls. Read all labels carefully.

Insects

Codling Moth.—The adult is grayish-brown with 19 mm (3/4 in.) spread and a golden-brown area near the apex of each front wing. The white or pinkish larva or "worm," 19 mm (3/4 in.) long, is found in the fruit near the core. A second type of injury, "stings," is caused by larvae eating through the skin and into the apple a short distance.

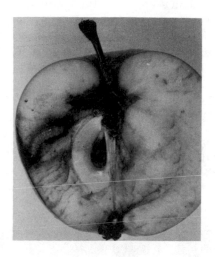

FIG. 1.25. CODLING MOTH LARVA

Bud Moth.—This pest causes some growers great trouble, although it may be of little importance in nearby orchards. The grayish-brown adult has a 13 mm wing spread with a grayish-white irregular area on each forewing. The brown larva which has a black head and is about 10 mm long bores into buds and continues to feed on opening leaves. The most serious injury is to the fruit, especially in the fall.

Fruit Tree Leaf Roller.—The adult is a rusty brown moth with nearly 2.5 cm (1 in.) wingspread. The larva, a slender geen caterpillar with black or dark brown head, feeds on the opening leaves and eats deep gouges in the young fruit.

Red Banded Leaf Roller.—The moth has a reddish oblique band on the front wings. The larva is about the same size as the fruit tree leaf roller but is pale in color with a yellowish head. The first brood feed on the underside of leaves and later, under the protection of a leaf or where apples are in clusters, on the small apples; second and sometimes third broods cause the most severe damage. The apples they attack have an irregular area of skin eaten from the fruit. All apple cutivars are subject to attack but Cortland and Rhode Island Greening are especially susceptible.

Apple Maggot.—The adult fly is slightly smaller than the house fly, with a black mark on each wing. The maggot is so small until the apple is overripe that it is seldom seen; when fully grown it is about 5 mm long. Infested apples have winding streaks of brown, corky flesh. The only evidence from the outside may be tiny pinprick-sized punctures. If the maggots wander through the flesh near the skin, brown depressed areas may appear. Heavily infested fruit is deformed. Some of the most susceptible cultivars are Yellow Transparent, Wealthy, and Fameuse. Northern Spy is moderately susceptible and McIntosh more resistant.

Plum Curculio.—The adult is a dark brown snout bettle about 6 mm long with a strong curved snout. The larva is a grub that develops in the skin. The female deposits the egg in a hole eaten in the skin of the fruit above a crescent-shaped cut, causing a typical D-shaped scar. The adult feeds on the fruit in the fall, eating cavities beneath holes in the skin.

FIG. 1.26. APPLE MAGGOT INJURY

Green Apple Aphid.—This is the most important aphid of apple. Aphids suck the sap from the leaves, particularly near the growing tips, and cause a sooty fungus on the fruit. Avoid excessive tree growth and remove watersprouts in early summer.

Rosy Apple Aphid.—Early injury is caused to the leaves and newly-set apples. Injured apples are dwarfed, wrinkled, and hang on the trees in clusters after normal fruit is ripe. Dormant sprays of dinitro cresol kill the eggs.

Woolly Apple Aphids.—These attack bark and roots, and are covered with a white waxy material. Clusters of them resemble little tufts of wool. They cause swellings on the wood which may rupture.

European Red Mite.—The adult is reddish-brown, unspotted, and just visible to the naked eye. The eggs are reddish and large compared with the adult mite. Mites suck the sap from the leaves and give a bronzed appearance. Tree growth and size of crop are reduced. Delicious, Yellow Transparent, Northern Spy, Wealthy, and Early McIntosh are highly susceptible. Certain insecticides that have destroyed beneficial parasites and predators have been a major factor in the increase of this pest.

The 2-Spotted Mite and 4-Spotted Mite.—These closely related species are similar in appearance, customs, and control. They are generally not so important as the European red mite.

Oystershell Scale.—The adult is about 3 mm long, the color of the tree bark, and resembles the shell of an oyster. If the infestation is severe, scales occur on the fruit and petioles.

Lecanium Scale.—This large, brown, convex scale is usually found on the underside of twigs, feeds on leaves and twigs, and secretes a honeydew on which an unsightly black sooty fungus grows.

San Jose Scale.—About the diameter of a pinhead, this flat, circular scale is ashy brown, with a raised central area.

Buffalo Tree Hopper.—The adult is about 9.5 mm long, greenish, and in form suggests a buffalo. The female injures young trees when she

FIG. 1.27. BUFFALO TREE HOPPER

deposits eggs in two rows under slits made in the bark. Heavily injured trees become dwarfed and stunted. Spray trees, cover crop, and fence rows; avoid alfalfa as a cover crop.

Round-headed Apple Tree Borer.—This is a large beetle with two white lines along the back. The larva, a white grub 2.5 cm (1 in.) long with a brown head, bores into the trunk near or just beneath the ground. Remove mountain ash, hawthorn, wild apple, and cherry within 550 m of the orchard.

Tarnished Plant Bug and Apple Red Bug.—These bugs suck the juices from leaves and fruit. The punctured areas of the fruit do not develop normally and depressions occur.

Apple Leafhopper.—This insect is of less general importance than it was before the use of modern insecticides. Both nymphs and adults cause white stippling on the foliage by sucking the sap.

Other Insects.—There are species that are usually controlled by the regular spray program but occasionally become serious pests in the apple orchard, e.g., tent caterpillar and cankerworm, which at times may reach outbreak proportions; green fruitworm, which sometimes eats large holes in the fruit; and shothole borer, which may spread from dead and dying wood to healthy trees in a heavy infestation. Rose chafer, casebearer, fall webworm, Japanese beetle, and apple skeletonizer are usually of minor importance.

Diseases

Apple Scab (*Venturia inaequalis*).—This is the most serious disease of apples in many areas. Besides damaging the current apple crop, heavy foliage infection weakens the trees and so reduces future crops. The fungus overwinters on old leaves and infects the opening leaves in spring.

Resistance to scab has been found in a number of different species and selections of apple (Lamb and Hamilton 1969).

Fire Blight (*Erwinia amylovera*).—A bacterial disease, fire blight is often serious on apples and pears (see Pears).

Apple Rust.—The disease infects the fruit, and sometimes the foliage. The fungus causing this disease (*Gymnosporangium juniperi virginianae*) spends part of its life on red cedar. Eradicate red cedar trees within 0.8 km or as an alternative, apply protective sprays used for scab control.

TABLE 1.30
APPROXIMATE HOURS OF WETTING REQUIRED FOR PRIMARY ASCOSPORE
LEAF-SCAB INFECTION IN THE ORCHARD

Temp (°C)	Light (hr)	Infection Moderate (hr)	Heavy (hr)	Temp (°C)	Light (hr)	Infection Moderate (hr)	Heavy (hr)
25.6	13	17	26	11.7	12	17	25
25.0	11	14	21	11.1	12	18	26
24.4	9.5	12	19	10.6	13	18	27
17–24	9	12	18	10.0	14	19	29
16.7	9	12	19	9.4	14.5	20	30
16.1	9	13	20	8.9	15	20	30
15.6	9.5	13	20	8.3	17	23	35
15.0	10	13	21	7.8	19	25	38
14.4	10	14	21	7.2	20	27	41
13.9	10	14	22	6.7	22	30	45
13.3	11	15	22	6.1	25	34	51
12.8	11	16	24	5.5	30	40	50
12.2	11.5	16	24	0.6–5.0	More than 2 days		

Source: Mills and LaPlante (1951).

Black Rot.—This is a fungus disease (*Sphaeria malorum*), known also as "canker," on the fruit as "blossom-end rot," and on the foliage as "leaf spot," "brown rot," or "frog-eye." Rhode Island Greening and Baldwin are highly susceptible.

FIG 1.28. APPLE SCAB

Sooty Blotch (*Gloeodes pomigena*) **and Fly Speck** (*Microthyriello rubi*).—These two often appear together on the fruit. Both are controlled by the regular summer spray program for scab. The common names of these diseases describe their appearance.

Brooks Fruit Spot (*Mycosphaerella pomi*).—Known also as "Phoma fruit spot," the disease causes some loss in certain areas, particularly to Jonathan, Baldwin, Grimes, Rome, Tolman, Sweet, and Stayman. The spots, which seldom exceed 4.8 mm, are often inconspicuous at picking time.

Powdery Mildew (*Podosphaera leucotricha*).—This is often a serious problem in the nursery and in areas where it is favored by climatic conditions. It results in dwarfed foliage and stunted growth (Pierson *et al.* 1971).

REFERENCES

ANDERSON, H.W. 1956. Diseases of Fruit Crops. McGraw-Hill Book Co., New York.

ARCHIBALD, J.A. 1966. Orchard soil management. Ontario Dept. Agr. Food Publ. *457*.

BAKER, C.E. 1949. Effects of different methods of soil management on the potassium content of apple leaves. Proc. Am. Soc. Hort. Sci. *42*, 7-10.

BATJER, L.P. 1954. Effect of 2,4,5-trichlorophenoxypropionic acid on maturity and fruit drop of apples. Proc. Am. Soc. Hort. Sci. *64*, 215-221.

BATJER, L.P. 1964. Progress report on the use of B-Nine on fruit trees. Wash. State Hort. Assoc. Proc. *60*, 27-28.

BATJER, L.P., FORSHEY, C.G., and HOFFMAN, M.B. 1968. Effectiveness of thinning sprays as related to fruit size at times of spray application. Proc. Am. Soc. Hort. Sci. *92*, 50-54.

BATJER, L.P., and THOMPSON, A.H. 1947. Studies with 2,4-dichloro-phenoxycetic acid sprays in retarding fruit drop of Winesap apples. Proc. Am. Soc. Hort. Sci. *49*, 45-48.

BATJER, L.P., and WESTWOOD, M.N. 1960. 1-Napthyl N-methyl carbamate, a new chemical for thinning apples. Proc. Am. Soc. Hort. Sci. *75*, 1-4.

BATJER, L.P., and WILLIAMS, M.W. 1966. Effects of N-dimethyl amino succinamic acid (Alar) on watercore and harvest drop of apples. Proc. Am. Soc. Hort. Sci. *88*, 76-79.

BECKENBACH, J.H., and GOURLEY, J.A. 1932. Effects of different cultural methods upon root distribution of apple trees. Proc. Am. Soc. Hort. Sci. *29*, 202-204.

BENSON, N.R., and COVEY, R.P., JR. 1976. Response of apple seedlings to zinc fertilization and mycorrhizal inoculation. HortScience *11*, 252-253.

BILLERBECK, F.W. 1955. Influence of 2,4,5-trichlorophenoxypropionic acid on the color of red-fruited apple varieties. Proc. Am. Soc. Hort. Sci. *61*, 170-180.

BISHOP, R.C., and KLEIN, R.M. 1975. Photo-promotion of anthocyanin synthesis in harvested apples. HortScience *10* (2): 126-128.

BLAIR, D.S. 1950. Apple growing. Can. Dept. Agr. Publ. *847*.

BLANPLIED, G.D. 1969. A study of the relationship between optimum harvest dates for storage and the respiratory climacteric rise in apple fruits. J. Am. Soc. Hort. Sci. *94*, 177-179.

BLANPLIED, G.D., SMOCK, R.M., and KOLLAS, D.A. 1967. Effect of Alar on optimum harvest dates and keeping quality of apples. Proc. Am. Soc. Hort. Sci. *90*, 467-474.

BLASBERG, C.G. 1953. Response of mature McIntosh trees to urea foliar sprays. Proc. Am. Soc. Hort. Sci. *62*, 147-153.

BOYNTON, D. 1954. Apple nutrition. *In* Fruit Nutrition. N.F. Childers (Editor). Horticultural Publications, New Brunswick, N.J.

BOYNTON, D. *et al.* 1950. Responses of McIntosh apple orchards to varying nitrogen fertilization and weather. Cornell Univ. Agr. Expt. Sta. Mem. *290*.

BRADT, O.A. 1955. Pruning methods for fruit trees. Rept. Hort. Res. Inst. Ontario, 22-30.

BRAMLAGE, W.J., and THOMPSON, A.H. 1962. The effects of early-season sprays of boron on fruit set, color, finish and storage life of apples. Proc. Am. Soc. Hort. Sci. *80*, 64-72.

BRASE, K.D. 1956. Propagating fruit trees. N.Y. State Agr. Expt. Sta. Bull. *773*.

BROWN, G. 1926. Apple thinning in the Hood River Valley. Oregon Circ. *76*.

BUKOVAC, M.J., and MITCHELL, A.E. 1962. Biological evaluation of 1-naphthyl N-methylcarbamate with special reference to the abscission of apple fruits. Proc. Am. Soc. Hort. Sci. *80*, 1-10.

BURRELL, A.B. 1940. Boron deficiency disease of apple. Cornell Univ. Agr. Expt. Sta. Bull. *428*.

CHANDLER, W.H. 1954. Cold resistance in horticultural plants. Proc. Am. Soc. Hort. Sci. *64*, 552-572.

COOPER, R.E., and THOMPSON, A.H. 1972. Solution culture investigation of the influence of Mn, Ca, B and pH on internal bark necrosis of apple trees. J. Am. Soc. Hort. Sci. *97* (1): 138-141.

CRANDALL, P.C. 1955. Relation of preharvest sprays of maleic hydrazide to the storage life of Delicious apples. Proc. Am. Soc. Hort. Sci. *65*, 71-74.

DAYTON, F.D. 1963. Distribution of red color in the skin of apple cultivars of McIntosh parentage. Proc. Am. Soc. Hort. Sci. *82*, 51-55.

DEVYATOV, A. *et al.* 1976. Sand as mulch in young apple orchards. Fruit Sci. Rept. (Poland) *3* (4): 41-42.

DEWEY, D.H., HERNER, R.C., and DILLEY, D.R. 1969. Controlled atmosphere for the storage and transport of horticultural crops. Mich. State Univ. Hort. Rept. No. *9*, 126-129.

DICKSON, G.H. 1939. Dropping of McIntosh apples. Sci. Agr. (Canada) *19*, 712-721.

DICKSON, G.H., and UPSHALL, W.H. 1948. Some effects of shortening the annual cultivation period in an apple orchard. Rept. Hort. Res. Inst. Ontario *1947*, 26-29.

DILLEY, D.R. 1976. Apple harvest and storage notes. 95th Ann. Rept. Mich. State Hort. Soc. 45-50.

DONOHO, C.W. 1968. Relationship of date of application and size of fruit to the effectiveness of NAA for thinning apples. Proc. Am. Soc. Hort. Sci. *92*, 55-62.

DRAKE, M. *et al.* 1966. Bitter pit as related to calcium level in Baldwin apple fruit and leaves. Proc. Am. Soc. Hort. Sci. *89*, 23-29.

DUNN, J.S. *et al.* 1976. Mechanical raspberry harvesting and the Lincoln canopy system. Am. Soc. Agr. Eng. Winter Meet. Paper *76-1543*.

EAVES, C.A. 1964. Effect of chemical and physical properties of lime on scrubbing efficiency. Mass. Agr. Expt. Sta. Bull. *422*, 16-21.

EDGERTON, L.J., and BLANPLIED, G.D. 1968. Regulation of growth and fruit maturation with 2-chloroethane phosphonic acid. Nature *219*, 1064-1065.

EDGERTON, L.J., and GREENHALGH, W.J. 1969. Regulation of growth, flowering and fruit abscission with 2-chloroethanephosphonic acid. J. Am. Soc. Hort. Sci. *94*, 11-13.

EDGERTON, L.J., and HATCH, A.H. 1969. Preparation of cherries and apples for mechanical harvesting. Proc. N.Y. State Hort. Soc. *114*, 23-27.

EDGERTON, L.J., and HOFFMAN, M. B. 1965. Some physiological responses of apple to N-dimethyl amino succinamic acid. Proc. Am. Soc. Hort. Sci. *86*, 28-31.

EDGERTON, L.J., HOFFMAN, M.B., and FORSHEY, C.G. 1964. Effects of some growth regulators on flowering and fruit set of apple trees. Proc. Am. Soc. Hort. Sci. *84*, 1-6.

FAUST, M., and SHEAR, C.B. 1968. Corking disorders of apples; a physiological and biochemical review. Botan. Rev. *34*, 441-464.

FISHER, A.G., EATON, G.W., and PORRITT, S.W. 1976. Internal bark necrosis of apple in relation to soil pH and leaf Mn. Can. J. Plant Sci. *57*, 297-299.

FISHER, E.G., and KWONG, S.S. 1961. Effect of potassium fertilization on fruit quality of McIntosh apple. Proc. Am. Soc. Hort. Sci. *78*, 16-23.

FISHER, D.V., and LOONEY, N.E. 1967. Growth, fruiting and storage responses of five cultivars of bearing apple trees to N-dimethylaminosuccinic acid (Alar). Proc. Am. Soc. Hort. Sci. *90*, 9-19.

FORSHEY, G.H. 1963. Potassium-magnesium deficiency of McIntosh apple trees. Proc. Am. Soc. Hort. Sci. *83*, 12-20.

FORSHEY, G.H. 1970. The use of Alar on vigorous McIntosh apple trees. J. Am. Soc. Hort. Sci. *95*, 64-67.

FRANCIS, F.J. *et al.* 1965. Detection of water core and internal breakdown in Delicious apples by light transmittance. Proc. Am. Soc. Hort. Sci. *87,* 78-84.

FRANKLIN, E.W. *et al.* 1970. Harvesting, storing and packing apples. Ontario Dept. Agr. Food Publ. *431.*

GARDNER, F.E. *et al.* 1939. Spraying with plant growth substances for control of preharvest drop of apples. Proc. Am. Soc. Hort. Sci. *37,* 415-428.

GERHARDT, F., and SCHOMER, H.A. 1954. Film liners for boxes of pears and apples. Pre-Pack-Age *7,* 14-17.

HALLER, M.H. 1954. Apple scald and its control. U.S. Dept. Agr. Bull. *1380.*

HAMMETT, L.K. 1976. Ethephon influence on storage quality of Delicious apples. HortScience *11* (1): 57-59.

HARDENBURG, R. 1963. Controlling CO_2 concentrations within sealed poly-ethylene-lined boxes of apples. Proc. Am. Soc. Hort. Sci. *82*, 83-91.

HARDENBURG, R., and SIEGELMAN, H. 1957. Effects of polyethylene box liners on scald, firmness, weight loss, and decay of stored apples. Proc. Am. Soc. Hort. Sci. *69*, 75-83.

HARLEY, C.P. 1941. Physiological factors associated with flower bud initiation in the apple. Proc. Am. Soc. Hort. Sci. *38*, 91-92.

HARLEY, C.P. *et al.* 1957. Effects of the additive Tween 20 on apple thinning by naphthalene acetic acid sprays. Proc. Am. Soc. Hort. Sci. *69*, 21-27.

HEENEY, H.B. *et al.* 1964. Zinc deficiency in Eastern Ontario orchards. Can. J. Plant Sci. *44*, 195-200.

HEINICKE, D.R. 1966. Characteristics of McIntosh and Delicious apples as influenced by exposure to sunlight. Proc. Am. Soc. Hort. Sci. *89*, 10-13.

HOFFMAN, M. B. 1939. Preharvest drop of mature McIntosh apples as influenced by applications of nitrogen fertilizers. Proc. Am. Soc. Hort. Sci. *37*, 438-442.

HOFFMAN, M.B., and EDGERTON, L.S. 1952. Comparison of NAA, 2,4,5-T and 2,4,5-TP for controlling harvest drop of McIntosh apples. Proc. Am. Soc. Hort. Sci. *59*, 225-230.

HOFFMAN, M.B. *et al.* 1955. Naphthaleneacetic acid and naphthaleneacetamide for thinning apples. Proc. Am. Soc. Hort. Sci. *65*, 63-70.

HOLUBOWICS, T., and BOE, A.A. 1970. Correlation between free amino acid content of apple seedlings treated with gibberellic acd and abscisic acid. J. Am. Soc. Hort. Sci. *95*, 85-88.

KENWORTHY, A.L. 1967. Plant analysis and interpretation for horticultural crops. *In* Soil Testing and Plant Analysis, Part II. M. Stelly (Editor). Soil Science Society of America, Madison, Wisconsin.

KETCHIE, D.O., and BEEMAN, C.H. 1973. Cold acclimation in apple trees under natural conditions during four winters. S. Am. Soc. Hort. Sci. *98* (3): 257-261.

KIDD, F., and WEST, C. 1925. The course of respiratory activity throughout the life of an apple. Gt. Brit. Dept. Sci. Ind. Rept. *1924,* 27-32.

KIDD, F., and WEST, C. 1932. Effects of ethylene and apple vapours on the ripening of fruits. Gt. Brit. Dept. Sci. Ind. Res. Food Invest. Board Rept. *1931,* 18–23.

KIDD, F., and WEST, C. 1945. Respiratory activity and duration of life of apples. Plant Physiol. *20,* 467-504.

KRAUS, E.J., and KRAYBILL, H.R. 1918. Vegetation and reproduction. Oregon Agr. Expt. Sta. Bull. *149.*

LAMB, R.C., and HAMILTON, J.M. 1969. Environmental and genetic factors influencing the expression of resistance to *Venturia inaequalis* in apple progenies. J. Am. Soc. Hort. Sci. *95,* 554-557.

LASKO, A.N., and MUSSELMAN, R.C. 1976. Effects of cloudiness on interior diffuse light in apple trees. J. Am. Soc. Hort. Sci. *101* (6): 642-644.

LATIMER, L.P., and PERCIVAL, G.P. 1947. Comparative value of sawdust, hay, and seaweed as mulch for apple trees. Proc. Am. Soc. Hort. Sci. *50,* 23-30.

LEUTY, S.J. 1971. Chemical thinning of tree fruits. *In* Fruit Production Recommendations. Ontario Dept. Agr. Food Publ. *360.*

LEVIN, I.R., ASSAF, R., and BRAVDO, B. 1972. Effect of irrigation treatment for apple trees on water uptake from different soil layers. J. Am. Soc. Hort. Sci. *97* (4): 521-526.

LIU, F.W. 1977. Varietal and maturity differences of apples in response to ethylene in CA storage. J. Am. Soc. Hort. Sci. *102* (1): 93-95.

LOONEY, N.E. 1968. Light regimes within standard size apple trees as determined spectrophotometrically. Proc. Am. Soc. Hort. Sci. *93,* 1-6.

LORD, W.J., GREEVE, D.W., and DAMON, R.A., JR. 1975. Evaluation of fruit abscission and flower bud promotion by ethephon and SADH. Proc. Am. Soc. Hort. Sci. *100* (3): 259-261.

LORD, W.J., SOUTHWICK, F.W., and DAMON, R.A., JR. 1967. Influence of N-dimethylamino succinamic acid on flesh firmness and on some pre- and post-harvest physiological disorders of Delicious apples. Proc. Am. Soc. Hort. Sci. *91,* 829-832.

LOTT, R.V. 1945. Terminology of fruit maturation and ripening. Proc. Am. Soc. Hort. Sci. *46,* 166-172.

LOTT, R.V., and RICE, R. 1955. Effect of preharvest sprays of 2,4,5-TP on maturation of apples. Illinois Agr. Expt. Sta. Bull. *588.*

LOUGHEED, E.C., CLARKE, K.A. and TESKEY, B.J.E. 1975. Harvesting, storing and packing apples. Ontario Min. Agr. Food Publ. *431.*

LOUGHEED, E.C., FRANKLIN, E.W., and SMITH, R.B. 1969. Ethylene analyses by automatic gas chromatography. Can. J. Plant Sci. *49,* 386-391.

LOUGHEED, E.C. *et al.* 1973. Firmness of McIntosh apples as affected by Alar and ethylene removal from the storage atmosphere. Can. J. Plant Sci. *53*, 317-322.

MACDANIELS, L.R., and HOFFMAN, M.B. 1941. Apple blossom removal with caustic sprays. Proc. Am. Soc. Hort. Sci. *38*, 86-88.

MACLAREN, D. 1971. Water trees with trickle irrigation. Can. Fruitgrower *27* (7): 6-7.

MAGNESS, J.R. 1939. Correlation of fruit color in apples to nitrogen content of leaves. Proc. Am. Soc. Hort. Sci. *37*, 39-42.

MARTSOLF, J.D. *et al.* 1975. Effect of white latex paint on temperature of fruit tree trunks. J. Am. Soc. Hort. Sci. *100* (2): 122-129.

MASON, J.L. 1964. Yield and quality of apples grown under four nitrogen levels. Proc. Am. Soc. Hort. Sci. *85*, 42-47.

MATTUS, G.E. 1966. Maturity standards for Delicious. Am. Fruit Grower *36* (8): 16.

MCDONNELL, P.F., and EDGERTON, L.J. 1970. Some effects of CCC and Alar on fruit set and fruit quality of apple. HortScience *5*, 89-91.

MEYER, B.S., ANDERSON, D.B., and BOHNING, R.H. 1960. Plant Physiology. Van Nostrand Reinhold Co., New York.

MICHELSON, L.F. *et al.* 1969. Response of McIntosh apple to differential placement of nitrogen and potassium fertilizers. HortScience *4*, 249-250.

MILLER, S.R. 1975. Color, firmness, starch content and persistence of 2-chloroethylphosphonic acid in McIntosh apples. Can. J. Plant Sci. *55* (4): 1001-1006.

MILLS, W.D., and LAPLANTE, A.A. 1951. Diseases and insects in the orchard. Cornell Univ. Agr. Expt Sta. Bull. *711*.

MURNEEK, A.E. 1947. Pollination and fruit setting. Missouri Agr. Expt. Sta. Bull. *379*.

MURNEEK, A.E. 1954. 2,4,5-T as a preharvest spray for apples. Proc. Am. Soc. Hort. Sci. *64*, 209-214.

NEBEL, B.R. 1940. Longevity of pollen in apple, pear, plum, peach, apricot and sour cherry. Proc. Am. Soc. Hort. Sci. *37*, 130-132.

OBERLEY, G.H., and FORSHEY, C.G. 1969. Cultural practices in the bearing apple orchard. Cornell Univ. Agr. Expt. Sta. Ext. Bull. *1212*.

OVERHOLSER, E.L., and OVERLEY, F.L. 1940. Effect of time of nitrogen on the responses of Jonathan apples. Proc. Am. Soc. Hort. Sci. *37*, 81-84.

PHILLIPS, W.R., and ARMSTRONG, J. G. 1967. Handbook on storage of fruits and vegetables. Can. Dept. Agr. Publ. *1360*.

PHILLIPS, W.R. *et al.* 1955. Effect of temperature near 32°F on the storage behavior of McIntosh apples. Proc. Am. Soc. Hort. Sci. *65*, 214-222.

PIERSON, C.F. *et al.* 1971. Market diseases of apples, pears and quinces. U.S. Dept. Agr., Agr. Handbook *376*.

POAPST, P.A., WARD, G.M., and PHILLIPS, W.R. 1959. Maturation of McIntosh apples in relation to starch loss and abscission. Can. J. Plant Sci. *39*, 257-263.

POLLARD, J.E. 1974. Effects of SADH, ethephon and 2,4,5-TP on color and storage quality of "McIntosh" apples. J. Am. Soc. Hort. Sci. *99* (4): 341-343.

PORRITT, W.W., and MEHERIUK, M. 1968. Application of chemicals for control of apple scald. Can. J. Plant Sci. *48*, 495-500.

PROCTOR, J.T.A. 1974. Color stimulation in attached apples with supplementary light. Can. J. Plant Sci. *54* (3): 499-503.

PROCTOR, J.T.A. *et al.* 1975. The penetration of solar global radiation into apple trees. J. Am. Soc. Hort. Sci. *100* (1): 40.

RAESE, J.T. 1976. Sprout control of apple and pear trees with NAA. HortScience *10* (4): 396-398.

RAESE, J.T., and WILLIAMS, M.W. 1974. Relationship between fruit color of G. Delicious apples and N content and color of leaves. J. Am. Soc. Hort. Sci. *99* (4): 332-334.

ROBERTS, R.H., and STRUCKMEYER, B. 1948. Notes on pollination. Proc. Am. Soc. Hort. Sci. *51*, 54-60.

SALSBURY, F.B., and ROSS, C. 1969. Plant Physiology. Wadsworth Publishing Co., Belmont, Calif.

SHANNON, L.M. 1954. Internal bark necrosis of Delicious apple. Proc. Am. Soc. Hort. Sci. *64*, 165-174

SHAW, J.K. 1943. Description of apple varieties. Mass. Agr. Expt. Sta. Bull. *430*.

SHEAR, C.B. 1972. Incidence of corkspot as related to Ca in the leaves and fruit of apples. J. Am. Soc. Hort. Sci. *97* (1): 61-64.

SHIM, K.K. *et al.* 1973. The upward and lateral translocation of urea supplied to roots of apple trees. J. Am. Soc. Hort. Sci. *98* (6): 523-525.

SHOEMAKER, J.S. 1926. Pollen development in the apple with special reference to chromosome behaviour. Botan. Gaz. *81*, 148-172.

SHOEMAKER, J.S., and TESKEY, B.J.E. 1962. Practical Horticulture. John Wiley & Sons, New York.

SHUTAK, V.G., and CHRISTOPHER, E.P. 1956. Storage scald control by orchard applications of mineral oil. Proc. Am. Soc. Hort. Sci. *67,* 80-81.

SIMONS, R.K. 1962. Anatomical studies of the bitter pit area of apples. Proc. Am. Soc. Hort. Sci. *81*, 41-50.

SIMONS, R.K. 1970. Phloem tissue development in response to freeze injury to trunks of apple trees. J. Am. Soc. Hort. Sci. *95*, 182-190.

SMITH, R.B., LOUGHEED, E.C., and FRANKLIN, E.W. 1969. Ethylene production as an index of maturity for apple fruits. Can. J. Plant Sci. *49*, 805-887.

SMOCK, R.M. 1953. Effects of climate during the growing season on keeping quality of apples. Proc. Am. Soc. Hort. Sci. *62*, 272-277.

SMOCK, R.M., and BLANPLIED, G. 1965. Effect of modified technique in CA storage of apples. Proc. Am. Soc. Hort. Sci. *87,* 73-77.

SMOCK, R.M., and BOYNTON, D. 1944. Effect of differential nitrogen treatments in the orchard cn the keeping quality of McIntosh apples. Proc. Am. Soc. Hort. Sci. *45,* 77-86.

SOLECKA, M.H., PROFIC, H., and MILLIKAN, O.F. 1969. Some biochemical effects in apple leaf tissue associated with the use of simazine and amitrole. J. Am. Soc. Hort. Sci. *94,* 55-57.

SOUTHWICK, L. 1938. Relation of seeds to preharvest McIntosh drop. Proc. Am. Soc. Hort. Sci. *36,* 410-412.

SOUTHWICK, F.W., and HURD, M. 1948. Harvesting, handling, and packing apples, N.Y. State Agr. Expt. Sta. Bull. *750.*

STILES, W.C. 1964. Influence of calcium and boron sprays on York spot and bitter pit of York Imperial apples. Proc. Am. Soc. Hort. Sci. *84,* 39-43.

TESKEY, B.J.E. 1968. Grafting fruit trees. Ontario Dept. Agr. Food Publ. *439.*

TESKEY, B.J.E. 1976. Compatibility and pollination of fruit trees. Ontario Min. Agr. Food Publ. *76-013.*

TESKEY, B.J.E., and FRANCIS, F.J. 1954. Color changes in skin and flesh of stored McIntosh apples sprayed with 2,4,5-TP. Proc. Am. Soc. Hort. Sci. *63,* 220-224.

TESKEY, B.J.E., LEUTY, S.J., and BRADT, O.A. 1977. Thinning tree fruits. Ontario Min. Agr. Food Publ. *77-021.*

TESKEY, B.J.E., and KUNG, S.D. 1967. Some effects of carbaryl on two apple cultivars. Can. J. Plant Sci. *47,* 311-318.

TESKEY, B.J.E., PRIEST, K.L., and LOUGHEED, E.C. 1972. Effects of Succinic acid, 2,2-dimethyl hydrazide (alar) and 2-chloroethylphosphonic acid (ethephon) on abscission and storage quality of McIntosh apples. Can J. Plant Sci. *52,* 483-491.

TESKEY, B.J.E., and SHANMUGANATHAN, M. 1971. Effects of low concentrations of 2-chloroethylphosphonic acid on CO_2 and ethylene evolution from N. Spy apples. Can. J. Plant Sci. *51,* 73-78.

TESKEY, B.J.E., and WILSON, K.R. 1975. Tire fabric waste as mulch for trees. J. Am. Soc. Hort. Sci. *100* (2): 153-157.

TESKEY, B.J.E. *et al.* 1973. Pruning and training fruit trees. Can. Dept. Agr. Publ. *1513.*

TOWNSEND, G.F. *et al.* 1958. Use of pollen inserts for tree fruit pollination. Can. J. Plant Sci. *38,* 39-40.

TUKEY, H.B. 1952. Uptake of nutrients by leaves and branches. Rept. 13th Intern. Hort. Congr. London, Sept. 8-15, Royal Hort. Soc.

UPSHALL, W.H. 1970. North American Apples: Varieties, Rootstocks, Outlook. Mich. State Univ. Press, East Lansing.

VERNER, L. 1938. Effect of a plant growth substance on crotch angles in young apple trees. Proc. Am. Soc. Hort. Sci. *35*, 415-422.

WANDER, I.W. 1943. Effect of heavy mulch in an apple orchard upon several constituents and the mineral content of the foliage and fruit. Proc. Am. Soc. Hort. Sci. *42*, 1-6.

WANDER, I.W. 1947. Calcium and phosphorus penetration in an apple orchard. Proc. Am. Soc. Hort. Sci. *49*, 1-6.

WAYWELL, C.G. *et al.* 1977. Guide to chemical weed control. Ontario Dept. Agr. Food Publ. *75*.

WEEKS, W.D., and LATIMER, P. 1938. Incompatibility of Early McIntosh and Cortland apples. Proc. Am. Soc. Hort Sci. *36*, 284-286.

WEEKS, W.D., and SOUTHWICK, F.W. 1957. Relation of nitrogen fertilization to annual production of McIntosh apples. Proc. Am. Soc. Hort. Sci. *68*, 27-31.

WEEKS, W.D. *et al.* 1958. Effect of varying rates of nitrogen and potassium on composition of foliage and fruit color of McIntosh. Proc. Am. Soc. Hort. Sci. *71*, 11-19.

WEIR, T.E., STOCKING, C.R., and BARBOUR, M.G. 1970. Botany. John Wiley & Sons, New York.

WILLIAMS, M.W., and BATJER, L.P. 1964. The site and mode of action of 1-naphthal N-methylcarbamate in thinning apples. Proc. Am. Soc. Hort. Sci. *85*, 1-10.

WOODBRIDGE, C.G. *et al.* 1971. Boron content of developing pear, apple and cherry fruit buds. J. Am. Soc. Hort. Sci. *96* (5): 613-615.

WRIGHT, R.C. *et al.* 1954. Commercial storage of fruits, vegetables, and florist and nursery stocks. U.S. Dept. Agr. Handbook *66*.

ZIELINSKI, Q.B. 1955. Modern Systematic Pomology. W.C. Brown Co., Dubuque, Iowa.

2

Pears

Together with all the other deciduous tree fruits discussed in this book, the pear belongs to the family Rosaceae. It is very closely related botanically to the apple with which it was placed in the same genus by Linnaeus. The pear cultivars commonly grown commercially and discussed in this text are the occidental pears *Pyrus communis*, the common pear of Europe and Western Asia. Like the apple, the pear fruit is classed botanically as a pome (see under Apples). Each of the leathery carpels contains 2 seeds and the receptacle and ovary wall become fleshy to constitute the edible portion of the fruit (Weir *et al*. 1971).

Pears are widely grown in the Eastern United States, in Ontario, and Nova Scotia; but about 90% of the pear production of the United States and Canada, 10.5 million hl (30 million bu), is west of the Rocky Mountains. California leads with 4.6 million hl, followed by Washington with about 2.5 million hl and Oregon with 1.8 million hl. The total western production is over 9 million hl (25 million bu). The North Central states contribute about 0.5 million hl of which nearly 255,000 hl are grown in Michigan. Nearly 255,000 hl of pears are produced in the North Atlantic states; about 211,000 hl in New York. Pear production in the South Atlantic section is about 281,000 hl while British Columbia, Ontario and the South Central region each produce about 0.5 million hl. Europe grows about 30 million hl of pears per year. Other important pear-growing countries are Australia, New Zealand, South Africa, Argentina, and Japan.

PROPAGATION

Pear cultivars are propagated in the nursery by techniques which, in general, have been described for the apple.

Rootstocks

The pear cultivars Anjou, Bartlett, Bosc, Comice, Seckel and Packham's Triumph were grown on 9 rootstocks and compared for tolerance to pear decline, tree size, density of bloom, yield, fruit weight and leaf nutrient content. The results indicate considerable variation among different cultivar/rootstock combinations. No single rootstock can be recommended for all situations, but each limiting condition should be considered before choosing a rootstock (Lombard and Westwood 1976).

European Pear.—The most widely used stock for standard trees is the European pear (*Pyrus communis*), commonly called French pear. It is vigorous, tolerates both considerable drought and excessive soil moisture, and gives satisfaction on a wide range of soils. It is also highly resistant to oak root fungus.

"French roots" now used are mostly seedlings from Bartlett seeds, obtained from canneries. Bartlett seeds give good germination and seedling vigor, and the seedlings have good affinity with nearly all cultivars. Certain other cultivars, such as Hardy and Winter Nelis, make good seedling rootstocks. The necessary afterripening period for the seed is 60–90 days at 0°–5°C. The optimum is 4.5°C.

Pyrus nivalis (snow pear) is widely distributed throughout Europe where it has limited use for perry (pear cider). In North America it has some possible value as a rootstock.

Seeds of species from warm winter climates require less chilling than those from colder climates, and the optimum chilling temperatures are higher (6.7°–10°C) than for the latter. Temperatures below freezing are relatively ineffective in breaking rest in all species.

Pear trees, generally, require 1200–1500 hr of chilling at or below 6.7°C to break rest. However, there are cultivar differences and influences transitted to the scion cultivar from the root. Also, the chilling requirement can be modified by the time at which it occurs, sunlight, fog, cloud, wind, and the tree condition. Dormancy may be prolonged in southern climates and result in a prolonged bloom period, dead flower buds, fewer flower buds, delayed foliation, and reduced vigor (Brown *et al.* 1967).

In Germany and Belgium, pear seedlings planted on old pear orchard sites that had been fumigated with DD and chloropicrin rooted and grew better than trees in untreated soil (Husbado 1969).

Japanese Pear.—Japanese pear (*P. serotina*) is more resistant to blight than European pear, but in California and parts of the northwest it promotes black end or hard end. This disease deforms and blackens the apex of the fruit, and thereby renders it unsalable. On soils that are not the best for tree growth, the loss from black end is often more pronounced than on open, fertile soils where the trees grow well. Inarching trees on Japanese roots with French seedlings to avoid black end has not proved worthwhile nor has the use of resistant cultivars as interstocks.

Other Oriental Pears.—Suitability of *P. calleryana* as a stock for cold areas is questionable; it is more subject to winter injury than French root. *P. ussuriensis* is very hardy but is not immune to blight and trees propagated on it are subject to black end of the fruit. *P. betulaefolia, P. serrulata, P. pyrifolia* (sand pear) and *P. pashia* are some others that have been tried as rootstocks for pears (Tables 2.1 and 2.2).

Quince.—*Cydonia oblonga* is used as a dwarfing rootstock for pear (see Chap. 3).

Old Home.—Old Home, which is very resistant to blight, may produce blight-resistant seedlings and therefore furnish a satisfactory stock. Old Home can be substituted for Hardy as the intermediate stock between

TABLE 2.1
PEAR DECLINE OF 6 PEAR CULTIVARS AS INFLUENCED BY ROOTSTOCK

	Average Pear Decline Rating						
Stock System	Anjou	Bartlett	Bosc	Comice	Packham's Triumph	Seckel	Average
Old Home	1.9	1.9	1.9	2.7	1.5	2.1	1.9
Old Home/Quince A	2.2	2.2	1.9	2.6	1.9	2.2	2.1
Old Home/Winter Nelis	1.7	1.9	1.9	2.6	1.7	2.0	1.8
Old Home/Bartlett Sdlg	1.4	2.0	1.6	2.7	1.9	2.0	1.8
Winter Nelis Sdlg	1.5	1.9	1.7	1.6	1.7	2.1	1.8
Bartlett Sdlg	1.3	1.9	1.4	1.7	2.0	2.3	1.8
P. calleryana	2.0	1.9	1.6	1.5	1.2	2.3	1.8
P. nivalis	1.4	2.5	2.0	1.8	1.6	1.9	1.9
P. betulaefolia	—	—	—	—	—	1.9	1.9

Source: Lombard and Westwood (1976).

quince roots and pear scion cultivars; thus, it furnishes resistance below ground as well as in the trunk and scaffold branches. Old Home does not give any increased resistance to blight to the cultivar grafted on it but it does ensure a resistant trunk and framework. Some decline and fire

TABLE 2.2
EFFECTS OF ROOTSTOCK AND CHILLING TIME ON AVERAGE
GROWTH OF BARTLETT AND *P. CALLERYANA* PEAR TREES

| | | | Average Shoot Growth[1] | | | |
| | | | Days After Planting | | | |
Scion/Rootstock	No. of Trees	(5°C) Chilling (hr)	30 (cm)	60 (cm)	90 (cm)	Total[2] (cm)
Bartlett/own root	2	372	0	0.5	0.5	0.5
	2	744	0	0.2	0.5	0.5
	2	1130	0	0.3	1.0	1.0
P. calleryana/own root	3	372	4.5	9.0	38.0	41.3
	2	744	4.0	32.0	43.5	48.0
Bartlett/*P. calleryana*	2	372	0	0	0	0
	2	648	0.2	1.0	6.5	11.0
	2	744	3.5	13.0	15.5	16.0
	2	1130	5.0	37.0	47.0	47.5

Source: Westwood and Chestnut (1964).
[1]The longest three shoots were measured on each tree.
[2]Average growth per shoot at the time terminal buds set.

blight clonal rootstock selections of Old Home × Farmingdale have shown more vigor than Old Home rootstocks while others are much less vigorous (Westwood *et al.* 1976).

Uncongeniality.—Kieffer seems to lack good congeniality with French or domestic pear seedlings. Premature autumnal coloration of the leaves is related to small size of tree and poor quality of the fruit. These characteristics may be due to nutritional difficulties brought about by a poor graft union.

A good plan is to eliminate all red-leaved, comparatively small, Kieffer trees from the nursery. Some trees with poor unions may show little evidence of stress in the nursery, but under less favorable conditions show signs of uncongeniality in the orchard. Replace such trees early in the life of the orchard.

Cuttings

Pear can be propagated with some degree of success from both hardwood and softwood cuttings under mist providing they are first treated with a rooting compound such as indole-3-butyric acid. Rooting is better when hardwood cuttings are callused while buds are still in the rest period. Removal of all buds on hardwood cuttings inhibits rooting and when buds show pronounced development their promotive influence on root formation is strong (Westwood and Ali 1966).

In British Columbia, treating roots of pear nursery trees with indole-3-butyric acid prior to planting stimulated production of new roots (Looney and McIntosh 1968).

In many countries, scion rooting of pear trees is one method of preventing decline of trees grafted on uncongenial rootstocks. In Israel the percentage of scion-rooted trees was improved by ringing the trees above the graft union, and further improved by covering the wound with lanolin impregnated with indole-3-butyric acid. Results were better still when the treated wound was covered with moist vermiculite. The effect of ringing on scion rooting was greatly diminished when the wound healed rapidly by the use of specific wound dressings rather than the lanolin mixture (Gur *et al.* 1968).

CULTIVARS

Pear cultivars vary greatly in size, form, quality, and other characters. Also, considerable variation may occur within a cultivar between areas.

Anjou.—Large, late pear of good quality; stores and ships well. Tree only fairly productive, good pollinizer. Origin: Belgium, 1823.

Bartlett.—Accounts for about 75% of the pear production in United States and Canada. Fruit of high quality for dessert, canning, and drying. Tree well adapted to a very wide variety of soil and climatic conditions. England (where it is called Williams), 1770.

FIG. 2.1. RUSSET BARTLETT PEARS

Russet Bartlett.—The same as Bartlett except for the russet skin color. The fruit can be left on the tree longer, and remains edible longer than Bartlett. Ontario, 1931; a bud sport of Bartlett.

Red Bartlett.—A bud sport of Bartlett which produces fruit of an attractive bright red color. In size, form, and flesh the fruit is typical of its parent. Washington state, 1938.

Bosc.—Large, bell-shaped, russeted fruit, covered with slight indentation; high quality for dessert and canning; matures 3–4 weeks after Bartlett. France, 1835.

Clapp (Clapp's Favorite).—Large, attractive, well colored; very good quality, early. Massachusetts, 1871.

Comice.—Large, roundish pyriform, often lightly russeted and blushed; late; quality and flavor good, especially in the Pacific Northwest. France, 1800.

Flemish Beauty.—Very attractive, high quality. Tree hardy. Susceptible to scab. Flanders, 1810.

Giffard.—An early pear of good quality; green with red dots and red marbling. France, 1840.

Hardy.—Late. Used for grafting on quince root as an intermediate stock, and to a small extent for its fruit which is of good quality. Tree fairly vigorous and productive, especially on quince root. France, 1820.

FIG. 2.2. BARTLETT PEARS SHOWING CHIMERA

Howell.—A large, yellow waxen pear of good quality; stores and handles well. Connecticut, 1830.

Kieffer.—Medium size; poor dessert, fair canning quality; matures 6–8 weeks later than Bartlett. Tree very productive, widely adaptable, hardy, and disease resistant. Pennsylvania, 1863.

Magness.—Medium size, good quality, a week after Bartlett. Blight resistant. Seckel × Comice. U.S. Dept. Agr. Maryland, 1960.

Moonglow.—Large, attractive, good quality, 2 weeks before Bartlett; stores 2 months. Tree blight resistant. U.S. Dept. Agr. Maryland, 1960.

Old Home.—Not grown for fruit but provides a blight resistant framework and a compatible union with quince. California, 1890.

Ovid.—Large; resembles Bartlett except for russet patches; fine-grained, tender, sweet; late. Tree resembles Bartlett. New York, 1878.

Seckel.—Small, brownish-green; high quality; ripens two weeks after Bartlett. Tree vigorous, hardy, and productive; succeeds best on dwarfing roots. Pennsylvania, 1784.

Winter Nelis.—Small, attractive, high quality; keeps very well in cold storage. Tree is difficult to prune and manage; widely distributed on the Pacific Coast. Belgium, 1880.

Identification from Tree Characters

Pear cultivars can be identified from distinct characteristics of the tree. This is of great advantage in the identification of nursery stock and of cultivars whose fruits are very similar. Some useful characteristics in this respect follow (Upshall 1926).

Red Shoots.—*Clairgeau and Duchess (a Comparison).*—Little difference in growth habits and bark characters; Clairgeau leaves are narrower, longer-pointed at the apex, less waved and twisted and with a less reflexed petiole. The surface of older leaves of Clairgeau is more netted than those of Duchess.

Clapp's Favorite and Flemish Beauty (a Comparison).—Both cultivars have dark reddish shoots and trunk. (A) Bark: Flemish more spreading; more and finer branches. Clapp generally has numerous spurs at the base of the head where Flemish would have branches. (C) Leaves: Flemish is duller green, surface not netted, smaller and shorter petioles.

Grit Cells

Stone or grit cells are a normal development in pears and some other fruits, e.g., quince. They are a genetic factor inherited quantitatively and particularly noticeable in *Pyrus pyrifolia* and *P. serotina*, both of which are known as the sand pear. Keiffer, in which stone cells are relatively prominent is the progeny of *P. serotina* and probably *P. communis*.

The grit cell, either singly or in clusters, is a brachysclereid; a sclerenchymatous cell whose wall is thick, highly lignified, and may be much stratified with clearly distinguishable lamellae. Mature grit cells are usually dead. Pear cultivars have been rated according to the number and size of the grit cells (Thompson *et al.* 1974).

CLIMATIC REQUIREMENTS

Although pears tolerate a wide range of climatic conditions, their commercial culture has been restricted mostly to areas that are particularly favorable for them. Pears are an important crop from south central California up into British Columbia, and from Michigan, Ontario, and New York in the north to the mideastern states.

High Temperature Effect

Prevalence of high winter temperatures limits the southern range of commercial pear growing. Most commercial pears require 900–1000 hr below 7°C to adequately break their rest period. Therefore, pears commonly grown in northern regions are not adapted for commercial production in areas where winter temperatures are mild.

Cultivars such as Hood, Pineapple, and Orient have a low enough chilling requirement to succeed well as far south as northern Florida, but the quality of the fruit leaves much to be desired. Better cultivars are being developed for the south.

Low Temperature Resistance

If pear trees are fully dormant, temperatures as low as −29°C usually do little injury. The wood and buds of pears seem somewhat more subject to injury from low temperatures than those of apples, but pear trees are more resistant to low temperatures than peach trees. In general, pear growing is hazardous where temperatures may be lower than −25° to −29°C.

Climate and Quality

A correlation exists between climate and quality of certain cultivars. Bartlett reaches its highest dessert quality and best shipping and storage qualities where temperatures for the two months preceding

FIG. 2.3. SEVERE SUNSCALD OR SOUTHWEST INJURY RESULTING FROM LACK OF PROTECTION FROM EXTREME FLUCTUATIONS IN DAY–NIGHT TEMPERATURES

harvest are high. When grown in the cooler sections, this cultivar tends to ripen faster after picking and to develop core breakdown while still firm and of prime eating appearance.

Bosc also reaches its highest dessert quality under relatively high temperatures. Other cultivars including Anjou, Hardy, Winter Nelis, and Easter Beurre are equally well suited to the cooler climates and to moderately hot conditions.

SOIL REQUIREMENTS AND MANAGEMENT

Pears grow well in a wide range of soils, provided these soils possess adequate moisture and are well drained. Although the trees tolerate heavy soils they grow best on deep fertile loams with a well-drained subsoil.

Good tree growth cannot be expected on shallow soil of low fertility or on poor soil with a wet subsoil. There should be at least 1 m (4 ft), preferably a minimum of 2 m (6 ft), of soil in which the roots can penetrate freely. A deeper soil can hold a much greater reserve of moisture and nutrients than a shallow soil. A plentiful supply of water, naturally held in the soil, is good insurance for healthy, productive trees.

Whether the original soil condition of an orchard is maintained, improved, or let deteriorate depends on the manner in which the soil is managed. No one system of soil management is best for all orchards. Cultural practices may change as the trees grow older and come into bearing.

In the East, soil-management and fertilizer practices with pears must be based on control of fire blight. As better controls for blight are devised the system may be revised, but meanwhile avoid any practice that promotes too succulent growth.

A soil-management system cannot be properly evaluated on a yearly basis. A good system may not show beneficial results for 2–3 years or more. Harmful effects of a poor system may not be evident for some time. Yield alone can not be used to measure results. A program that causes swift breakdown of organic matter may temporarily produce crop increases through sudden release of nitrates and other plant nutrients.

Efficient soil-management practices should provide favorable conditions for growth and fruiting over the life of the orchard and protect the soil against physical and chemical deterioration arising from erosion and other causes.

Permanent Cover

The use of grass and legumes as a permanent cover for pear orchards is widespread cultural practice. Permanent sods of bluegrass, redtop, fescue, orchard grass, and alfalfa, alone or in mixtures, are satisfactory in bearing orchards. Ladino, red clover, birdsfoot trefoil, and Korean lespedeza are good mixtures where medium-sized legumes are desired. Sods reduce the leaching of water-soluble nutrients, especially N from the soil.

In Ohio, a bluegrass sod without mulch was generally preferred. N fertilizer was best limited to intermittent applications. A suggested application for bearing orchards was 225 kg per ha (200 lb per acre) of 10-10-10 fertilizer over the entire orchard floor every 2–3 years (Patterson 1957).

In the Wenatchee and Yakima Valleys in Washington, and the Hood River Valley in Oregon, where abundant water is available for irrigation, pear orchards are largely maintained in permanent sod; alfalfa and sweet clover are most widely used. The area between the trees is seeded, preferably before the trees reach full bearing age. The crop is allowed to grow throughout the season, matting down to form a dense soil cover during late summer. Usual practice is to disk such orchards in spring, working the ground thoroughly to incorporate the vegetation. Prep-

**FIG. 2.4. PEARS THRIVE WHEN GROWN IN A PERMANENT SOD
COVER ALONG WITH A HEAVY ORGANIC MULCH**

aration is then made for summer irrigation. Disking also checks the growth of alfalfa during spring when the trees bloom and their growth is most rapid. Thus, competition between cover crop and trees is avoided during this critical period. After disking, however, the alfalfa soon begins to grow and a heavy soil cover is obtained by midsummer.

The permanent cover crop or sod system of orchard management represents the minimum cost for cultivation and provides a steady supply of organic matter for the soil. The soil is shaded and is several degrees cooler throughout the season than with clean cultivation. The incorporation of organic matter, minimum cultivation, and penetration of the cover crop roots help keep the soil in good condition. One drawback of this system of management is that somewhat more moisture is required than under clean summer cultivation. Also, the heavy growth of vegetation tends to harbor certain insect pests. Occasional seasons of clean cultivation may be required.

Cultivation

In some areas, particularly in the west, clean culture is practiced throughout the summer. The orchards are usually plowed or disked in spring and the soil is worked down thoroughly. Shallow cultivation is usually given after each irrigation. Where irrigation is not used, disking or other shallow cultivation is frequently practiced throughout spring and summer to check vegetation that would compete with the trees for moisture. Clean cultivation tends to deplete the organic matter in the soil unless positive measures are taken to counteract the ill effect.

In soils of relatively heavy texture, which are often used for pears, timing of cultivation is particularly important, especially in nonirrigated orchards. If such soil is worked too early it dries into hard clods and remains in poor physical condition throught the summer. Soil that is too dry is also difficult to work properly. If a winter and spring cover crop is growing in nonirrigated orchards, work down the cover crop before the soil becomes too dry.

TABLE 2.3
EFFECT OF SYSTEMS OF SOIL MANAGEMENT ON TRUNK CIRCUMFERENCE (IN CM)
OF BARTLETT TREES

1 Lot Treatment	1930	1934	1939	1944	1949	1954	1959	1961
			No. Years After Establishment of Sod Cover					
			5	10	15	20	25	27
Cultivated plus N	5.3	19.6	36.8	46.2	53.3	60.7	65.8	68.6
Sod mulch plus N	5.3	19.3	32.2	40.4	50.0	59.7	66.5	69.6
Cultivated, no N	5.3	18.5	36.1	44.4	50.3	57.6	62.7	65.0
Sod mulch, no N	5.6	18.8	30.7	38.4	47.2	56.6	63.2	66.3

Unfavorable effects of cultivation result chiefly from rapid destruction of the soil organic matter by oxidation or loss from erosion. Sandy loam soils that contain large amounts of organic matter and other favorable types of soils often may be cultivated for many years without obvious ill effects, especially if annual cover crops are grown. Many clay or clay loam soils show the destructive effects of cultivation more quickly.

In areas where permanent cover crops are not used, seed with over-wintering crops in late summer. Widely used are rye, buckwheat, wheat, oats, and barley in the east, and vetch, horsebeans, native legumes, and grain in the west. In many orchards, abundant weeds develop in the fall after cultivation ceases, and this native vegetation helps maintain the organic supply.

Mulching

Pears, like apples, thrive when mulched. The same benefits are obtained and the same possible disadvantages must be considered. Mulching generally consists of applying straw, hay, sawdust, strawy manure, corn cobs, pea straw, pomace, or other organic material on the soil beneath the tree from near the trunk out to or beyond the branches. A mulch over the entire orchard floor is highly beneficial but because of the cost and lack of availability of mulch material this practice is seldom followed.

Maintain the mulch deep enough to smother competing vegetation. For the first 2–3 years apply more N than normally used to offset the temporary tie-up of N in breakdown of mulch material. After this, a new biological balance is established among the soil microorganisms and the nitrate N begins to accumulate beneath a mulch. With a straw mulch, apply about 25% more nitrogen fertilizer than needed by trees in sod alone; heavy mulches of sawdust or shavings may make necessary a 50% or more temporary increase in N application. The amount of N fertilizer required and the time when it may be reduced or omitted entirely varies

FIG. 2.5. TREES MULCHED WITH WASTE TIRE FABRIC AND PROTECTED AGAINST RODENTS BY PLASTIC GUARDS

with the soil, kind and amount of mulch material used, and rapidity of its decomposition. A legume or manure mulch may promote excessive growth, thereby increasing the danger of blight.

The ultimate results of a mulching program are a more favorable supply of nitrate N, the movement of K and other nutrients into the soil, improved physical condition of the soil with greater moisture retention, and less fluctuation in soil temperatures.

In Michigan, alfalfa and grass sod cover retarded growth and fruit production of pear trees for the first 12 years after the cover was established. From then on, the trees made greater trunk circumference increase and produced heavier annual crops and larger fruit than trees under clean culture (Table 2.3) (Toenjes 1955).

In the west where irrigation is a standard part of the cultural program, pear orchards are maintained under a semipermanent sod or clean cultivation system and mulching is not practiced.

Many pear growers who have followed a good mulching program over an extended period find that it has been profitable through better tree growth, increased yield of improved quality fruit, savings in fertilizer costs, and beneficial effects on the soil. If the mulch material is applied in late fall and a regular mouse control program followed, the mouse hazard need be no greater with a mulch than in an orchard in sod alone.

FERTILIZERS

In commercial pear orchards a wide but satisfactory range of nutrients in the leaves can be expected. This range includes the extremes of so-called incipient deficiency and heavy feeding. N fertilizers produce the greatest response in pears. It would appear, from tissue analysis, that the pear tree is relatively tolerant to massive applications of N, and that varying levels of N fertilization have little effect on fruit ripening, firmness, and other attributes of fruit quality. However, because succulent growth is least resistant to blight, avoid too rapid growth and use N fertilizers with caution on pears. For the different kinds of N fertilizers and their rates and methods of application, see Chap. 1.

The best guide to fertilizer requirements of pear trees is the appearance and condition of the trees themselves. Soil tests can give some indication but the tree root system is so ramifying that a representative sample of rooting medium is difficult to obtain. Tissue tests indicate what nutrients the tree is actually taking up. But, differences in leaf composition with respect to N, P, K, Ca, and Mg may be brought about by differences in cultivar, soil management, date of sampling, and season in unfertilized pears grown on the same rootstock under uniform conditions of soil type, pruning, moisture supply, and pest control. "Normal" growth and yield may occur accompanied by considerable variation in composition. Not only the general level of the nutrients mentioned but their ratios may be affected (Hewitt 1967; Proebsting 1953).

In most sections of Oregon and Washington, and in some parts of California, annual applications of fertilizers high in N stimulated growth of the trees and improved production. On some deep, fertile, alluvial soils, particularly where the orchards were maintained under summer cultivation, little response was obtained from fertilizers. Where fertilizers proved of value, those high in N were generally best, particularly annual moderate applications in late fall (Kinman and Magness 1949).

Phosphorus.—Deficiency symptoms of P deficiency in pear orchards are severe burning of the margins and tip halves of the leaf blades early in the growing season, decrease in leaf size, failure of fruit to develop

FIG. 2.6. THE PEAR TREES ARE FED BY APPLYING FERTILIZER TO THE
SOD COVER

properly, very short terminal growth, scaly bark, and dying back of new
growth.

Potassium.—Sometimes K deficiency is encountered in pears but they
seem better able than apples to obtain enough K to avoid leaf scorch in
soils moderately low in this element. However, K deficiency can be so
severe as to inhibit growth and development of tree and fruit.

Magnesium.—Generally the Mg content of the tissues reflects the
amount in the soil. Mg increases in the bud as it enlarges and rapidly as
the flower opens. The floral parts may contain more Mg than the leaves.
Mg deficiency shows in the older leaves at the base of some of the
current season's growth. Oblong islands of dark purplish-brown to
almost black tissue are arranged in rather orderly fashion between the
parallel veins on both sides of the midrib, extending from tip to base of
the leaf. Some chlorosis at the margins of these interveinal islands is
present with severe deficiency. Leaf analysis indicates very low Mg and
early defoliation take place beginning with the basal leaves. On acid soils
a liming program using dolomitic limestone tends to increase Mg
availability. $MgSO_4$ (MgO 18%) is a water soluble source of Mg and can be
applied to the soil or as a foliar spray for temporary relief of Mg
deficiency.

Boron.—Lack of B, indicated by corky spot and browning and rosetting of the leaves can be corrected by broadcasting borax under the trees at the rate of 33.8 kg per ha (30 lb per acre) or 0.1–0.2 kg (1/4 to 1/2 lb) per tree in early fall. Such treatment usually prevents B deficiency for 3

FIG. 2.7. "MEASLES" ON PEAR TWIG CAUSED BY BORON DEFICIENCY

years. If the soil pH is above 6.5, when necessary spray the foliage at the rate of 236 g (2 lb) of borax in 100 litres (100 gal.) of water per 0.4 ha (acre) in 3 applications (calyx, and first and second cover sprays).

"Blossom blast" can be controlled by a spray of 236 g (2 lb) of boric acid per 100 litres (100 gal.) of water applied a week before full bloom. In this disorder, the blossoms may wilt, die, become dry, fail to absciss, and may persist until the following year. Sometimes leaves also die and the limbs may not put out new buds within a few weeks after the blast occurs, and so die before the next year. Blast in an orchard is random in distribution, rarely affecting more than 10–20% of the trees, and usually is confined to 1–2 leaders per tree. It is often confined to trees that have previously shown the disorder. Affected orchards usually are located on heavy soils, and blast appears more frequently in years when above-average rainfall and below-average temperatures occur prior to and during bloom.

Copper.—Cu deficiency in pears has been reported in widespread areas. The exanthema and dieback symptoms are similar to those of apple.

Manganese and Iron.—Deficiencies of Mn and Fe occur on calcareous soils of high pH. The typical interveinal leaf chlorosis symptoms have been described by many workers. Foliar sprays of manganese oxide are effective in controlling Mn deficiency (Boynton 1966).

Zinc.—Zn deficiency symptoms in pear are similar to those of apple and occur under the same conditions (Bould *et al.* 1949).

Irrigation

In areas such as the western valleys where annual precipitation is less than 29 cm, irrigation is essential. But even where the annual precipitation is between 63 and 90 cm, supplementary water at times may be very beneficial. The distribution of precipitation throughout the year and throughout the growing season may be more important than the total amount of annual precipitation.

An adequate supply of good water free of sediment and precipitates, for pest control as well as for irrigation, is a prime requisite in the choice of the orchard location. Water supply may be obtained from a stream, lake, well or farm pond. In the semiarid valleys of the west, water is brought to the orchards by gravitational flow from higher altitudes by means of flumes.

Amount of Water.—Pronounced diurnal variations, directly related to climatic conditions, have been noted in the water potential of both leaves and fruit of pear. Most water loss from the fruit occurs through the pedicel into the xylem of the tree. At night the plant water potential reflects the soil moisture status. In the heavier soils, a moisture supply representing about 50% or more of the maximum available moisture in the upper 3 ft is essential for maximum growth of fruit, shoots, and trunk. A minimum irrigation requires about 5 cm (2 in.) of water or 513 kl per ha (54,000 gal. per acre). Lighter soils, or those with a gravel substratum, need more frequent and heavier applications. A loam should be wet 1.5–1.8 m (5–6 ft) deep (about 950 kl [100,000 gal.] of water per hectare [acre] per application). Besides the rainfall, 76–89 cm (30–35 in.) of irrigation may be applied annually in some regions. One method of determining when and how much water to apply is to study the average monthly rainfall records for the area and to irrigate with 5–7.6 cm (2–3 in.) of water when rainfall is deficient. Another method is to examine the soil frequently with an auger. Excessive irrigation is harmful and can be disastrous (Klepper 1968).

Rate of growth of pear fruits in California was reduced by exhaustion of readily available moisture in the top 1.2 m (4 ft) of soil, and by high temperatures and severe evaporating conditions. Wetting the leaves increased the rate of growth temporarily, even in the unirrigated plot where the trees had been without readily available moisture in the top 1.2 m (4 ft) of soil for over a month (Hendrickson 1941).

Reduced water supply to the tree during summer, as indicated by reduced rate of fruit enlargement, resulted in decreased proportion of water to total solids in the fruit at harvest. This decrease was correlated with a higher pressure test of fruit during all except the end of the harvest period. Reduced water supply to the tree during summer also resulted in less rapid development and less core breakdown of fruit upon ripening after cold storage (Ryall and Aldrich 1937).

Methods.—Irrigation in furrows has been a common method of supplying water to pears and other fruit trees. Ususally 6–8 furrows about 15 cm (6 in.) deep are made between every 2 rows of trees. If the water contains injurious salts they may accumulate in the soil between narrow furrows with wide ridges. Broader furrows with uniformly even bottoms avoid this salt accumulation and make more surface soil available for root growth. Basins are sometimes formed by making levees around single trees or large enough to include several trees on the same elevation. Such basins allow more water to be applied at one time to a broad surface. Little concentration of salts and nutrients occurs with this method.

Sprinkler irrigation is now preferred to ditch and furrow method in most areas. It is less troublesome; water can be applied at the rate at which it can penetrate the soil; and puddling of the soil and salt accumulation can be avoided. Sprinkler irrigation sometimes can be used to successfully combat frost injury to blossoms and provide some measure of environmental control (Lombard *et al.* 1968).

Drip or trickle irrigation has the potential to deliver a specific amount of water to each tree according to its exact needs. It thus could eliminate both water waste and tree stress due to water shortage. Branch lines carry the water from the main down the orchard row and deliver a small, constant, precise amount of water to each tree by an emitter system. The plastic lines can be buried and remain in place.

POLLINATION AND FRUIT SET

All pear cultivars are commercially self-unfruitful, but some inconsistency exists. Bartlett, for example, is considered self-fruitful in some areas, as in the Sacramento Valley in California. The degree of self-fruitfulness varies with the season and location. At best, however, the set does not usually exceed 1–3% when Bartlett is planted in large blocks without provision for cross-pollination. Self-fertilization frequently results in parthenocarpic fruit. Cross-pollination is advantageous to fruit setting of practically all pear cultivars.

Pollinizers

Compatibility.—Pear cultivars, like apples, may be either diploid or triploid. Diploid cultivars, which include all commercial cultivars grown on this continent, have viable pollen and, except in a few combinations, can successfully pollinize one another. Boussock is a triploid. Seckel and Bartlett are incompatible.

Courtesy of Canada Dept. Agr. Res. Sta., Harrow, Ont.

FIG. 2.8. THIS KIEFFER PEAR TREE HAS BEEN FRAMEWORKED TO ANJOU IN ORDER TO PROVIDE A POLLINIZER FOR BARTLETT AND TO GET A BETTER CULTIVAR AT THE SAME TIME

Age of Bearing.—Most pear cultivars begin flowering at about the same age (6–8 years). Seckel and Cayuga are relatively late, and Bosc is 2–3 years behind Bartlett in coming into production in some areas. This may necessitate the use of bouquets or "imported" pollen in the orchard until the late cultivar begins to flower.

Bloom Period.—The bloom period must overlap sufficiently for adequate cross-pollination between cultivars. Some early bloomers like Kieffer, Howell, and Duchess may not be good pollinizers for Bartlett in some regions in the east. Likewise, late bloomers such as Winter Nelis and Wilder may not be the best pollinizers for Bartlett. However, on the Pacific Coast, Bartlett is successfully used as a pollinizer for most other important cultivars. In the Sierra Nevada foothills, Bartlett is almost entirely self-unfruitful and Winter Nelis is a good pollinizer for it. Cross-pollination of Anjou is desirable, and Bartlett, Easter Beurre, and White Doyenne are effective pollinizers.

Resistance to Blight.—A great hazard to pear production, especially in the east, is fire blight. Avoid pollinizers that are particularly susceptible to blight (Van der Zwet et al. 1974).

Planting Plan.—Pear trees have been commonly planted 7.5 × 7.5 m (25 × 25 ft) apart. However, the trend in new plantings is toward a higher density of trees per ha by closer spacing in the tree rows. Provide for cross-pollination by planting a cultivar not more than 15 m (50 ft) from a pollinizer. Interplanting of commercial cultivars in alternate rows provides for cross-pollination. If more trees of one cultivar than of the other are desired, plant 2–3 rows to each row of pollinizer. If the pollinizer is solely or chiefly for the benefit of the other cultivar, plant single trees at intervals throughout the orchard. One tree only is sufficient to pollinate 8 others if bees are active to carry the pollen (see also Pollination of Apples).

Agents of Pollination.—Pear pollen, like that of other fruit trees, is heavy and wind plays no appreciable role in pollination. Honeybees are the chief agents of cross-pollination. However, pear blossoms hold little nectar and what they do have is not very attractive to honeybees. Therefore, bees working in pears are chiefly after pollen. Quite commonly, the amount of pollen produced by pears is inadequate. Thus, pollination problems may arise unless adequate pollen and a high concentration of bees are provided in the orchard during bloom. Supply at least 2 strong colonies of bees per ha.

Fruit Set

Spraying to Increase Set.—In Washington and California in a Bartlett orchard where cross-pollination was not provided, both gibberellic acid and 2,4,5-TP resulted in an increased set but not so where cross-pollination was provided. The treatment resulted in varying amounts of fruit breakdown at the stem and calyx ends and in inferior keeping quality. Aqueous sprays of gibberellic acid and of 2,4,5-TP and its amine salt greatly increased the parthenocarpic set of Bartlett (Griggs and Iwakiri 1961).

Most results with B sprays applied to pears in bloom indicate that fruit set is increased only when a B deficiency is present. In Oregon, B sprays were applied to Anjou trees over a 4-year period, when about 80% of the blossoms were open, at concentrations of 52, 104, 208, and 416 ppm. No responses in improved fruit set or any other factors were found in these trees having no deficiency symptoms such as blossom blast and dieback. All four years were quite favorable for fruit set in Anjou. Marked benefit resulted from B applications when deficiency symptoms were present (Degman 1953).

Daminozide indirectly causes greater fruit set by reducing terminal growth and promoting fruit-bud initiation. In some cases fruit size is adversely affected (Batjer *et al.* 1964; Griggs 1968).

Fruit Thinning.—When blossoms or young fruits are thinned to one fruit per cluster, ultimate percentage fruit set is generally increased. Usually it is the king blossom that sets in this case.

Ringing.—Scoring, girdling or ringing has some positive effect on fruit set in the current year, if done prior to 3 weeks after bloom. When this treatment is combined with early fruit thinning the resulting fruit set increase is greater than if either treatment is used alone. The greatest effect of either treatment, however, is on fruit bud differentiation which mostly occurs sometime within 20 days after bloom.

Pruning.—Thinning out of the old fruit spurs tends to increase the ultimate percentage fruit set. A light-to-medium pruning is more effective than a heavy pruning and a heading-back type results in more fruit set than does thinning-out pruning (Westwood and Bjornstad 1974).

PRUNING

Pruning is an important cultural operation in the production of large pears typical of the cultivar in shape and quality. It is also one of the most difficult problems confronting the grower. The influence of the growth condition and the differences in characteristics of growth and production among cultivars may all affect the best type of pruning in a given orchard. Careful study of cultivar characteristics and the influences of local conditions and treatments on them is necessary in evolving the best pruning system.

Pruning at Planting Time

If the nursery tree is an unbranched whip, cut it off at the height desired for the top main branch, usually at 102–122 cm (40–48 in.). With vigorous trees more branches form along the trunk the first year. Select the scaffold branches from these after the first growing season.

Orchard Pruning After One Year

At the end of the first year, select 4–5 vigorous branches, 15–30 cm (6–12 in.) apart, along the trunk and well distributed around it for the main framework, and remove the other limbs. Make the central branch

**FIG. 2.9. THESE WELL-TRAINED YOUNG PEAR TREES ARE
NOW READY FOR SUMMER PRUNING TO DEPRESS
EXCESSIVE VEGETATIVE GROWTH**

dominant, since this forms the leader and stronger crotches result.
Unless the tree has made excessive growth, little or no heading back of
these branches is necessary. Such a distribution of branches gives a
stronger tree and reduces the danger of sunburn injury and of blight to
the tree as a whole.

Pruning in the Formative or Prebearing Period

The pruning that the tree should receive at this stage varies somewhat
with its growth. The less pruning the more quickly the tree comes into
bearing. Consequently, once the main branches are selected, do a minimum of pruning until the tree is bearing. Usually a light thinning out of
branches suffices. If growth is very strong, heading back the branches
lightly for 1–2 years gives a more compact tree that stands more stiffly
in the wind. Generally, avoid heading back from the second year until the
trees are in heavy bearing.

Usually, 4–5 main branches are sufficient to build a good tree. With
cultivars susceptible to blight and where blight is likely to be serious,
leave more framework and secondary branches, since infection with this

disease may make it necessary later to remove some of them.

As the trees approach bearing age, corrective pruning may be necessary to obtain a satisfactory fruiting condition. Most young trees make upright growth and do not branch freely. Spreading the tree by mechanical means may encourage formation of more fruit buds, but once the tree is in moderately heavy production the weight of fruit is usually sufficient to open the tree.

Pruning Mature Pear Trees

In pruning the young tree, a basic purpose is to build a suitable framework with the least possible cutting and to provide for maximum size and production at an early age. After the tree is in full bearing, however, the purpose of pruning is to maintain the fruiting wood in vigorous condition in order that the trees may produce regular crops of fruit of good size and quality. Pruning tends to reduce the number of fruiting points, but it stimulates thriftier growth of those that remain. In general, practice the minimum pruning that maintains the spurs and fruiting wood in vigorous growing condition. This means keeping the tree open to light and air by the annual removal of some old fruiting spurs and old branches. Young shoots generated as a result of this type

Courtesy of Bill Luce (1977)

FIG. 2.10. GROWING PEARS IN THE HEDGEROW SYSTEM IN THE YAKIMA VALLEY

of pruning should be retained if they are located where they will have space and light and will not interfere with other fruiting branches. If too long (more than 1/2 m) shoots should be headed back by 1/3 or more.

If too much new growth occurs, prune more lightly and/or do more of the pruning in early summer. Prune most heavily in that part of the tree showing the least amount of growth; prune least heavily where most new growth is taking place. Pear shoots (or sprouts) have been successfully controlled with a spray application of napthaleneacetic acid (Raese 1975). Pruning practices for bearing trees should be correlated with soil-management and other cultural practices.

Fruiting Habits.—Pears bear their fruit from terminal buds on short spurs on 2-years and older wood. Fruit buds occasionally terminate the growth of 1-year-old wood, but such terminal buds seldom set fruit and are of little importance to the fruiting habit of the tree.

In a young, fast-growing tree, spurs may become well-developed on 2-year-old wood and as the tree increases in age continue to develop. They do not die out after 1, 2, or 3 years of fruiting as in the plum.

Pruning, therefore, develops into a method of keeping these spurs in a healthy and vigorous state with an ample supply of sunlight and air, prevention of overbearing, and encouragement of some new growth. On young trees the long, 1-year growth may be shortened and thinned out, giving the spurs full opportunity to develop. Too severe cutting, however, may produce wood growth at the expense of spur development.

On older trees, the annual growth becomes less rapid. Branches which are 12—14 years old bear a mass of fruit spurs. Each fruit bud bears 4—5 flowers and as many leaves. A spur continues to produce fruit as long as it is healthy and vigorous; but old spurs that produce small fruit or become unproductive should be removed in the annual pruning. This promotes new growth and better fruit down in the tree.

Cultivar Differences.—Some cultivars strongly tend to develop and retain spurs throughout the tree and to produce little new growth except at the terminals of branches and near points of pruning. Hardy and Lawson produce great numbers of spurs but tend to produce little new wood along the older branches. Other cultivars having this tendency, though less pronounced, include Bosc, Clairgeau, Flemish Beauty, and to a still lesser extent, Anjou and Comice. Such cultivars require rather heavy and detailed pruning. Head back the branches and new shoots to maintain the vigor of fruiting branches and to induce growth of new branches; and thin out old fruiting branches and spurs to maintain vigor on those that are left.

Other cultivars tend to produce twigs and vigorous shoots, not only from the terminals of branches but along the sides of branches as well if conditions are favorable for wood growth. Development of spurs in these cultivars is much less marked. Bartlett and Winter Nelis have this

FIG. 2.11. THE ESPALIER SYSTEM OF GROWING PEARS IN
FRANCE

growth habit. Their shoots generally must be thinned in order to keep
the trees adequately opened to the light and to maintain sufficient vigor
in the new growth. Moderate annual pruning is desirable to maintain the
trees in the best fruiting condition. In general, the pruning should be

FIG. 2.12. PEARS IN ITALY TRAINED TO THE PALMETTE
SYSTEM

detailed and distributed over the whole tree, since the greatest response
in increased vigor comes in the part of the tree adjacent to the pruning
cuts.

Pruning and Fertility.—If soil-fertility and soil-management practices are such that fairly vigorous trees are maintained, good results occur with only a limited amount of cutting. If the pruning is reduced too much, however, it is a difficult to obtain adequate fruit size even with fairly heavy fruit thinning. Overpruning, on the other hand, reduces the bearing surface and leaf area of the trees so much that yield is likely to be decreased. A balanced program of moderate annual pruning and good soil management generally results in maximum fruit production.

Pruning and Blight.—In all areas where blight is serious, the danger of inducing vigorous growth that is very susceptible to blight must be considered. Practice only the heading back and the thinning out necessary to enable the tree to make a moderately thrifty growth and produce fruit of the desired size; avoid unnecessary cutting that results in heavy new growth.

FRUIT THINNING

Many pear cultivars tend to set excessively heavy crops which the tree cannot develop to good marketable size. This may occur particularly with Winter Nelis, Bosc, and Bartlett and sometimes with Anjou. If medium-sized to large fruit is desired, thin part of the developing crop from such trees in order to have a large leaf area per fruit.

Many cultivars, such as Bartlett and Bosc, tend to set the fruit in clusters, often 3–5 fruits per spur. If the set of fruit on the tree as a whole is excessive, reduce these clusters to 1–2 fruits each. But, if the set of fruit on the tree as a whole is not excessive, fruit on these clusters reach satisfactory size and quality without thinning. In California, thinning of Bartlett is seldom necessary.

It is difficult to give specific advice for thinning pears. The number of fruits a tree will carry and develop to good marketable size varies with its vigor and with the growing conditions. With nearly all cultivars 30–40 good leaves per fruit are essential for building materials that go to make the fruit. These leaves, however, need not be directly adjacent to the fruit. With extremely heavy sets of fruit, thinning to reduce the amount of fruit in proportion to the leaf system is essential to obtain fruit of best size and quality.

The response of pears to chemical thinning agents tends to be inconsistent from season to season and from one area to another. Naphthaleneacetic acid (NAA) applied shortly after bloom can be reasonably effective in thinning pears but better results have been realized with naphthaleneacetamide (NAD) immediately after petal fall, followed by NAA five days later. Results with 1-naphthyl-N-methylcarbamate (carbaryl) have not been very satisfactory to date (Teskey *et al.* 1977).

Prediction of the full bloom date for pears by recording the accumulative heat units gives some assistance in planning spray and thinning programs. The methods of accumulating heat unit data based both on daily maximum and daily mean temperatures were examined for the period prior to bloom of apple, pear, cherry, peach and apricot for 20 years. The accumulation of daily maximum temperatures above a base resulted in as low or lower variability than any other method examined. Coefficients of variation were used in determining suitable start dates and base temperatures for each kind of fruit at Summerland, B.C., and for apple at Kentville, N.S. A method of predicting the full bloom date was examined. In all cases some improvement over using the average date of bloom was obtained (Anstey 1966).

HARVESTING

Pears should be picked when fully mature but still hard and green. If left on the tree until ripe, pears break down and turn brown internally. If picked while immature, they will not develop full flavor and if stored, shrivel sooner than mature pears. After being picked, mature fruit ripens in 5–7 days at about 25°C.

Indices of Maturity

Pressure Test.—Softness of flesh is fairly accurate index of maturity of pears. However, pears develop somewhat differently in various areas

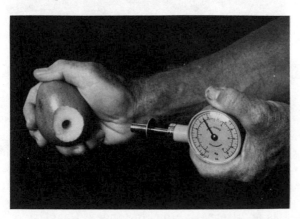

FIG. 2.13. A PRESSURE TESTER FOR DETERMINING THE STAGE
OF MATURATION OF PEARS AND APPLES

so that the pressure representing the best stage to pick a cultivar in one area is not the best for all areas. In general, pressures of 9–9.5 kg (20.0–20.5 lb) indicate picking maturity for Bartlett.

In British Columbia, Bartlett pears picked 1–2 weeks beyond date of commercial harvest, at pressure of 6.1–7.2 kg (13.5–16.0 lb) ripened with good quality for both eating and canning after 6 weeks at −1° to 0°C. Keeping quality as measured by length of time for core breakdown to develop after removal to the ripening room indicated that lengthening the period of cold storage decreased storage life to a greater degree than increasing the maturity at harvest. Following harvest, Bartlett held at 20°C underwent a period of inactivity during which no softening occurred for 3 days with fruit picked at optimum maturity. With fruit picked later, however, softening commenced immediately (Fisher and Porritt 1955).

Percentage of Sugar.—Percentage of total sugars in the fruit is another useful index of maturity. About 13% is commonly the minimum refractometer reading accepted by canners of Bartlett.

In many Bartlett orchards in California, firmness and soluble solids were higher in fruit from nonirrigated than from irrigated trees. Both

FIG. 2.14. BLACK END, A PHYSIOLOGICAL DISEASE OF PEAR

soluble solids and total sugars in the juice were higher in the "dry" fruit, but the ratios between the two were similar in "dry" and in "wet" fruit. The results had practical application in that soluble solids, easily measured by a refractometer, indicated sugars and that the same growth factors retarding softening increased sugar content. When, because of restricted soil moisture, fruit was delayed in softening, soluble solids could be used as an alternative picking index. Quantity of winter rain had little effect since summer irrigation was still necessary to maintain an available supply of soil moisture throughout the growing season.

Ease of Separation from Spur.—When pears reach optimum maturity for picking they usually can be separated from the spur by a slight twisting pull. Begin picking at the first indication of sound fruit dropping from the tree. Generally, if a large percentage of spurs come off with the fruit, pears are not ready to be picked.

Chemical Control of Preharvest Drop

Dropping of fruit before or during harvest causes much loss in some years. This loss can be greatly reduced by the use of hormone sprays on some cultivars, including Bartlett and Bosc.

Naphthaleneacetic acid (NAA) has effectively prevented fruit drop when 5 ppm were applied at the start of the harvest drop. In Washington, 2 ppm of naphthaleneacetic acid, applied about 2 weeks before harvest, has controlled drop of Bartlett.

NAA at 10 ppm controls fruit drop when applied to pears 3–4 days before picking time, or at the first evidence of sound fruit dropping. Combinations of naphthaleneacetic acid and naphthaleneacetamide give about the same results as either material alone (Table 2.4).

TABLE 2.4
NUMBER OF PEARS THAT DROPPED BEFORE AND AFTER TREATMENT

	No. of Trees	Before Treatment	After Treatment
Naphthaleneacetic acid	17	39.3	135.6
Naphthaleneacetamide	13	45.3	73.0
Fruitone[1]	19	38.3	132.0
Control	16	33.0	198.3

[1]Naphthaleneacetic acid plus naphthaleneacetamide.

A period of more than 3–4 days was necessary for expression of the influence of the sprays (NAA, NAD, Fruitone). The drop, after treatment with chemicals of the strength of 10 ppm and 5 ppm, was reduced to about 50% of that from untreated trees, except that in one case where naphthaleneacetamide was used the drop was reduced to about 33% of that in the checks. Strickland *et al.* (1941), encountering much heavier drop of pears (41% of the drop at commercial harvest time), were able to reduce the loss to 25% of that in the checks by applying 20 ppm in 3 applications at 2-week intervals up to harvest time.

Succinic acid-2-dimethyl hydrazide (daminozide) has not, to date, produced satisfactory results as an agent in the control of preharvest drop of pears.

The effect of any one of these chemical thinning materials depends largely on the health and vigor of the tree. Trees with weakened foliage, or otherwise unthrifty, do not react satisfactorily to any foliar spray treatment. Response to hormone treatment also varies with the cultivar, season, and other factors.

STORAGE

Pear cultivars differ as to the length of time they can be stored successfully (Table 2.5).

TABLE 2.5
LENGTH OF TIME AT −1° TO −6°C FOR SAFE STORAGE OF CERTAIN CULTIVARS OF FALL AND WINTER PEARS AT SHIPPING POINT AND AFTER SHIPMENT TO MARKET

Storage Treatment and Variety	Storage Period Months	End of Storage Period
Stored immediately after		
Hardy	2–3	Sept.–Nov.
Comice	2–3	Nov.–Dec.
Bosc	3–3.5	Nov.–Dec.
Kieffer	2–3	Sept.–Nov.
Anjou	5–6	Mar.
Winter Nelis	6–7	Mar.–May
Easter Beurre	5–7	Mar.–May
Stored after 12-day transit period (precooled)		
Anjou	4–5	Mar.
Hardy	2–3	Sept.–Nov.
Comice	2–3	Nov.–Dec.
Bosc	2–3	Nov.–Dec.
Winter Nelis	6–7	Mar.–May

Source: Wright *et al.* (1954).

Bartlett pears when harvested at 7–9 kg pressure can be stored successfully for two to three months at −1°C and a relative humidity of 90–95%. Unless fully mature, pears may freeze at −2°C. They are often stored by canneries for as long as four months, then removed from storage and ripened quickly at 23°–25°C.

Differences occur in keeping qualities of pears from different production areas. Those from relatively cool areas generally have a shorter life. If picked when too immature, pears may wilt and scald; core breakdown is common in pears picked too late.

Temperature and Humidity

Pears to be stored after shipment may need to be precooled. The most desirable temperature for storage of pears is $-2°$ to $-1°C$. In general, pears undergo the same postharvest changes as apples, but these changes, particularly in Bartlett, are more rapid; therefore, speed in handling and quick cooling are even more essential. For best results extra refrigeration capacity with an air temperature as low as $-5°C$ may be required for the initial cooling. Give close attention to fruit and air temperatures to prevent freezing (Phillips and Armstrong 1967).

Pears lose moisture more rapidly than apples so hold the humidity at 90–95%.

Ripening

Temperature.—The best ripening temperature for all pears after storage is about $18°C$. At $22°-27°C$ they usually fail to soften. The skin of pears is sensitive to fumes from chlorine solutions or sulfur dioxide used in plant sanitation. After using these materials, ventilation avoids fruit damage.

Volatiles Evolved.—Ethylene is associated with the ripening of fruits. Bartlett pears produced a maximum of 0.87 mg of ethylene per kilogram of fruit per 24 hr after 84 days at $0.5°C$. This was more than 14 times the rate for Anjou. The nonethylene (odorous) volatiles were evolved at a maximum rate of 33.2 and 33.4 mg for Bartlett and Anjou, respectively. Rates of emanation of volatiles from pears and apples were highest at $12.8°C$. The nonethylenic group of volatiles was evolved at a continuously increasing rate with the progress of senescence (Gerhardt 1954).

Effects of 2,4,5-T and Ethylene.—Postharvest applications of 2,4,5-T (200 ppm) and ethylene (20–30 ppm) effectively accelerate the rate of respiration of preclimacteric Bartlett and hasten the softening of the flesh. The riper fruits, treated before start of the respiratory rise, were more responsive to ethylene than to 2,4,5-T, and the difference in response between the two treatments was in about the same order as the difference in CO_2 production. Effects of both compounds were greatly reduced when applied to the fruit at $20°C$ following a period of cold storage. Ethylene did not affect respiration or volatile production when applied for limited periods to fruit at $1.7°C$ (Blanplied and Hansen 1968; Uota and Dewey 1953).

Shipping tests were conducted with early season Bartlett to determine the effects of postharvest treatment with ethylene or 2,4,5-T on rate of ripening at transit temperatures. Holding tests were also made to determine the effects of similar treatments on Bartlett subsequently held at 10°C for a simulated transit period. Both ethylene and 2,4,5-T increased the rate of softening at transit or simulated transit temperatures as compared with untreated lots. Ethylene treatment prior to shipment or holding at 5°C resulted in more rapid development of yellow skin, but 2,4,5-T treatment usually caused a blotchy appearance due to irregular disappearance of the green color. Treatment with ethylene prior to shipment had possibilities for hastening the softening and coloring of a minimum maturity Bartlett from early districts (Ryall and Dewey 1955).

Bartlett Pears.—Do not pick Bartlett for storage until the ground color begins to lighten and the lenticels have corked over. When this cultivar begins to show yellow in storage, it has been stored too long and will fail to ripen or soften on removal. Remove the fruit while it is still light green. The maximum storage period for canning and local fresh markets is about 90 days, and at terminal markets, 50–60 days. Storage breakdown may occur if preharvest sprays are used to control early dropping and if picking is delayed beyond optimum maturity.

Storage studies with N.Y. grown Bartlett indicated that core breakdown in fruits ripened after storage was influenced by crop load on the tree, harvest date, storage temperature, the use of CA, and folded polyethylene box liners, as well as the length of storage period, and poststorage ripening temperature. Core breakdown was controlled by storage in air at −1°C (Blanplied 1975A).

Fall and Winter Pears.—The length of time to safely store fall and winter pears depends on the cultivar and when the fruit is picked, and also on whether it is shipped before or after storage. Also, wide differences in keeping quality often occur in pears from various producing areas. If Bosc, Flemish Beauty, and Comice pears are held in cold storage beyond their season, they do not ripen satisfactorily or they may not ripen at all.

Chief decays of fall and winter pears in storage are gray mold rot (*Botrytis*) and blue mold rot (*Penicillium*). Gray mold rot can spread from decaying to sound, healthy fruit; hence, it is often called "nest rot." Losses from this rot can be reduced by the use of paper wrappers impregnated with copper. In the Pacific Northwest, blue mold rot (pinhole rot) is sometimes more important on pears, particularly Winter Nelis, than gray mold rot. Losses from blue mold can be reduced by careful picking and handling, prompt storage at −1° to 0°C and the use of paper wrappers to prevent direct contact between diseased and sound fruit.

Some control of decay is obtained by treating pears with a 0.4–0.6% solution of sodium chloro-2-phenylphenate. Anjou pears are subject to a

FIG. 2.15. PEAR LEAF BLISTER MITE ON APPLE LEAF (LEFT) AND PEAR LEAF (RIGHT)

scald that can be controlled by oiled paper wraps; but oiled wraps are not effective against scald on other pear cultivars.

Kieffer pears, if they are sound, firm, and still green when stored, and are held under the conditions advised for other fall and winter pears, should keep well for 2–3 months. If intended for storage, handle all pears with care in picking and packing because even slightly bruised or rubbed places may turn black and seriously damage the sales value. In all cultivars, a ripening temperature of 15°–18°C is essential for attainment of maximum quality for either dessert or canning (Wright *et al.* 1954).

Controlled-atmosphere (CA) Storage

Several pear cultivars, including Bartlett and Anjou can have their storage life and shelf-life lengthened by controlling the percentages of O_2 and CO_2. At 0°C storage, O_2 is maintained at 5% and CO_2 at 2%. Storage costs, however, are greater with CA storage, and the operator can not remove fruit from the room at will because the desired atmosphere would be lost and several weeks are required to regain it.

Film box liners for pears in storage produce an effect similar to that of CA storage. The sealed plastic film depresses the general rate of metabolism of the fruit by raising the $CO_2:O_2$ ratio within the liner and thereby contributing to longer storage life. Use film liners only for that part of the crop intended for late storage, and pack under film seal only sound fruit washed with an effective fungicide (Gerhardt 1955).

Incidence of pithy brown rot core (PBC) of Bosc pear held in controlled atmosphere storage was increased by delayed harvest, prolonged storage, high CO_2, and by low O_2 levels. At 2–3% O_2, PBC was prevalent even when CO_2 was maintained at 0.2% and below. PBC was not related to levels of CO_2 in the flesh because pears tolerated CO_2 at high O_2 concentration (Blanplied 1975B).

Treatment of Anjou pears with high CO_2 atmosphere for a short period immediately following harvest prolonged storage life, retarded ethylene production, delayed the climacteric rise in respiration, reduced loss of malic acid, suppressed increase in protein N, retained firmness, quality, and the capacity to ripen after long storage. Treatment with 12% CO_2 for 2 or 4 weeks provided the best results without injury (Wang and Mellenthin 1975).

INSECTS AND DISEASES

Control of insects and diseases is by biological and chemical means. The latter depends on the kind and concentration of pesticide used and the careful timing and thoroughness of its application. Familiarity with the local performance of the insects and diseases is necessary. Follow local control schedules.

FIG. 2.16. PLANT BUG INJURY

Insects

Pear Psylla.—The adult is about the size of a winged aphid. Both adults and nymphs suck juice from leaves, fruit stems, and young shoots.

Differences in resistance to pear psylla exist among pear species and cultivars (Westigard and Westwood 1969).

Pear Thrips.—These minute, brown insects with gray, fringed wings injure the opening blossoms and give the leaves a silvery blistered appearance. The nymphs are similar to the adults in general form and feeding habits.

Mites.—Arachnids, including red spiders and pear blister mites, often cause much injury to pear leaves. They are most numerous in orchards in dry, hot sections where irrigation is not practiced or where it has been neglected and the soil has become dry. An integrated mite control program in Washington, involving the use of virus inoculum, predators, and sanitation, is yielding promising results (Johnson *et al.* 1970).

Pear Slug.—The adult is a black and yellow sawfly. Injury is done, particularly on young trees, by the slimy, black or greenish-black, sluglike larva. Nearly all insecticides are effective against it.

Plant Bugs.—Several species, including the tarnished plant bug, may attack the pear. Injured fruits show depressed, russeted areas or they may be knotty and deformed. Hard tissues extend from the injury to the core. Control with orchard sanitation and insecticide sprays.

Other Insects.—Oriental fruit moth, scales, buffalo tree hopper, fruit tree leaf roller, and codling moth may damage pears.

Diseases

Fire Blight.—This most destructive disease of pear is caused by the bacterium *Erwinia amylovora*. It is transmitted by bees, aphids, and other insects. It usually appears first as a blossom blight and spreads later to shoots. Blighted blossoms and leaves of blighted shoots turn brown and remain on the tree. From blighted blossoms and shoots the disease-producing bacteria may enter the trunk and main limbs, causing cankers in which they may live over the winter and act as a source of infection for the next year.

During the dormant season remove all infected parts 7–10 cm below obvious infection. Disinfect the cutting tools and wounds, particularly in the growing season, to prevent transmitting the disease (see Pruning). Use a dye-colored mixture of cyanide of mercury and bichloride of mercury (each 1 to 500) in sufficient glycerine to prevent rapid evaporation. Helpful preventive measures consist of removing all suckers and watersprouts from the tree and employing cultural practices that oppose too vigorous wood growth. Antibiotics, such as streptomycin, applied during bloom, give some promise of control (Patterson 1957).

In Yakima Valley of Washington, extensive fire blight developed just prior to harvest. It was found that some of the fruit was blighting from fire blight and some from "sprinkler rot" (*Phytophthora*). If no ooze or no target-like spot is present to identify the causal organism, the fruit would have to be placed in a moist chamber for 24–48 hr to see if the fire blight ooze or the white mycelium of *Phytophthora* would appear (Covey 1970).

Resistance of pear to fire blight is lowered by nutritional deficiencies. N does not increase susceptibility directly but only as it might result in too lush growth. A breeding program for fire blight resistance must involve cultivars of low commercial quality because all of the better quality sorts have relatively low resistance to blight. A comprehensive evaluation of cultivars for their degree of resistance to fire blight indicates considerable differences among cultivars in this respect (Van der Zwet *et al.* 1974).

Scab.—This fungus disease appears as dark, moldy patches on both fruit and leaves. It often causes heavy reduction in yield and serious defoliation in sections having considerable spring rainfall. Sulfur or captan give good control.

FIG. 2.17. PEAR SCAB (*VENTURIA PIRINA*)

Black End.—This disease makes fruits hard, rounded, and often black over the blossom end as they approach maturity. The trouble occurs almost exclusively on trees propagated on oriental stock. No satisfactory control measure is known, but the trouble may be avoided by using French pear as rootstocks (Hayashi and Wakisaka 1957).

Pear Decline.—This disorder is prevalent in western North America and develops at any time in the life of a tree. The decline and death of the tree may be rapid (quick decline) or lingering over several years (slow decline). It appears to be an induced incompatibility in which the scion causes necrosis of the sieve tubes of the rootstock. Pears on *P. serotina* and *P. ussuriensis* are highly susceptible; those on imported *P. communis* are intermediate. Domestic Bartlett seedlings are highly resistant. The role of pear psylla in the incidence of the disease has been substantiated under orchard conditions (Batjer 1960).

REFERENCES

ALLEN, F. W., and CLAYPOOL, L.L. 1948. Modified atmospheres in relation to storage life of Bartlett pears. Proc. Am. Soc. Hort. Sci. *51*, 192-204.

ANSTEY, T.H. 1966. Prediction of full bloom date for apple, pear, cherry and apricot from air temperature data. Proc. Am. Soc. Hort. Sci. *88*, 57-66.

BATJER, L.P. 1960. Relation of pear decline to rootstocks and sieve-tube necrosis. Proc. Am. Soc. Hort. Sci. *76*, 85-97.

BATJER, L.P. *et al*. 1964. Effects of N-dimethyl amino succinamic acid and vegetative and fruit characteristics of apples, pears and sweet cherries. Proc. Am. Soc. Hort. Sci. *85*, 11-16.

BLANPLIED, G.D. 1975A. Core breakdown of New York Bartlett pears. J. Am. Soc. Hort. Sci. *100*(2): 198-200.

BLANPLIED, G.D. 1975B. Pithy brown core occurrence in Bosc pears during controlled atmosphere storage. J. Am. Soc. Hort. Sci. *100*(1): 81-84.

BLANPLIED, G.D., and HANSEN, E. 1968. Effect of O_2, CO_2 and C_2H_4 on ripening of pears. Proc. Am. Soc. Hort. Sci. *93*, 813-816.

BOULD, C. *et al*. 1949. Zinc and copper deficiency of fruit trees. Univ. Bristol Ann. Rept. (1948), 37-38.

BOYNTON, D. 1966. Pear nutrition. *In* Fruit Nutrition. N.F. Childers (Editor). Horticultural Publications, New Brunswick, N.J.

BROOKS, H.J. 1964. Responses of pear seedlings to N-dimethylamino-succinamic acid. Nature *203*, 1303.

BROWN, D.S. *et al*. 1967. Effect of winter chilling on Bartlett pear and Jonathan apple trees. Calif. Agr. *21*, 10-14.

COVEY, R.P. 1970. Fire blight detection sometimes difficult. Better Fruit Veg. *64*, (8): 22-23.

COYIER, D.L., and MELLENTHIN, W.M. 1960. Effect of lime-sulfur-oil sprays on Anjou pear. HortScience *4*, 91.

DAVEY, A.E., and HESSE, C.O. 1953. Experiments with sprays in the control of pre-harvest drop of Bartlett pears. Proc. Am. Soc. Hort. Sci. *61*, 218-222.

DEGMAN, E.S. 1953. Effect of boron sprays on fruit set and yield of Anjou pears. Proc. Am. Soc. Hort. Sci. *62*, 167-172.

DENNIS, F.G. 1968. Growth and flowering responses of apple and pear seedlings to growth retardants and scoring. Proc. Am. Soc. Hort. Sci. *93*, 53-61.

DEWEY, D.H., and UOTA, M. 1953. Post-harvest applications of chemicals for ripening canning Bartlett pears. Proc. Am. Soc. Hort. Sci. *61*, 246-250.

FAUST, M., SHEAR, C.B., and BROOKS, H.J. 1969. Mineral element gradients in pears. J. Sci. Food Agr. *20*, 257-258.

FISHER, D.V., and PORRITT, S.W. 1955. Late harvesting and delayed cold storage of Bartlett pears. Proc. Am. Soc. Hort. Sci. *65*, 222-230.

FRANCIS, J.F. 1970. Anthocyanin in pears. HortScience *5*, 42.

GERHARDT, F. 1954. Rates of emanation of volatiles from pears and apples. Proc. Am. Soc. Hort. Sci. *63*, 248-254.

GERHARDT, F. 1955. Film box liners for storage of apples and pears. U.S. Dept. Agr. Circ. *965*.

GRIGGS, W.H. 1968. Effects of succinic acid 2-dimethyl hydrazide (Alar) sprays used to control growth in Bartlett pear trees. Proc. Am. Soc. Hort. Sci. *92*, 155-166.

GRIGGS, W.H., and IWAKIRI, B.T. 1961. Effects of gibberellin and 2,4,5-TP sprays on Bartlett pear trees. Proc. Am. Soc. Hort. Sci. *77,* 73-89.

GRIGGS, W.H., and IWAKIRI, B.T. 1969. Effect of rootstock on bloom periods of pear trees. Proc. Am. Soc. Hort. Sci. *94*, 109-111.

GRIGGS, W.H. *et al.* 1951. Effect of 2,4,5-Trichlorophenoxypropionic acid applied during bloom on fruit set, shape, size, stem length, seed content, and storage of pears. Proc. Am. Soc. Hort. Sci. *58*, 37-45.

GUR, A.R., SAMISH, R.M., and MAIMON, J. 1968. Stimulation of scion rooting in pear trees. Proc. Am. Soc. Hort. Sci. *93*, 83-87.

HANSEN, E., and MELLENTHIN, W.M. 1967. Chemical control of superficial scald on Anjou pears. Proc. Am. Soc. Hort. Sci. *91*, 860-862.

HARLEY, C.P. 1947. Magnesium deficiency in Kieffer pear trees. Proc. Am. Soc. Hort. Sci. *50*, 21-22.

HAYASHI, S., and WAKISAKA, I. 1957. Studies on yuzyhada disorder of *Pyrus serotina*. J. Hort. Assoc. Japan *26*, 178-184.

HENDRICKSON, A.H. 1941. Factors affecting the rate of growth of pears. Proc. Am. Soc. Hort. Sci. *39*, 1-7.

HEWITT, A.A. 1967. Effects of nitrogen fertilization on some fruit characteristics in Bartlett pear. Proc. Am. Soc. Hort. Sci. *91*, 90-95.

HUSBADO, P. 1969. Planting in old fruit orchards. Gartneryrket *59*, 381-386.

JOHNSON, D.E. *et al.* 1970. Summer mite control. Western Fruit Grower *24*, No. 7, 9-14.

KINMAN, C.F., and MAGNESS, J.R. 1949. Pear growing in the Pacific Coast States. U.S. Dept. Agr. Bull. 1739.

KLEPPER, B. 1968. Diurnal pattern of water potential in woody plants. Plant Physiol. *43*, 1931-1934.

LEWIS, L.N., and KENWORTHY, A.L. 1962. Nutritional balance as related to leaf composition and fire blight susceptibility. Proc. Am. Soc. Hort. Sci. *81*, 103-115.

LOMBARD, P.B., and WESTWOOD, M.N. 1976. Performance of six pear cultivars on Clonal Old Home, double rooted, and seedling rootstocks. J. Am. Soc. Hort. Sci. *101*(3): 214-216.

LOMBARD, P.B. *et al.* 1968. Overhead sprinkler system for environmental control and pesticide application in pear orchards. HortScience *1*, 95-96.

LOONEY, V.E., and MCINTOSH, D.L. 1968. Stimulation of pear rooting by pre-plant treatment of nursery stock with indol-3-butyric acid. Proc. Am. Soc. Hort. Sci. *92*, 150-154.

LUCE, W.E. 1977. Tree wall Anjou, Bartlett pruning, training described. The Goodfruit Grower *28*(3): 10-11.

LUTZ, J.M., and HARDENBURG, R.E. 1968. The commercial storage of fresh fruits and vegetables, and florist and nursery stocks. U.S. Dept. Agr., Agr. Handbook *66*.

MITCHELL, J.W., and LIVINGSTON, G.A. 1968. Methods of studying plant hormones and growth regulating substances. U.S. Dept. Agr., Agr. Handbook *336*.

MODLIBOWKSA, I. 1965. Effects of 2-chloroethyl-tri-methylammonium chloride (CCC) and gibberellic acid on growth, fruit bud formation and frost resistance in one-year-old pear trees. Nature *208*, 503-504.

MODLIBOWSKA, I. 1966. Effect of GA and CCC on growth, fruit bud formation and frost resistance of blossoms of young pears. E. Malling Res. Sta. Rept. *49*, 88-93.

PATTERSON, J.V. 1957. Growing pears. Ohio Agr. Expt. Sta. Bull. *359*.

PHILLIPS, W.R., and ARMSTRONG, J.G. 1967. Handbook on the storage of fruits and vegetables for farm and commercial use. Can. Dept. Agr. Publ. *1260*.

PROEBSTING, E.L. 1953. Factors affecting the concentration of N, P, K, Ca, and Mg in pear leaves. Proc. Am. Soc. Hort. Sci. *61*, 27-30.

RAESE, J.T. 1975. Sprout control of apple and pear with NAA. HortScience *10*(4): 396-398.

RANDHAWA, G.S. *et al.* 1949. Undersize fruit in Kieffer pear orchards. Sci. Agr. (Canada) *29*, 482-489.

ROGERS, B.L., and THOMPSON, A.H. 1968. Growth and fruiting response of young apple and pear trees to annual applications of succinic acid 2,dimethyl hydrazide. Proc. Am. Soc. Hort. Sci. *93*, 16-24.

ROGERS, H.T. 1977. What you do when the canal runs dry. Am. Fruit Grower *97*(4): 13-32.

RYALL, A.L., and ALDRICH, W.W. 1937. Efects of water supply to the tree upon water content, pressure test, and quality of Bartlett pears. Proc. Am. Soc. Hort. Sci. *35*, 283-288.

RYALL, A.L., and DEWEY, D.H. 1955. Post-harvest treatments to hasten ripening of early season Bartlett pears for fresh market. Proc. Am. Soc. Hort. Sci. *61*, 251-256.

RYALL, A.L., and PENTZER, W.T. 1974. Handling, transportation and storage of fruits and vegetables, Vol. 2. AVI Publishing Co., Westport, Conn.

RYUGO, K., and DAVIS, L.D. 1968. Yuzuhada, a physiological disorder of oriental pear and its possible relation to hard-end of Bartlett. HortScience *3*, 15-17.

STRICKLAND, A.G. *et al.* 1941. Spraying with plant growth substances. Fruit World Market Grower (Australia) *42*, 7-10.

TESKEY, B.J.E., LEUTY, S.J., and BRADT, O.A. 1977. Thinning tree fruits. Ontario Min. Agr. Food Factsheet *77-021*.

TESKEY, B.J.E. *et al.* 1973. Pruning and training fruit trees. Can. Dept. Agr. Publ. *1513*.

THOMPSON, J.M. *et al.* 1974. Inheritance of grit cells in fruits of *Pyrus communis*. J. Am. Soc. Hort. Sci. *99*(2): 141-143.

TOENJES, W. 1955. Response of Bartlett pear trees under sod mulch and clean cultivation systems of soil management. Mich. Agr. Expt. Sta. Quart. Bull. *37*, 363-374.

UOTA, M., and DEWEY, D.H. 1953. Respiration and volatile emanation of Bartlett pears as influenced by post-harvest treatment with ethylene and 2,4,5-T. Proc. Am. Soc. Hort. Sci. *61*, 257-265.

UPSHALL, W.H. 1926. Nursery stock identification. Ontario Dept. Agr. Food Bull. *319*.

VAN DER ZWET, T. *et al.* 1974. Fire blight resistance in pear cultivars. HortScience *9*(4): 340-341.

WALLACE, T. 1956. Soils and manure for fruit. Brit. Min. Food Fisheries Bull. *107*.

WANG, C.Y., and HANSON, E. 1970. Differential response to ethylene in respiration and ripening of immature Anjou pears. J. Am. Soc. Hort. Sci. *95*, 314-316.

WANG, C.Y., and MELLENTHIN, W.M. 1975. Effect of short-term high CO_2 treatment on storage of "d'Anjou" pear. J. Am. Soc. Hort. Sci. *100*(5): 492-495.

WEIR, T.E., STOCKING, C.R., and BARBER, M.G. 1971. Botany—An Introduction to Plant Science. John Wiley & Sons, New York.

WESTIGARD, P.H., and WESTWOOD. M.N. 1969. Host preference and resistance of *Pyrus* species to the pear psylla. J. Am. Soc. Hort. Sci. *95*, 34-36.

WESTWOOD, M.H., and BJORNSTAD, H.O. 1974. Fruit set as related to girdling, early cluster thinning, and pruning of pear. HortScience *9*(4): 342-344.

WESTWOOD, M.N., and ALI, N. 1966. Rooting of pear cuttings as related to carbohydrates, nitrogen and rest period. Proc. Am. Soc. Hort. Sci. *88*, 145-150.

WESTWOOD, M.N., and CHESTNUT, N.E. 1964. Rest period chilling requirement of Bartlett pear as related to *Pyrus colleryana* and *P. communis* rootstocks. Proc. Am. Soc. Hort. Sci. *84*, 82-87.

WESTWOOD, M.N., and LOMBARD, P.B. 1968. Effect of seeded fruits and foliar applied auxin on seedless fruit set of pear. HortScience *3*, 168-169.

WESTWOOD, M.N. *et al.* 1968. The possibility of wind pollination in pear. HortScience *1*, 28-29.

WESTWOOD, M.N. *et al.* 1976. Performance of Bartlett pear on standard and Old Home × Farmingdale clonal rootstocks. J. Am. Soc. Hort. Sci. *101*(2): 161-164.

WOODBRIDGE, C.G., and LASHEEN, A.M. 1960. The nutrient status of normal and decline Bartlett pear trees in the Yakima Valley. Proc. Am. Soc. Hort. Sci. *75*, 93-99.

WRIGHT, R.C. *et al.* 1954. Commercial storage of fruits, vegetables and nursery stocks. U.S. Dept. Agr., Agr. Handbook *66*.

3

Dwarfed Apples and Pears

With rising costs of production there has been greater interest by fruit growers in recent years in smaller trees. Although dwarf fruit trees have been grown in Europe for centuries, until relatively recently they have evoked little attention in North America.

Merits of dwarf trees are that they are more easily and more economically pruned, sprayed, thinned, and harvested, and usually they come into fruiting at a much earlier age than standard trees. These advantages more than compensate for the extra initial cost involved in the purchase and planting of more dwarf trees per hectare. Small trees require less space and, if not crowded too much, will produce a very high percentage of well colored fruit (Zeiger and Tukey 1960).

DWARFING ROOTSTOCKS

Dwarf trees generally are produced by propagating the desired cultivar on dwarfing rootstocks. Various degrees of dwarfness are attained by the use of different rootstocks. A positive explanation of how certain rootstocks cause dwarfness is not yet known although many possibilities have been investigated (Dickson and Samuels 1956).

Malling.—The Malling Research Station in England commencing about 1912 developed and typed a number of dwarfing rootstocks; hence, the name Malling stock (M). Malling (M) apple rootstocks are numbered in Arabic numerals. Each number refers to a specific rootstock which produces a tree with definite characteristics of size, bearing age, and so forth, but the Malling number of the rootstock bears no relation to the ultimate size of the tree grafted on it. Most M rootstocks are propagated vegetatively from layers, suckers, or cuttings. In Table 3.1 several M rootstocks are compared with each other and with French crab seedlings.

TABLE 3.1
EFFECT OF ROOTSTOCK ON BRANCH SPREAD AND ACCUMULATED YIELD TO 14TH YEAR

Scion Cultivar	Rootstock	Spread (m)	Yield per Tree (kg)	No. of Trees	Yield per Hectare to 14th Year (t)
McIntosh	M.1	5.6	434	35	33
McIntosh	M.2	5.5	376	35	29
McIntosh	M.4	—	329[1]	35	25[1]
McIntosh	M.8	3.6[2]	96[2]	70	15[2]
McIntosh	M.9	3.2	144	145	46
McIntosh	Fr. crab seedling	5.8	488	27	29
Delicious	M.8	4.6[2]	125[2]	70	19[2]
Delicious	M.9	—	126	363	101
Delicious	Fr. crab seedling	5.2	335	27	20
Northern Spy	M.9	4.3	209	115	67
Northern Spy	Fr. crab seedling	5.7	443	27	26

Source: Upshall (1956).
[1]At 11 years.
[2]At 12 years.

M.9 is the most dwarfing of the commercial M series for apples. It produces a tree only 20–40% of standard size. All of the crop can be picked from the ground. The trees come into bearing very early, often in the second or third year after planting. A mature tree on M.9 may produce up to 0.1 kl of apples, and because of the close planting possible (3–4 m) the trees may give high yields per hectare. Fruit on M.9 is

FIG. 3.1. APPLE CULTIVARS BENCH GRAFTED TO M.9 ROOTSTOCKS AND LINED OUT IN THE NURSERY ROW

usually of high quality and well colored, partly as a result of excellent light conditions. The root system is comparatively small and the root tends to be brittle. For this reason, a tree on M.9 roots must be supported throughout its life.

M.7 has given variable size effects with different cultivars but generally produces a tree 40–60% of standard size. Yield per hectare can be higher than with M.9 and the trees come into bearing earlier than standards. The trees may require some support particularly in their early years in exposed locations. High budding and deep planting should make staking unnecessary on good sites. This practice will tend to minimize suckering also.

M.2 at the end of 14 years in the orchard is about 65% of standard size, but has very little spread of branches. The trees come into bearing about 1 year before standards and are a little more productive for their size. They do not require support.

M.1 is similar to M.2 but is slightly smaller and a little more productive for its size. It does not require support.

M.4 rootstock gives a very productive tree about 75% of standard size. The root system is not large enough to give the tree proper anchorage;

FIG. 3.2. DELICIOUS/M.7 IN THEIR EIGHTH YEAR; REGULAR
(LEFT) AND SPUR-TYPE (RIGHT)

hence, the tree must be supported. With such large trees artificial support is very difficult and costly.

M.11 is a clonal rootstock widely grown in northern Europe, especially north Germany, because of its hardiness, but extensive trials in Poland and Canada have shown that M.11 is similar to M.16 in hardiness and is less winter hardy than Antonovka, Hibernal, and Robusta No. 5.

In New York at the end of 10 years, trees on M.1 and M.13 were medium-sized and likely to reach 3/4 the size of trees on a standard rootstock. Trees on M.7 were the smallest of those tested, followed in order by those on M.5, M.4, and M.2; these four rootstocks were regarded as semi-dwarfing. M.13 and M.16 induced no growth-restricting influence on the cultivars grown upon them. The smaller trees gave a smaller total yield per tree over the 10-year period. However, on a per-hectare basis, yield from the smaller trees equaled or surpassed that from standard trees during the early years (Brase 1953).

TABLE 3.2
TREE SIZE OF 4 CULTIVARS OF APPLES ON 4 EAST MALLING ROOTSTOCKS AT THE END OF 12 GROWING SEASONS

Scion	Rootstock	No. of Trees	Trunk Circumference (cm)	Tree Spread (m)	Tree Height (m)
Delicious	M.1	16	59.0	4.3	4.6
Delicious	M.13	9	55.4	4.0	5.5
Gallia	M.1	19	48.0	3.5	4.1
Gallia	M.13	17	49.9	3.4	4.6
Gallia	M.16	6	53.8	4.0	4.9
Gallia	M.12	6	50.1	4.0	4.9
Golden Delicious	M.1	9	49.0	3.7	4.6
Turley	M.1	10	67.5	5.2	4.9
Turley	M.13	17	60.2	5.3	4.9
Turley	M.13	18	59.9	4.1	3.8

Source: Tukey *et al.* (1954).

In certain apple growing regions the Malling stocks are not fully hardy. Complete killing may not occur but the performance and longevity of the trees on these rootstocks are severely reduced because of the accumulated injury to the root system. M.9 and M.1 rootstocks may be more susceptible to winter injury than the other rootstocks of this series.

M.26 rootstock is a cross of M.16 and M.9 and produces a tree slightly larger and sturdier than does M.9 and is well adapted for high density planting. Like M.9, it is precocious but it may have a little better

TABLE 3.3
AVERAGE SIZE AND CUMULATIVE YIELD OF MCINTOSH TREES AT EIGHT YEARS OF AGE

Rootstock	No. of Trees	Spread (m)	Cross Section Area (cm²)	Cumulative Yield (kg/Tree)
Antonovka	16	4.0	103.5	96
M. Robusta No. 5	16	3.6	80.6	80
M.1	16	3.6	72.9	98
M.2	16	3.6	75.3	91
M.7	16	3.2	47.8	58
M.12	16	4.3	112.2	107

Source: Blair (1955).

anchorage than M.9 in some cases. No positive differences in hardiness can be stated for M.26 as compared with the other M rootstocks, but, although virus-free, a greater degree of susceptibility to fire blight is suspected. In a high altitude test in Tennessee both M.26 and M.9 succumbed to winter injury while Mm7 showed the greatest resistance (Mullins and Gilmore 1977). This rootstock has a greater degree of compatibility with Delicious and its spur types than does M.9 but Golden Delicious on M.26 is a smaller tree than when grown on M.9 (Carlson 1970A).

Malling–Merton.—To gain resistance to woolly aphid, crosses were made of the Malling series with Northern Spy, a cultivar known to be aphid resistant. The resistant rootstocks have been designated as the Malling-Merton (MM) series. Four selections of the MM series have shown promising characteristics other than their resistance to woolly aphid (Hutchinson 1969A).

MM.104 is an easily propagated, well anchored clone which produces a tree the same size as that on M.4 or a little larger. Trees on MM.106 are comparable to those on M.7, that is, about 50% standard size. Trees on MM.109 are slightly larger than those of M.2, while MM.111 rootstock

FIG. 3.3. MCINTOSH/M.9: (LEFT) SUPPORTED BY SINGLE
STAKE; (RIGHT) TRAINED TO A TRELLIS

gives a tree about the same size as that on M.2, that is, about 65% of standard size. More than 10 years of observations at Guelph, Ontario, revealed no lack of hardiness nor any differences in this respect among the four MM rootstocks mentioned here (Teskey 1969B; Tydeman 1953).

A2.—Originating at Alnarp, Sweden, A2 is reported as being hardy in northern Europe but tests at Ottawa, Canada, have not shown it to possess more hardiness than the M or MM stocks. Generally A2 gives a tree similar in size and production to M.2 but it may be useful as an understock for spur-type cultivars.

Virus Tested Rootstocks.—Although still infected with some of the latent viruses such as stem-pitting, chlorotic leaf and spy epinasty, the virus tested rootstocks M.2A, M.7A and M.9A are free of the most economically important viruses such as apple mosaic, rubbery wood and chat fruit.

The East Malling-Long Ashton (EMLA) clones of M.2, M.7, M.9, M.26, MM.106, and MM.111 are entirely virus free.

M.9 Crosses.—In this series of rootstocks, M.9 is a parent in each case. The range of vigor is from more dwarfing to more vigorous than M.16. More dwarfing than M.9 is 3426, while 3436 produces a little larger tree than does M.9. M.27 (formerly 3431), a cross of M.13 × M.9 is a very dwarfing apple rootstock that does not sucker. Trees on M.27 are about one-half the size of trees on M.9 (Rogers 1957).

Robusta No. 5.—This is a selection of *Malus robusta* developed at Ottawa, Canada, and is useful where a high degree of hardiness is required. It is vigorous and stools well, although the stem is inclined to be somewhat rough. In very favorable areas, Robusta No. 5 has only a slight dwarfing effect but under more severe conditions it is similar to M.2. Robusta No. 5 has been used successfully in some of the colder apple regions as the trunk and framework for the less hardy scion cultivar. In areas of fluctuating winter temperatures Robusta No. 5, which has a low

FIG. 3.4. THE ROOT SYSTEM AND VERY HARDY FRAMEWORK OF THIS TREE ARE ROBUSTA 5; THE SCION CULTIVAR IS MCINTOSH

rest period, may break dormancy and stimulate activity in the scion cultivar, resulting in winter injury in the form of sunscald, bark splitting and cambium injury.

Pear Rootstocks

There has been much less commercial interest in North America in dwarfed pear than in dwarfed apple and consequently considerably less research has been done with pear dwarfing rootstocks.

Pear trees are dwarfed by using the quince rootstocks *Cydonia oblonga*. Quince A is one of a series of Angers quinces selected by East Malling. It produces a tree 30–60% of standard size. Under most conditions the quince root hastens the time of coming into bearing but this is not a constant effect. The per-hectare (per-acre) yield of pears on Quince A is generally less than for standard trees (Table 3.4). Virus tested stock is available.

Quince C gives a tree about 50% the size of those on Quince A and is slightly more productive for its size (Upshall 1948).

Bosc and most strains of Bartlett show signs of incompatibility with Quince A when budded directly to it. It is thus better to use an interstock such as the cultivars Duchess, Old Home and Hardy, which are compatible with both the Quince A root and the scion cultivar. A strain of Quince A which is fully compatible with Bartlett is now available.

TABLE 3.4
COMPARISONS BETWEEN STANDARD (FRENCH PEAR ROOTSTOCKS) AND
DWARF (QUINCE A ROOTSTOCKS) PEARS AFTER 13
YEARS IN THE ORCHARD

	Area of Cross Section of Trunk (cm^2)		Year of First Crop		Accumulated Yield (kg/Tree)	
	Standard	Dwarf	Standard	Dwarf	Standard	Dwarf
Orchard No. 1						
Clapp Favorite	116	41	7	5	72	41
Bartlett	47	28	4	4	52	48
Vermont Beauty	122	64	6	6	32	49
Anjou	109	41	7	7	64	24
Orchard No. 2						
Clapp Favorite	181	63	5	4	36	4
Bartlett	85	32	4	3	31	31
Vermont Beauty	120	83	6	4	10	39
Anjou	190	71	6	4	11	30

Source: Upshall (1948).

The Quince A and Quince C root systems lack strength, and pears on these roots have a strong tendency to break at the union. For this reason, dwarf pear trees must be supported thoughout their lives.

Dwarf pear trees have sometimes been planted deep in the orchard in the belief that deep planting (union 15 cm below soil surface) would eventually cause rooting above the union, and such scion-rooted trees would then grow to standard size after having gained the advantage of precocity. Results of deep planting have been extremely variable. At Vineland, Ontario, Clapp Favorite, Anjou and Bartlett on deep-planted dwarfing roots outgrew trees on shallow-planted dwarfing roots after 5 years. Shallow-planted and deep-planted Vermont Beauty on Quince A roots after 13 years were exactly the same size but the yields were 50% higher for the shallow trees. In three other cultivars the differences in yield were insignificant (Upshall 1946).

Interstocks, Knotting, and Bark Inversion

Interstocks.—Dwarfing effect on a scion cultivar can be induced by the use of a stem piece instead of a dwarfing rootstock, thus avoiding the disadvantage of a weak root system. For example, a strong seedling is selected for the understock. About 15−30 cm above the crown the understock is grafted to a dwarfing stock, e.g., M.9. When the M.9 stem piece or interstock has developed 7−15 cm long, the desired cultivar, e.g., Delicious, is grafted onto it. Thus, the dwarfed tree would be comprised of three parts: the seedling rootstock, a stem piece of M.9, and the top of Delicious. With such double-worked trees the interstock, if too short, may become completely overgrown by the understock and/or scion cultivar.

A variation of the above method involves a stem piece consisting of the bark only of the dwarfing stock. With this method the understock forms the stem. A piece of the bark 7 cm or more long is removed from the stem and is replaced by the same sized piece of bark from the dwarfing stock. This method often gives unsatisfactory results.

An interstock is required with Bartlett to overcome the lack of congeniality between Bartlett scion and quince rootstock. Old Home and Hardy are commonly used but some pear cultivars may produce larger fruit with interstocks other than Old Home or Hardy (Tehrani and Hutchinson 1969).

Knotting and Bark Inversion.—By tying a loose knot in the stem of a young seedling, it is possible to check vegetative growth and hence induce fruiting earlier than might otherwise occur. The bearing of fruit by young trees, in turn, has a definite dwarfing influence. However, there is no evidence that the practice of knotting actually reduces the juvenile period of the tree so treated. A knot tied in the stem of the rootstock or interstock does tend, however, to promote earlier fruiting of a clonal apple cultivar.

FIG. 3.5.   A TEN-YEAR-OLD SPARTAN APPLE ON M.9
ROOTSTOCK

Inversion of a ring of bark on the trunk also dwarfs the tree and induces earlier fruiting of the clonal cultivar. With this method, a ring of bark 5–10 cm long is removed from the trunk and replaced in an inverted position. Bark inversion has the same effect as ringing, the use of dwarfing rootstocks or interstocks, or knotting the stem of the tree. All of these treatments check the normal flow of nutrients in the phloem.

Bark inversions avoid the hazards of ringing while producing similar results. The effect lasts several years if bark inversion is done on young trees. The dwarfing effect of bark inversion, however, is not so permanent as the use of dwarfing rootstocks or dwarfing interstocks but the method is inexpensive and can be repeated if vegetative growth becomes excessive (Dickson and Samuels 1956).

Burr-knot

Burr-knot was described in the early 1800s. At first thought to be the result of a bacterium, burr-knots are now known to be an aggregation of dormant root initials common to some apple and pear cultivars. They may appear on any portion of the tree and sometimes reach the size and appearance of a large bird's nest. Woolly aphids and fire blight are sometimes associated with burr-knot but neither is a causal organism. Several clonal rootstocks have shown susceptibility to this phenomenon. In some cases burr-knots may contribute to trunk twisting and fluting of the scion and may promote fire blight infection (Rom 1970).

TABLE 3.5
CROPPING DATA FOR GOLDEN DELICIOUS IN THE MEADOW SYSTEM (1974)

Rootstock	Cycle No.	No. Blossom Clusters/Tree	Fruits/Tree No.	Wt (kg)	Equivalent to: Tonnes/ha	100 Fruit Wt (kg)
MM.111	1st	12.0	9.9	0.98	70.3	9.90
M.9	1st	16.9	11.5	1.08	77.5	9.39
MM.109	1st	15.7	12.3	1.13	81.0	9.19
M.26	1st	14.3	10.7	1.16	83.2	10.84
MM.106	1st	9.0	7.1	0.72	51.7	10.14
MM.106	2nd	17.2	—	1.34	96.1	—

Source: Luckwill *et al.* 1974.

PROPAGATION

Dwarfing stocks are propagated vegetatively from clonal rootstocks with the result that known combinations produce trees with a predictable behavior.

Mound Layering

In establishing a stool bed select, if possible, a fertile loam soil well supplied with organic matter and a high moisture-holding capacity. With a heavier soil the incorporation of sawdust is beneficial. Plant the desired rootstocks preferably in the spring 30—45 cm apart in rows 2—3 m apart. The following spring before bud break, cut off the top growth of the stocks at ground level to force shoot growth below the cut. When the shoots have become about 30 cm high mound them to a height of 5 cm in the center of the row with a 1-bottom plow or other device. Repeat the mounding operation in July so that the final mound is 30—38 cm high. Leave 15-25 cm of the tips of the shoots showing above the mound.

In the fall when the rooted stools are mature but before the ground is frozen, pull the mound down with rakes and remove the rooted shoots from the parent stem. Mound the rootstocks again sufficiently to

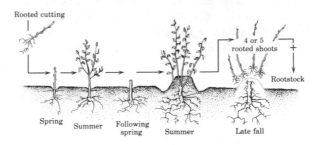

Modified from Carlson (1955)

FIG. 3.6. PROPAGATION OF CLONAL ROOTSTOCK BY MOUND
LAYERING

provide winter protection. Store the rooted shoots over the winter to be planted in the nursery row in the spring, or bench graft and then store them for spring planting in the nursery row. Place the bud at least 15–20 cm above the crown in order to ensure against scion rooting.

A simple modification of the common mound layering method is to place a galvanized steel strip of 0.5 cm mesh screen over the cut stump before covering with soil. The shoots grow through the screen and, as their diameter expands in growth, the screen has a girdling effect. The girdling promotes rooting above the structure and facilitates the separation of the rooted shoots from the mother plant. A variation of this technique consists of bending the juvenile tree over and placing the screen above it. The young shoots from the bent-over tree (instead of the stump) grow up through the screen (Hogue and Granger 1969).

The use of intermittent mist may make the propagation of clonal rootstocks easier and faster. Treating the cuttings with a rooting compound, such as indolebutyric acid, in conjunction with mist sprays, may give best results (Hartman and Hansen 1955).

PRODUCTION

There appears to be little difference among scion cultivars of apples as to their performance on any given dwarfing rootstock. Some variability has been noted in the effects of interstocks on certain cultivars but such differences are inconsistent and may be due more to environmental or other factors than to the cultivar itself. Certain very vigorous cultivars such as Northern Spy are somewhat difficult to control on very dwarfing rootstocks such as M.9 while Golden Delicious trees on M.9 may be more dwarfed than desirable.

With pears, fire blight is a very important factor to consider in the choice of cultivars. In this respect, as in the others, pear cultivars are the same whether on dwarf or standard seedling roots. Certain cultivars, e.g., Clapp Favorite, Bartlett, Flemish Beauty, Sheldon, and Bosc, do not agree well with quince rootstock and should be topworked on an inter-stock which does agree with quince, such as Hardy, Old Home, Anjou, or Duchess.

In general, fruit on dwarf trees tends to be of somewhat higher quality than that on standard trees. This is probably the result of good light conditions for leaves and fruit, and ease of spraying.

FIG. 3.7. SHOWING A CROSS-SECTION OF A BURR-KNOT ON DWARFED APPLE TREE

Site and Soil

Dwarfing rootstocks possess sufficient hardiness for most apple and pear growing areas. However, in the coldest fruit growing districts Malling and quince rootstock are not fully hardy. Of the Malling stocks M.9 and M.1 seem to be the least resistant to low temperature injury; nevertheless, they are as hardy as any of our commercial cultivars in North America. Locate trees on M.9 and M.7 where they will receive some shelter from strong winds.

Soil requirements for dwarfing rootstocks are generally the same as for seedling roots, but the smaller types, being less extensive and more shallow rooted, need a soil higher in fertility and more retentive of moisture than that required by seedling roots.

Planting

Trees on dwarfing rootstocks can be planted on the square, rectangle, hexagonal, or contour, as with standard trees. In the northern regions where winter injury is a factor to consider, plant dwarf fruit trees in early spring. In more southern areas good success is obtained with fall-planted trees.

Depth of Planting.—Set the trees a little deeper than they were in the nursery but make sure that the graft union is well above ground level. If the union becomes covered with soil or mulch for only a short time, rooting takes place from the scion wood. Such a scion-rooted tree continues growth until it is standard size. Inspect trees in a cultivated orchard frequently, and immediately remove any scion roots which may have developed.

Courtesy of Hort. Res. Inst., Vineland, Ont.

FIG. 3.8. HOWELL PEAR ON QUINCE C ROOTSTOCK THE
SECOND YEAR AFTER PLANTING

Spacing.—Correct spacing depends on the ultimate size of the trees. This, in turn, is determined by the rootstock or interstock used, the cultivar, the environmental conditions, and nature of the soil. Trees on the very dwarfing rootstocks, M.9 and Quince A and C, can be spaced as close as 3 × 3.5 m (10 × 12 ft) or 2 × 4 m (8 × 14 ft) to form a hedgerow. Trees on very vigorous roots, such as M.12 and M.16, require the same spacing as standard trees.

Spacing for the semidwarf types, such as M.4 and M.7, may be about 6 × 6, or 5.5 × 6.5 m (20 × 20, or 18 × 22 ft); for moderately vigorous trees, like those on M.2 and M.1, proper spacing is about 9 × 9 m (30 × 30 ft). Harvesting, spraying, and other orchard operations should be considered when deciding on the best spacing and arrangement for fruit trees.

Pollination

The pollination requirements of fruit cultivars are the same whether the tree is standard or dwarf. All apple and pear cultivars should be considered as commercially self-unfruitful. Therefore, the plan of an orchard should include pollinizer cultivars. Base the spacing of pollinizers on the assumption that the effective work range of a honeybee in the orchard is about 25 m.

Support

Trees on M.9 and Quince A and C roots require support throughout their lives since the root systems are small and tend to be brittle. Unsupported trees with a load of fruit may uproot or break just below the bud union.

One method of support is tying the tree to a steel or treated-wood post which is placed before planting. This method is simple and allows for cultivation in both directions. Another method of support is by means of a wire fence or trellis to which the trees are trained somewhat similar to the Kniffin system of training grapes. Or, the wires may be used simply to support the tree without any attempt being made to train the branches to them. Use No. 9 gage wire on posts placed 7 m (24 ft) apart. Anchor the end posts firmly and include a device for loosening and tightening the wires for summer and winter temperatures.

Soil Management

Prepare the soil well for fruit trees. The smaller dwarf trees require somewhat more care and attention than larger trees.

Cultivation.—Cultivate the soil though the growing period for at least the first 4–5 years of the tree's life to avoid weed competition for moisture and nutrients. Seed the orchard to a cover crop in early summer. Indications are that this method of limited cultivation should be continued throughout the lives of very dwarf trees such as those on M.9

and Quince A unless a heavy mulching system is used.

Mulching.—A heavy mulch of hay straw, or other organic matter, improves soil structure, retains moisture under the tree, and supplies plant food. Mulching also reduces the amount of soil temperature fluctuation. Soil beneath a deep mulch may average 12°C or more cooler in the summer than does adjacent unmulched soil. Several of the East Malling rootstocks appear to prefer a relatively cool root temperature (Nelson and Tukey 1956).

Much winter injury is the result of high summer temperatures and droughty conditions in the rooting medium. Mulching is greatly preferred to the use of herbicides, since with the latter the soil is left bare to

FIG. 3.9. OVERHEAD IRRIGATION OF DWARFED APPLE AND PEAR TREES IN THE OKANAGAN VALLEY; (TOP) CONDUCTING PIPES AND SPRINKLER HEADS ABOVE THE TREES; (BOTTOM) UNDERGROUND PIPES WITH SPRINKLER HEADS ON RISERS

become subject to drought and high temperatures. Also, herbicides add neither organic matter nor nutrients for the tree but do create the possibility of herbicidal injury.

Dwarf fruit trees are attractive to mice. Proper precautions to prevent mouse injury include keeping the ground surface free of mulch, grass, and weeds within at least 60–90 cm (2–3 ft) of the trunk.

Fertilizer.—In general, fertilizer recommendations for standard fruit trees are applicable to trees on dwarfing roots, allowing modifications for differences in size. Dwarf trees may need a little more feeding for their size than does the same cultivar on standard roots. A complete fertilizer high in nitrogen worked into the soil under the tree or broadcast and covered with mulch is generally beneficial. Often, when a mulch is used no inorganic fertilizer supplement is needed. Base specific recommendations, however, on observations of growth, vigor, and production.

Pruning and Training

A combination of summer and dormant pruning is a recommended practice for dwarfed fruit trees. Dormant pruning stimulates vegetative growth; summer pruning tends to depress vegetative growth and increase fruit production. Thus a correct balance between growth and fruit production can be maintained by carefully regulating the balance of summer to dormant pruning.

Dwarf trees in the orchard should not be headed much lower than standard trees since early production pulls the branches down quickly. In the home garden the head of the tree can be somewhat lower than is desirable under orchard conditions.

The same principles apply to the pruning of dwarfed trees as for standard treees. However, much less pruning relative to size is required for the smaller trees and emphasis is on the continuous renewal of the fruiting wood. Since the pruning operation is so much easier with the smaller trees, more attention can be given to making the proper cuts.

In training trees to wires, the trunk, after the initial cutting back at planting time, is allowed to reach the top wire (1.5 m above ground) as quickly as possible. To obtain arms at the desired points, disbudding or deshooting methods are very useful. Remove all buds or spurs except those in suitable positions for developing arms. This can be done most easily just as the buds are bursting. Each year loosely tie the new terminal growth along the wire. Once the arms have been selected little further pruning is necessary other than the removal or cutting back of the too vigorous upright-growing shoots and some thinning out where the branches become too crowded.

FIG. 3.10. A "MEADOW' PLANTING, LONG ASHTON, ENGLAND:
IN FULL BLOOM (LEFT); PARTIALLY HARVESTED (RIGHT). THE
SCION CULTIVAR IS TWO YEARS OLD

Dwarf trees other than those trained to wires are trained to the modified central leader system and pruned as are standard trees. For very dwarf trees such as those on M.9, select 4–6 branches that make wide angles (not less than 35°) with the trunk and are spaced 10–20 cm apart spirally along the central leader.

Train the tree to a modified central leader with the main laterals spaced and spiralled (except in the case of the trellis-trained tree) a-round the central stem. Vary the spacing between branches from 10–40 cm depending on the ultimate size of the tree. Avoid crotch angles less than 35°, especially with small trees in areas subject to deep snow. Narrow angled branches can suffer much breakage under the weight of melting snow.

Once the tree is trained, practice a light annual pruning with the object of controlling excessive growth and maintaining a constant renewal of fruiting wood. This will entail the heading back of too vigorous shoots as well as the thinning out of the older and weaker wood.

Fruit Thinning

With many cultivars, especially of apple, an annual fruit thinning program is necessary if fruit of good size and quality is to be harvested. For annual bearing in cultivars that have a strong biennial bearing tendency, thin within three weeks after full bloom.

Very small trees may be thinned efficiently by hand with the worker standing on the ground, but fruit on dwarf trees can be thinned by chemical sprays in the same way as for standard trees (Chap. 1 and 2).

Dwarf and semidwarf trees are precocious and often "fruit out," i.e., lose their leaders and fail to gain proper size if allowed to fruit too soon.

This should be guarded against in the leaders particularly. Defruiting can be accomplished successfully by spraying with NAA and oil, or NAA and carbaryl.

Insects and Diseases

Cultivars on dwarfing rootstocks are subject to attack from the same insects and diseases that attack standard trees. With smaller trees, however, it is possible to control these pests at less cost than with standard trees and to do so more effectively and with greater control of drift.

REFERENCES

BLAIR, D.S. 1955. Apple investigations. Can. Agr. Res. Sta. Rept. (Smithfield, Ontario), 13-14.

BRASE, K.D. 1953. Ten years results with size-controlling rootstocks. N.Y. Farm Res. *19*, 4.

CAMPBELL, A.I. 1969. The effect of some apple viruses on the susceptibility of two clonal rootstocks to collar rot, *Phytophthora cactorum*. J. Hort. Sci. *44*, 69-73.

CARLSON, R.E. 1955. Cultural practices in propagating dwarfing rootstocks. Mich. Agr. Expt. Sta. Quart. Bull. *34*, 492-497.

CARLSON, R.E. 1970A. Compact fruit tree. Dwarf Fruit Tree Assoc., Mich. State Univ. *3*, 95-96.

CARLSON, R.E. 1970B. A proven apple rootstock—EM.7. Fruit Var. Hort. Dig. *25*, 11-12.

CHANDLER, W.H. 1937. Zinc as a nutrient for plants. Botan. Gaz. *98*, 625-646.

DICKSON, A.Q., and SAMUELS, E.W. 1956. Mechanism of controlled growth of dwarf apple trees. J. Arnold Arbor. Harv. Univ. *37*, 307-313.

HARTMAN, F.O., and HANSEN, C.J. 1955. Rooting of softwood cuttings of several fruit species under mist. Proc. Am. Soc. Hort. Sci. *66*, 157-167.

HOBBIES, E.W. 1958. Pillar Trees. *In* The Fruit Year Book. Royal Hort. Soc., London.

HOGUE, E.J., and GRANGER, R.L. 1969. A new method of stool bed layering. HortScience *4*, 29-30.

HUTCHINSON, A. 1969A. A 13-year study with Malling-Merton and other apple rootstocks. Ann. Rept. Agr. Res. Inst. Ontario *299*.

HUTCHINSON, A. 1969B. A 26-year study of M.7 apple rootstocks in comparison with M.1, M.2, French crab seedlings. Ann. Rept. Agr. Res. Inst. Ontario *299*.

HUTCHINSON, A. *et al.* 1969. Rootstocks for fruit trees. Ontario Dept. Agr. Food Publ. *334.*

LUCKWILL, L.C. *et al.* 1974. The meadow orchard system. Long Ashton Res. Sta. Rept., May 1975.

MARTIN, G.C., and WILLIAMS, M.W. 1967. Comparison of some bio-chemical constituents of EM.9 and EM.16 bark. HortScience 2, 154.

MULLINS, C.A., and GILMORE, T.R. 1977. Tree survival of three Malling rootstocks in Tennessee. Fruit Var. J. *31*(2): 40-41.

NELSON, S.H., and TUKEY, H.B. 1956. Effects of controlled root temperature on growth of East Malling rootstocks. J. Hort. Sci. *31*, 55-63.

QUINLAN, J.D. 1969. Mobilization of ^{14}C in the spring following autumn assimilation of $^{14}CO_2$ by an apple rootstock. J. Hort. Sci. *44*, 107-110.

ROGERS, W.S. 1957. Advances in rootstock research. J. Royal Agr. Soc. (England) *118*, 64-75.

ROM, R.C. 1970. Burr-knot observations on clonal apple rootstocks in Arkansas. Fruit Var. Hort. Dig. *24*, No. 3, 66-68.

SHOEMAKER, J.S., and TESKEY, B.J.E. 1962. Practical Horticulture. John Wiley & Sons, New York.

TEHRANI, G., and HUTCHINSON, A. 1969. Interstocks for dwarf pear. Res. Inst. Ontario Rept. *21.*

TESKEY, B.J.E. 1969A. Eighteen years observations of EM.9 rootstocks at Guelph, Ontario. Ontario Dept. Agr. Food Agdex *211 /36.*

TESKEY, B.J.E. 1969B. Malling-Merton rootstocks performance at Guelph, Ontario. Ontario Dept. Agr. Food Agdex *211 /36.*

TUKEY, H.B. 1964. Dwarfed Fruit Trees. Macmillan Co., New York.

TUKEY, R.B. *et al.* 1954. Twelve years performance of Malling rootstocks. Proc. Am. Soc. Hort. Sci. *64*, 146-150.

TYDEMAN, H.M. 1953. Description and classification of Malling-Merton and Malling 25 apple rootstocks. East Malling Res. Sta. Ann. Rept., 55-63.

TYDEMAN, H.M. 1954. Description of certain M.9 crosses. East Malling Res. Sta. Ann. Rept., 86-88.

UPSHALL, W.H. 1946. Dwarf apple and pear trees. Can. Hort. Home Mag., June.

UPSHALL, W.H. 1948. Dwarf apple and pear trees. Ontario Dept. Agr. Food Bull. *456.*

UPSHALL, W.H. 1956. Notes on Malling rootstocks for apples. Hort. Res. Inst. Ontario Rept. *13.*

ZEIGER, D., and TUKEY, H.B. 1960. An historical review of the Malling apple rootstocks in America. Mich. Agr. Expt. Sta. Bull. *226.*

4

Peaches

Botanically, the fruit of the peach (*Prunus persica*) is a drupe. It develops entirely from a superior ovary, and consists of the outer portion, or skin (exocarp), the middle portion (mesocarp) of the ovary wall, which becomes fleshy, and the inner portion (endocarp), which becomes hard and forms the stone.

Horticulturally, the production of peaches in the United States and Canada averages about 2.3 million kl (65 million bushels) per year. Production of peaches in North America comprises 80% of the world's production. Because of the soft fruit, the peach plays a minor role in fresh fruit trade, but the product is suitable for canning.

Peaches are produced commercially in about 2/3 of the states in the United States. The volume varies from year to year but comparative production for the 10 leading states is roughly as follows (in millions of litres [millions of bushels]): California more than 1050 (30); South Carolina, 211 to 247 (6 to 7); Georgia, 106 to 141 (3 to 4); Pennsylvania, 70 to 106 (2 to 3). New Jersey, Michigan, North Carolina, and Virginia, 35 to 70 (1 to 2) each; Arkansas and Alabama, 18 to 35 (1/2 to 1).

California produces about 50% of the total peach crop of the United States; about 63% is canned, 27% is sold as fresh fruit, 6% is dried, and 4% is frozen.

In Canada, peaches are produced commercially only in British Columbia (mostly in the Okanagan Valley), and in Ontario (in the Niagara district and several southwestern counties); the total averages about 88 million litres (2 1/2 million bushels) annually.

Great increases in real estate values have led to the abandonment of many established peach orchards and the planting of new ones in areas of lower overhead. The desire for lower production costs has brought about an interest in labor-saving decices and practices, such as chemical thinning of the fruits, and in high-density orchards.

AGE OF BEARING, ORCHARD LIFE, AND YIELD

Ordinarily, a northern peach orchard requires 5 to 6 years to reach production. Then the trees may continue to be profitable until they are 18 years old or more.

Peach orchards are shorter lived in the Deep South than in the North, often lasting less than half as long (averaging about 8 years), but they come into bearing earlier, often in the fourth year.

Yields in mature orchards may average 106 to 176 litres (3 to 5 bushels) per tree. Yields vary from year to year according to freeze damage and other factors from a crop failure to a very heavy crop, and are consistently more uniform in some districts and on certain sites than others.

PROPAGATION

A peach cultivar (variety) is not reproduced true to its name from seed. The seedlings differ greatly, even though there has been no cross-pollination. New cultivars originate from seed, but it may be necessary to plant thousands of seeds and wait for a long period of testing before anything better, or as good as the parents, is found.

All nursery peach trees consist of two parts: the rootstock, and the top, which is the desired cultivar.

Rootstocks

Some years ago most of the seed used for peach rootstocks came from seedling peach trees growing wild in the Carolinas, Tennessee, Kentucky, and elsewhere in the south. Such seed became known as "naturals." The ancestors were brought to this country by the early settlers. The supply of "naturals" has dwindled, and has given way to seed of Lovell, Elberta, Halford, Suncling, and others largely obtained from processing plants.

Certain peach rootstocks, such as Rutgers Red Leaf, Siberian C, and Nemaguard are grown primarily for seed, not for their edible fruit.

Bearing trees of Loring, Redhaven, and Babygold 5 on Siberian C seedlings have defoliated earlier than those on the other seedling rootstocks. Early cold acclimation on scions in the fall and scion cold hardiness in midwinter were enhanced more by Siberian C seedlings than by those of the other seedling rootstocks. Bud survival and fruit set of Redhaven and Babygold 5 scions were greater on Siberian seedlings than on the other seedling rootstocks following an outdoor stress of $-23.3°C$ ($-10°F$) in January. The cold hardiness of phloem, cambium, and xylem stem tissues were closely correlated with each other in the fall, but were not correlated with cold hardiness of flower buds on the same shoots. Seedlings of Siberian C appeared to enhance early scion dormancy and they increased scion bud hardiness by as much as 4.7°C (40.5°F) in the fall and 1.3°C (34.3°F) in mid-February (Layne *et al.* 1977).

Because of nematode (eelworm) injury in certain areas, attention has been given to rootstocks that are highly resistant or immune to this pest. The U.S. Dept. of Agriculture some years ago introduced Shalil from India, Yunnan from China, and Bokhara from Russia. Seed of these is in very limited supply, is comparatively expensive, and the pest-resistance quality is not as high as desired.

Elberta, "naturals," Lovell, and other commonly used seedling root-stocks are susceptible to both *Meleidogyna incognita* and *M. javonica* nematode species. Fumigation of the soil before planting with such as EDB, DD, or Telone at manufacturer's directions in bands 1.8 to 2.4 m (6 to 8 ft) wide might be worthwhile on some sites.

Morphological reactions of Nemaguard to *M. javonica* probably are based on ability of these rootstocks to restrict the nematode's food supply, consequently inhibiting its normal development. These resistant stocks are not recommended for cold areas (Malo 1967).

Afterripening of the Seed

In Florida, sow afterripened seed of Nemaguard in late January or early February for budding in June. Remove the seed from the pit in autumn and store it in damp peat or perlite at 1.7° to 7.2°C (35° to 45°F) for 40 to 60 days before planting. If the seeds are not removed from the pits, stratify and keep the pits in cold storage for up to 100 days. If plantings are allowed to become dry before sprouting occurs, seeds may again become dormant (Sharpe 1966).

Lovell has a long afterripening requirement (120 days at 0.6° to 10°C [33° to 50°F]; its optimum is 7.2°C [45°F]). Viability of its seeds decreases with age during dry storage; pits stored for more than one year give low germination. Loss in viability during dry storage is somewhat less with "natural" pits than with Lovell (Brase 1948).

After 140 days of afterripening, seeds which must be excised from the stony pericarp—even if sound—appear to be of doubtful planting value. A high percentage of pits which have split open after completing the required afterripening period may contain either decayed or nonviable seeds which on planting will fail to grow.

Once the seed has afterripened it is very susceptible to disease. Contamination is more likely to occur if seeds must be disturbed and transferred from the afterripening medium to the soil. In fall planting the individual pits are not in contact with each other as they are during stratification. Treatment of the seed (seed separated from the enclosing stony pericarp) with Fermate or Spergon increases the stand.

Fall Planting of Seed

Common nursery practice is to plant the seed in the field in the fall, about 5.1 cm (2 in.) deep, in rows about 1.2 m (4 ft) apart. Seed so planted germinates the following spring. Poorly drained soil results in a poor stand of seedlings. Guard against destruction of the seed by mice.

Fall planting eliminates storage afterripening of the seed and planting the sprouting embryos in the spring. The stony pericarp surrounding the embryo may not crack at the same time, so it may be necessary to inspect the afterripening seed lots several times during early spring and to sort out ones that have germinated. Delay in planting the sprouted embryos may result in many seedlings with "goose necks" that are worthless as budding stock.

In Ontario, in 9 out of 12 years, nonshelled Elberta pits, bedded in soil for a year, then planted in nursery rows, gave fair to good commercial stands of seedlings the following spring. In the other 3 years most of the viable pits germinated in the bed the first spring, thus leaving wholly or largely worthless pits. A good stand of seedlings was obtained the following spring (in mid-April) by cracking the pits of the previous year's crop which had been put in a soil bed as soon as the flesh was removed (Upshall 1942).

Whenever good 1-year-old pits are available, nurserymen prefer to use them because of good commercial stands without resort to cracking. The 1-year-old pits germinate quicker and more uniformly, and commonly result in a 60–75% stand of seedlings.

Germination and Vigor

Pits from high-chilling cultivars ripening less than 30 days before Elberta usually give better stands of seedlings than pits from earlier cultivars.

Nurserymen often obtain peach pits for rootstocks from canneries, where the pits remain in piles for varying lengths of time. Cull fruits from which the pulp has disintegrated are also often used as a source of seed. However, pulp disintegration may adversely affect the viability of the seed. Reduced viability may occur with seed that remains in rotting pulp for 10 days or in fermented peach juice for 7 days (Haut and Gardner 1935).

Handling Seedling Rootstocks

Two common methods of handling peach rootstock seedlings are (1) transplanting the seedlings from a shallow germination bed to the nursery row soon after they have emerged in spring, and (2) planting the pits directly in the nursery row in fall or early spring.

A marked depression of growth may occur after transplanting. At budding time, transplanted seedlings often are smaller and produce smaller nursery trees. That the latter may catch up to the larger ones after 2 years in the orchard is not much encouragement to nurserymen, since a No. 2 tree does not command top price. Since many nurseries dig peach trees in the fall and give them some winter protection, the possible winter injury to larger trees is not likely to deter nurserymen from trying to obtain a high percentage of No. 1 trees (Upshall 1940A).

Budding

Budding is done during the first year of seedling growth. June is the best time in the south, and August to early September in the north. The T-bud is the standard method, using buds from either bearing or nursery trees. After union of the bud and seedling, the top of the seedling is cut off just above the point on it where the bud of the desired cultivar was inserted. In northern nurseries this is done when growth has begun in the spring following late summer budding. Doing this work in the dormant season may result in buds developing slowly or in failure to grow.

With June budding, the seedling top may be cut off after the bud has "taken." Usually, however, forcing of June buds is necessary and, for this purpose, some leaves are left on the stock.

Early Versus Late Removal of Nursery-row Laterals.—Young peach trees in the nursery row in the north commonly make several lateral branches low on the trunk. Usual practice in Ontario is to remove laterals up to 46 cm (18 in.) from the ground either in June (early removal) or early September (late removal) (Upshall 1940B).

In June in the north, break out the laterals by hand by a pull to one side; the wounds heal quickly. The leaves from the axils of which the laterals have arisen are left intact. They usually become large and aid in thickening of the trunk. The tree should reach a good height by early July when the tarnished plant bug may cause damage by stopping the growth at terminal points.

Removal of laterals every week during June is unnecessary for there is not enough difference between 1-week and 2-week intervals to warrant the nearly double labor required. Usually 3 prunings suffice but the periods between them should not exceed 2 weeks, or else some of the laterals may be hard at the base and their removal by hand may damage the trunk. Adequate budwood usually will be available from 1 branch per tree and this later need should be considered at the time of removal in June, i.e., leave 1 branch lower than is necessary to make an acceptable nursery tree. This early removal of laterals promotes a smooth clean trunk.

If the laterals are left until early Septembber, use a knife or shears to completely remove all leaves on the lower 46 cm (18 in.) of the trunk. Since the low branches are often large by September, their removal at this time leaves an objectionable wound. If the pruning is delayed until late September or October the wounds may remain open to canker infection until spring. The wounds heal well enough to avoid canker if the pruning is done in early September.

Those who favor comparatively late removal of laterals claim that the largest tree results when the lower laterals are left intact during the summer. These branches provide plenty of good budwood. No real difference in tree size results when lower laterals are removed in early September, or when they are broken out weekly, or every second week, up to a point 46 cm (18 in.) above the union. Retaining all low laterals until digging time gives larger trees, but in practice such unpruned trees are hard to handle and, if pruned, invite canker infection at the wounds.

Combinations of ethephon and Dupont-WK surfactant were effective in stimulating leaf abscission of 20 cultivars on 5 species of tree fruit nursery stock. Species and cultivars varied considerably in sensitivity to mixtures of these chemicals, but one to three applications at weekly intervals of 200 to 400 ppm ethephon plus 1 to 2% D-WK were generally effective. These treatments caused little xylem, phloem, or bud damage except to Redhaven peach and Early Italian prune. Rome apple was sensitive but was not damaged by a concentration of 200 ppm ethephon plus 1% D-WK. D-WK stimulated leaf abscission when used alone at 1 to 2% but acted more slowly than when combined with ethephon. Ethephon alone at 200 to 400 ppm was usually ineffective (Larsen 1973).

Nursery Tree Grades.—Nursery peach trees are graded commercially according to height and diameter as follows:

Caliper		Minimum Height	
(cm)	(in.)	(m)	(ft)
1.7	11/16 and up	1.4 and up	4 1/2 and up
1.4−1.7	9/16−11/16	1.2 and up	4 and up
1.1−1.4	7/16−9/16	0.9 and up	3 and up
0.8−1.1	5/16−7/16	0.6 and up	2 and up

CULTIVARS (VARIETIES)

Groups or Races of Peaches

South China.—Characteristics of this race are small fruit, usually a beaked end (Peen-to is saucer-shaped), white flesh, a sweet honeylike flavor and low acidity, but an astringent skin, dull fruit color, soft flesh of poor carrying quality, and a low chilling requirement, e.g., Jewel, Okinawa.

Spanish.—The fruit is rather small, pubescent, low in quality, and clingstone or freestone. This race, occasionally seen in Florida, seems to have contributed little to the development of our modern cultivars.

Persian.—It includes cultivars originating as importations from Persia by way of Italy and Great Britain, e.g., the once highly regarded Early Crawford and Late Crawford. High quality and a good amount of red skin color feature the more popular cultivars.

North China.—Chinese Cling was imported from China in 1850 and first fruited in America at Columbia, S. Carolina. It is one of the parents of Elberta (Chinese Cling × Early Crawford) and of a number of white-fleshed cultivars, e.g., Belle of Georgia, and Greensboro. The fruit is dull colored, and white- or yellow-fleshed. The leaves are large. Most commercial cultivars are found in this group, in the Persian group, or in crosses of them.

Purposes, Preferred Color, and Trends

The main outlets for peaches are fresh fruit shipments, processing, local stores, and roadside markets. Certain cultivars are suitable for all markets; some, however, are more desirable for one specific purpose.

The preferred flesh color is yellow. Only a few commercial cultivars are now white-fleshed.

The popularity of cultivars changes. For example, in 1950, 45% of the peach production of South Carolina consisted of Elberta. By 1958, it made up 21% of the crop, and by 1965, it had dropped to 11%. Golden Jubilee made up 1 1/2% of the volume in 1958, and today is just about nonexistent in this state. Dixigem decreased from 4 1/2% in 1950 to 3 1/4% in 1958, and to less than 1% in 1965. Redhaven continued to increase from 2% in 1958 to 7 1/2% in 1964. Redskin, which had not been planted in 1950, accounted for 2 1/2% of the production in 1958, and 2% in 1965 (Van Blaricom and Sefick 1967).

Selected Cultivars

The selected cultivars are listed in order of first picking. However, since more than one picking is usual, cultivars of each group may overlap within their group, and often with those in the one-week gap between groups. Also, a given cultivar will be picked earlier in the south than the same cultivar in the north. State and province names indicate the source of origin.

Very Early (7–9 Weeks Before Elberta).—Most of these cultivars, especially the older ones, are clingy at the pit, soft, comparatively small, and adapted primarily for quick, limited use.

Earlyvee.—Yellow; highly colored; fair quality; very tender skin; semifree. Comes into bearing early; requires heavy thinning; for roadside or local market only. (Ontario)

Candor.—Bright red blush over 75% of the fruit; rich yellow ground color; semifree. (North Carolina)

Earlired.—Bright attractive red color; sizes well when thinned; 3–4 days after Earlyvee. (Maryland)

Collins.—Not as bright as Earlired, but less subject to split pits, and somewhat hardier in bud. (New Jersey)

Maygold.—Not adapted to the north. Fairly low chilling requirement. Yellow; small to medium; soft. (Georgia)

Others.—Cardinal (Georgia), Marcus (North Carolina).

Early (5–6 Weeks Before Elberta).—Mostly sold on fresh fruit markets; lack the requirements for a superior processed product. They are, however, somewhat larger, with somewhat meatier flesh and richer flavor than the previous group. Some of them tend strongly to split pits.

Jerseyland.—Fruit medium size, round, considerable red color, usually freestone when ripe, fairly firm. (New Jersey)

Coronet.—Firm, medium size; freestone; pit tends to cling when slightly immature; good freezer. (Georgia)

Royalvee.—Very productive; above average hardiness; needs thinning for size; good quality; semifree. (Ontario)

Garnet Beauty.—A bud sport of Redhaven which it resembles but the stone is not free.

Harbelle.—Medium firm; bright red over yellow ground color; rich flesh color; above average quality; freestone. (Ontario)

Sunhaven.—Medium-large; round, uniform. (Michigan)

Sentinel.—Medium size; fairly firm; attractive red blush over a yellow ground color. (Georgia)

Others.—Dixigem (Georgia), Redcap (Georgia), Marigold (New Jersey) Prairie Dawn (Illinois).

Early Midseason (3-4 Weeks Before Elberta).—These cultivars are larger than early ones, with more substance to the flesh; usually freestone; mostly acceptable to processors.

Redhaven.—Above medium size (heavy crops require thinning); solid red skin color; colors while immature, making it difficult to estimate proper picking stage. When well-grown and properly handled it is a superior cultivar in the east and midwest. Ranked second in California in 1969 in fresh market shipments but declined in popularity there, as under high temperatures the fruit tends to soften rapidly. (Michigan)

Golden Jubilee.—Elberta-type fruit; color rather poor for fresh market (red cultivars preferred for this purpose); bruises rather easily; freestone. (New Jersey)

Envoy.—Medium size, attractive fruit of good quality; freestone; ripens evenly, requiring only two pickings; tree not particularly strong, but crops consistently. (Ontario)

Ranger.—It has considerable resistance to bacteriosis and for this reason has become popular in the Ridge and Sandhill areas of South Carolina. (Maryland)

Loring.—Large; firm; attractive; freestone. Possibly more tender in bud than Redhaven, and is more susceptible to peach canker. (Missouri)

Veteran.—Has produced good crops under conditions when many cultivars had a light crop; high quality for both fresh and processing markets, but lacks appeal when compared with redder cultivars. (Ontario)

Velvet.—Medium to large; attractive all-over red color. (Ontario)

Harmony.—Notably winter hardy; highly productive; good size; medium firm; better than average quality. (Ontario)

Southland.—High quality; freestone; considerable red around cavity. (Georgia)

Sunhigh.—Good color, flavor, and shape. (New Jersey)

Others.—Hiley, (white), Fairhaven, Keystone, Washington, Redglobe, Fireball, July Elberta, Triogem, Halehaven.

Midseason (1-2 Weeks Before Elberta).—As with most other fruits, midseason cultivars usually possess some superiority over earlier ones in size, firmness, and certain other characteristics.

Sullivan Early Elberta.—Large; freestone; pit slightly larger than in Elberta; uneven ripening may be a problem for the frozen market. (Georgia)

Vanity.—Fruit medium to large; attractive red color. (Ontario)

Early Elberta (Stark's or Gleason).—Fruit brighter yellow than Elberta; requires thinning; tree not as rugged as Elberta.

Suncrest.—Captured the lead in California in 1969 in fresh fruit shipments. (U.S. Dept. Agr.)

Others.—Belle of Georgia (white), Madison, Cresthaven (Michigan).

Elberta Season.—At one time Elberta was by far the leading freestone peach. It is still extensively planted, but has lost considerable favor to earlier and redder cultivars, particularly for the basket trade. Its faults include mediocre quality, poor appearance, and dropping of the fruit before full maturity. It has size, firmness, and adapts to various soils. It is a good processor, but may also be largely replaced for this purpose.

Fay Elberta.—Ranked third in 1969 in California in fresh fruit shipments.

Dixieland.—Freestone; makes good canned and frozen products.

Blake.—Large; firm; roundish; highly red colored; freestone; canned product often shows serious brown and purple discoloration; tree none too hardy in the north. (New Jersey)

Redskin.—Fruit medium size; good quality; freesone; better color and quality than Elberta, which it is largely replacing in some areas as a basket peach. (Maryland)

Others.—Olinda (Ontario), Jefferson (Virginia), Jerseyqueen (New Jersey).

Late (2-4 Days After Elberta).—*Shippers Late Red.*—Large; firm, brilliant red at pit cavity. (Michigan)

Rio Oso Gem.—Large; firm; red pit cavity. (California)

Others.—Afterglow.

Firm, Nonmelting Yellow Clingstones.—This type is used in the manufacture of baby food purées.

Babygold 5, 7, 8, 9.—The range in season is from 2 weeks before to about 1 week after Elberta, in the order listed. No. 5 is probably the best of the four. (New Jersey)

Suncling.—More resistant to brown rot and makes a better tree than Babygold 7, with which it matures. Fruit medium size; blushed red where exposed to sun.

Pollination

Most peach cultivars are self-fruitful, so single cultivars can usually be planted in solid blocks. Exceptions, such as J.H. Hale in which little or no effective or viable pollen is produced, have been eliminated from new commercial orchards.

LOCATION AND SITE

Many growers have made serious and costly mistakes in selection of site, soil, cultivars, and so forth. Profitable fruit growing depends greatly on how well the project is planned from production and marketing standpoints before planting. Too much emphasis cannot be placed on careful study of the location and site from every aspect. Orchards in fringe areas or on marginal soils often are the most seriously damaged by cold or other factors.

Commercial peach production may often be a more hazardous enterprise than apple production owing to greater susceptibility to damage from spring frosts and low temperatures. The producer has little or no control of winter temperatures and only limited control over spring frosts. The chief decision is not how to reduce damage from cold in established peach orchards but rather the selection of sites which will avoid such losses.

Air Drainage

A commercial peach planting may be a failure if the site selected lacks at least some elevation above broad areas of nearby land. A slight slope or rolling land allows better air drainage that lessens hazard of frost damage. Cold air drains away from high land into lower levels. If the low land is represented by a narrow valley without an air drainage outlet, this area can quickly fill with cold air which may spread over the entire region.

Climatic history of a particular area and the performance of orchards or individual trees are useful in evaluating a particular location or site for a future peach orchard. A variation in elevation of 6.1 to 9.1 m (20 to 30 ft) which affords adequate air drainage may give frost protection. Avoid obstacles to air drainage, e.g., a solid high hedgerow or a windbreak along the lower side of the planting restricts air movement to an adjacent low area, and this condition may cause loss of fruit on trees in several rows. Similarly, an orchard on a slope may suffer damage because certain buildings restrict air movement.

In general, a northeastern exposure is desirable. It may prevent too early bloom, and there is less possibility of winter injury.

Tempering Effects of Water

Large, deep, open bodies of water, such as the Great Lakes, have a tempering effect on temperatures on sites close to them, particularly on their south or east sides. Comparatively small inland lakes, unless numerous, do not provide sufficient water volume to have much effect on air temperatures. Some successful peach orchards overlook wide river valleys which provide good conditions for drainage of the cold air layer away from the orchard site.

Soil and Soil Drainage

Poorly drained sites result in a spotty planting, short periods of production, low yields per acre, and high cost of production per bushel.

A high water table or a hard impervious soil layer prevents deep rooting. Injury from "wet feet" is caused by water coverage of part or all of the root system for more or less extended periods. The results are flagging of the succulent shoots and leaves, yellowing and premature dropping of the foliage, and in severe cases cessation of growth and death of the tree. Good drainage helps avoid the ailment. However, in many peach areas the occurrence of wet pockets in the soil, low spots in otherwise well-drained orchards, or a temporary rise in the water table due to heavy rains and swollen streams, all make the injury difficult to avoid entirely.

Heavy rains that cause standing water, and a waterlogged soil in a young peach orchard may result in death of the trees in a few days if the temperatures are 26.7°C (80°F) or above. An explanation is a breakdown due to to shortage of oxygen to the roots and release of a hydrolyzing enzyme that acts as a cyanogenic glucoside in the tree and releases hydrogen cyanide (Lowe 1970).

Growth and yield may be increased by blocked compared with open tile. This is probably related to moisture stress at certain critical periods. Peaches are particularly sensitive to compaction. Reduction in K uptake under compaction may reduce growth and yield. There is favorable growth and yield response to planting trees directly over the tile compared with between tile lines. Tree location in relation to tile lines may therefore be important (Cline 1967).

During periods of excessive rain, in a loose topsoil underlain by a compact subsoil fruit trees may be blown over because of a shallow root system. During dry periods, the trees may make short top growth which results in low yields and fruit of poor size and quality.

Deep, naturally well-drained soils are best (Table 4.1). A soil that permits deep root distribution produces trees that are longer lived, more productive and better able to withstand such adverse conditions as

TABLE 4.1
SOIL EFFECTS IN PEACH PRODUCTION AND LONGEVITY

| | Yield per Tree | | | | Trees Remaining After | |
| | 5–6 Years | | 13–14 Years | | 6 Years | 14 Years |
Soil Description	(litres)	(bu)	(litres)	(bu)	(%)	(%)
Shallow and imperfectly drained soils	25.7	0.73	19	0.54	80	19
Soils intermediate in depth and drainage	31.3	0.89	39.8	1.13	73	62
Deep and well-drained soils	37.7	1.07	62	1.76	91	81

Source: Archibald (1966).

drought, excessive rain, and low winter temperatures than does a shallow soil. It is in dry years that fruit size will likely be greater on trees on a deep soil than on a shallow soil (Archibald 1966).

Plum roots are notably resistant to "wet feet" and are sometimes used as stocks for peach trees in heavy soils. Peach and plum are not particularly compatible, however, and poor unions usually result in weak, short-lived trees.

Peach and apricot were more sensitive to waterlogging than was plum. All three species became more sensitive as temperature was increased between 17° and 27°C (62.6° and 80.6°F). A scion of a more tolerant species did not overcome the sensitivity of the roots. Both cyanogenic fluoride content and the proportion of it that was hydrolyzed during waterlogging was higher in peach than in plum roots. The rate of cyanogenesis increased with both temperature and time (Rowe and Catlin 1971).

Root System

Over 90% of the small or feeding roots of peach trees is located in the upper 46 cm (18 in.) of the soil. With young trees, an even greater percentage of the total tree roots is located within this soil depth. This condition may account for the detrimental effect of some cover crops in the early life of the trees. It may also explain the reduced growth of trees on land which is even slightly eroded. Roots of five-year-old peach trees may penetrate 3.1 m (10 ft) on some soils. Even though the number of feeding roots at the lower depths is small, this may be of great value during drought and for anchorage during strong winds.

The roots spread slightly farther out than the spread of the branches. Fertilizer should be spread under the tree and slightly beyond the perimeter of the tree.

PLANTING THE ORCHARD

There are two types of nursery peach trees, the one-year-old and the June-budded. The former is commonly planted in the north, and the latter in the south.

Year-old Nursery Trees

A one-year-old nursery tree has completed one full year's growth in the nursery since it was budded, and is nearly always branched. The largest grade is usually the highest priced tree. It is the size preferred by many growers, but medium-sized trees often give good results.

June-budded

June-budded trees usually are smaller than one-year-old trees, and are mostly whips or with only a few branches. After three years trees of the largest size may still be slightly larger than the other sizes of nursery stock. But this small difference may not affect the yield in later years.

Differences after two years in the orchard is attributed to practices used in digging, transplanting to the orchard, and the method and severity of pruning during the first few years (Savage 1955).

Exceptionally uniform planting material may increase in variability whereas variable planting material may become more uniform (Moore 1968).

When to Plant

In all regions planting is best done during the dormant period of the trees, ranging from late fall in the south to early spring in the north. In regions with mild winters the optimum time of planting during the dormant period is usually during or close to a rainy period.

When temperatures below zero accompanied by high winds are likely to occur and where soil without a snow cover freezes deeply, fall-planted trees may dry out considerably. Very late fall planting is not generally wise in northern regions.

Plant in the north as soon as the soil is dry enough to work in the spring, so that the trees may make new root growth and become established while soil temperatures and moistures are favorable. The percentage of survival often depends on the degree of maturity of the trees when dug in the nursery and the method of storage and spring delivery. Buds of late-spring transplanted trees force leaves quickly, and if the new root system cannot absorb enough moisture and nutrients for growth the trees may be killed by spring or summer growth. Do spring planting at least 2 to 3 weeks before the leaf buds will make much growth.

Planting Plans and Procedure

Where the land is level or slopes not more than 1.5 m in 30.5 m (5 ft in 100 ft), the square system is commonly used. Sometimes the spacing between rows is made greater than that in the row to aid in spraying and certain other operations.

For commercial production, do not set peach trees closer than 6.1 × 6.1 m (20 × 20 ft) in the square method. Even when planted at this spacing the trees often crowd each other when eight years old. To allow for full development of the trees, 7.3 × 7.3 m (24 × 24 ft), 6.1 × 7.6 m (25 × 25 ft), 6.9 × 6.9 m (22.5 × 22.5 ft), or even 7.6 × 7.6 m (25 × 25 ft) or more may be preferred to 6.1 × 6.1 m (20 × 20 ft) of fertile soil capable of producing vigorous trees. The wider spacing permits cultivation and use of power driven machinery without severely breaking the trees and scarring the branches.

TABLE 4.2
NUMBER OF TREES PER HECTARE (ACRE) FOR VARIOUS PLANTING DISTANCES

Spacing (m)	(ft)	Trees (per ha)	(per Acre)	Spacing (m)	(ft)	Trees (per ha)	(per Acre)
3 × 6.1	10 × 20	81	200	6.1 × 7.6	20 × 25	35	87
4.6 × 4.6	15 × 15	78	194	6.8 × 6.8	22.5 × 22.5	35	87
4.9 × 4.9	16 × 16	69	170	7.3 × 7.3	24 × 24	31	76
4.6 × 6.1	15 × 20	58	144	7.6 × 7.6	25 × 25	28	70
5.5 × 5.5	18 × 18	54	134	7.9 × 7.9	26 × 26	26	65
6.1 × 6.1	20 × 20	44	108	8.5 × 8.5	28 × 28	22	55

With the advent of the "in-and-out" power hoes some 3 × 6.1 m (10 × 20 ft) or 4.6 × 6.1 m (15 × 20 ft) plantings have been made, with the expectation that more trees per hectare (acre) will result in increased yields per hectare (acre) in the early life of the orchard. There has been little experience on how to handle these trees in later life, and pruning and other practices need to be developed. This is true also for developments in mechanized harvesting.

Handling Trees on Arrival from the Nursery.—Many trees are so weakened by drying out before planting that they can start only poorly, or not at all. Make every effort to avoid such damage.

The time when trees are received from the nursery may be unsuitable for immediate planting because of soil or weather conditions. Nursery trees may be held temporarily by "heeling in" as described for apple trees.

Setting the Trees.—Before peach trees are planted, prune off any broken or diseased roots. Root development takes place more readily on smooth than on broken and torn surfaces. While planting, keep the roots from drying out.

Digging the holes by hand, although costly, is likely to result in the best stand of trees. Where large hectarages (acreages) are to be planted, plow a furrow on the line of the staked tree rows. A subsoil plow may be run through the bottom of the furrow to aid in digging holes. Dig the holes large enough in width and depth to accommodate the tree roots without crowding them. Set the tree 2.5−5 cm (1−2 in.) deeper than it stood in the nursery, with the bud union showing just above ground. Distribute the roots in the hole to their natural positions. They then anchor the tree properly and continue to do so when the tree needs more anchorage. If all roots grow in one direction, anchorage is mainly from one side only.

In the Deep South, it is customary to apply water in a "water cup" of soil made around the young tree. Apply a handful of complete fertilizer on the soil near the tree in late spring or early summer after the tree has developed foliage, or apply a few forkfuls of strawy manure around the base of the tree soon after planting.

Since topsoil contains more organic matter and mineral nutrients than subsoil, firm topsoil around the root system to bring the particles into contact with the roots. Complete the filling in with some loose soil.

Filler Trees.—Planting peach trees in an apple orchard may result in the harvesting of several crops before the apple trees come into good production, thus reducing the cost of bringing an apple orchard into bearing. The main drawbacks of the peach tree as a filler are: the necessity of using different spray programs; the peach trees compete with the apple trees; the employment of certain practices, such as the use of cover crops, to increase or maintain soil fertility may be impractical; good peach sites eventually become occupied by apples.

Setting solid blocks of peach trees simplifies operations, results in orchards of longer life, and gives a greater return on the initial investment. If filler trees are planted between permanent trees, nothing should interfere with the timely removal of the fillers.

Avoid growing plums either as interplants or in close proximity to peach orchards, in areas where peach yellows or phony disease are serious problems. The plum is a favorite host of the leafhoppers which transmit these diseases.

Intercrops.—It is often economically desirable in orchards on level ground to grow an intercrop during the first 2−3 years. The trees may benefit from the tillage given such a crop and from the fertilizer applied to it. Avoid intercrops such as strawberries, tomatoes, and other solanaceous plants that may be host to verticillium wilt; also long-term perennials, e.g., raspberries, rhubarb, etc., that will interfere with orchard operations; and root crops, e.g., late carrots, the harvesting of which amounts to a late cultivation of the orchard soil. In the second and third years allot less space to the intercrop than in the first year. Unless the trees are spaced at least 7.6 m (25 ft) apart, do not plant an intercrop after the third year. In fact, growing intercrops in peach orchards is the exception rather than the rule.

Peach Replants.—The practice of replanting old peach sites to peaches is becoming more frequent with the decrease in suitable new sites. Often the trees die in their first and second years, or they seem to grow well for a few years and then decline in vigor and yield.

Many of the peach tree losses in replanted peach orchards are due, in part, to a fungus disease caused by mushroom root rot (*Clitocybe tabescens*). This rot shows a whitish color under the dead bark of the base of the trunk and characteristic small mushroomlike growths. It often occurs on land where oak trees grew previously (Savage 1954).

Favorable response sometimes is obtained from the addition of relatively low rates of high-Mg lime to the soil. A readily available source of N may give the trees a rapid start, perhaps, in turn, overcoming an inhibiting effect.

Replant trees set out in fumigated soil have fewer nematodes associated with their roots than nonfumigated orchards.

TRAINING AND PRUNING PEACH TREES

In general, heavy pruning of young peach trees delays their coming into full production, but perhaps not as much as with the apple.

Pruning has at least four major functions in the peach orchard: (1) training the tree to a strong framework, thereby preventing undue losses from breakage of limbs; (2) shaping the tree within a reasonable size for convenient harvesting and other orchard operations; (3) thinning the crops, thereby influencing size of fruit and yield; and (4) helping to maintain adequate shoot growth on mature trees. Pruning of peach trees is required every year. Peach trees are usually trained to an open center.

Pruning Peach Trees at Planting Time

Peach trees usually should be pruned at the time they are set in the orchard. This pruning is done better immediately after than before plantings. Postpone the pruning of fall-planted trees in northern regions until late winter or early spring.

Pruning at planting time has a twofold purpose: (1) to reduce the top of the tree to balance the loss of roots and the shock of transplanting, and thereby encourage tree survival; and (2) to promote the selection of a suitable framework.

June-budded Trees.—June-budded trees are dug the same year they are budded and planted in fall or winter months in the south. With trees less than 0.6 m (2 ft) high, no heading back may be necessary at planting time.

Cut back unbranched trees over 0.6 m (2 ft) tall (whips) to 61–76 cm (24–30 in.), depending on the original height. If there are laterals, but these are not strong or high enough to make good scaffold branches, cut them back to 1–2 buds at planting time and reduce the height of the tree to 61–76 cm (24–30 in.).

If there are well distributed branches with strong buds on the tree, select 3–4 of these branches, preferably spaced spirally around the trunk, and eliminate the others. Cut back the selected ones to 2–3 live

buds. If the branches are sturdy they need not be cut back quite so
heavily.

Year-old Trees.—Northern nursery peach trees are sold as 1-year-old
stock. They are usually branched.

Most peach trees that are grown close together for a year in the nursery
have few, if any, laterals which are desirable for framework branches.
Usually the laterals are weak, broken at the base, or lack uniformity in
size and spacing.

Trees of medium to large size are good to plant. Weak trees are objectionable, and overvigorous ones are not necessary. Cut back the strongest
laterals to 1−2 bud stubs (almost to a whip). Do not attempt to select
leading branches at this time. During the first growing season remove
suckers originating below the point of nursery budding and also watersprouts on the trunk below the desired height or head level.

An alternative, but little used, method to the preceding one is deshooting. Buy tall trees, 1.2 m (4 ft) or more high and 14−17 mm (9/16−11/16
in.) in diameter. Cut them back to 91−112 cm (36−44 in.) and the laterals
to 1−2 buds. When the shoots from these buds become 5.1−12.7 cm (2−5
in.) long, select 3−5 of them to make well-spaced scaffold branches and
remove the others. In this method more emphasis is placed on selecting

FIG. 4.1. PEACH TREE WITH ABOUT 50% OF THE NEW WOOD
REMOVED IN PRUNING

the main framework branches during the year of planting than is the
case when only dormant pruning is practiced. The lowest shoot is
commonly left about 51 cm (20 in.) above ground.

In the Deep South, where fully grown trees are maintained lower and
are shorter lived than those of the north, a guide is as follows: at
planting time, cut back the trees to a single stem 61 cm (24 in.) high. If

laterals have formed, cut the lower ones off flush with the stem, but allow 2.5–5.1 cm (1–2 in.) stubs to remain on the upper ones, to ensure leaving buds for new shoot development. After the tree sprouts in spring, select three evenly spaced, vigorous shoots to be the main scaffolding. Remove or cut back other shoots and remove all low growing watersprouts and suckers. In the first winter, cut back each main branch about 1/3 to an outward-growing lateral. Remove watersprouts and limbs that are too near the ground. Keep the trees growing low, for easy picking in later years. In early spring each year, rub off watersprouts from main branches within 0.6 m (2 ft) of the head or center of the tree. Continue the training procedure for the second and third winters (Sharpe 1966).

After the third winter, pruning consists of removing over-crowded branches, removing watersprouts, heading back terminal growth to check excess height, and keeping the center open. Fruiting laterals need to be thinned and removed depending on vigor and cultivar habits, in order to reduce fruit load.

Pruning Peach Trees After One Year in the Orchard

After 1-year-old nursery peach trees have grown a year in the orchard, in late winter select 3 strong, well-placed, wide-angled, outward-growing laterals for the scaffold branches and remove all others, including any strong or overlapping central growths. If 2 branches of equal size tend to divide the tree and form a Y at the trunk, remove 1 of them. Cut off the main leader just above the top scaffold branch. Head back the selected branches lightly to an outward-growing lateral. Shorten any branch that is growing out of proportion to the rest to keep the tree symmetrical.

Aim to develop a good open-center tree. The scaffold branches must be able to sustain heavy loads, so select those that form a strong, wide-angled union with the trunk. They must be well spaced for strength and to allow room for secondary branching.

Pruning in the Formative or Prebearing Period

Dormant Pruning.—The framework of most trees after 2 years in the orchard commonly consists of 3 main scaffolds plus secondary branches. A first good move after sizing up a tree is to make a few selected cuts in the center of the tree and to remove badly placed branches. Remove limbs and watersprouts that grow up through or across the center of the tree. Then, cut back branches to outward-growing laterals to continue the spreading and open-center process.

Killing of the tips of the shoots by the Oriental fruit moth results in the development of laterals back of the injury. Some of these laterals grow desirably outward, but much of this type of injury and subsequent growth makes the tree denser and requires more thinning out than is normally done.

After 3 years in the orchard, the framework of the trees should be well formed, and only light to moderate pruning (mostly thinning-out) is needed to keep the center open and fruiting branches well spaced. Most of the first crops are located on the lower 1/3−1/2 of the tree.

Pruning reduces yield in direct proportion to its severity, but may increase size of fruit. In its other effects on the tree, heavy pruning acts like an application of N by delaying maturity and reducing color but often increasing size. The following experiment in Ontario, Canada, on 3 cultivars which was started 1 year after planting and which at the end of the sixth year was still in progress, illustrates some of the effects just mentioned.

Light pruning involved the removal of low drooping branches, and dead, dying, and very weak wood (Table 4.3). Heavily pruned trees received the light pruning as above plus a heavy heading to an outward-

TABLE 4.3
PRUNING YOUNG PEACH TREES THE FIRST 6 YEARS IN THE ORCHARD

	Light Pruning		Heavy Pruning	
Prunings per tree (avg)	0.5 kg/year	1.1 lb/year	1.3 kg/year	2.9 lb/year
Terminal growth (6th year)	35.6 cm	14 in.	56 cm	22 in.
Yield (avg. 5th and 6th years)	18 kg/tree	40 lb/tree	8.2 kg/tree	18 lb/tree
Amount picked at 1st picking (%)	52		20	
Fruit color	Good		Poor	

Source: Upshall (1940B).

growing lateral on 2-year wood. This treatment reduced tree height but at the expense of bearing area.

Heavy heading increased terminal growth over light pruning by 20 cm (8 in.) in the sixth year but reduced average yields by 10 kg (22 lb) per tree for the fifth and sixth years. The crop under both treatments was a failure through frost injury in the fourth year, was normal in the fifth year, and light in the sixth year through poor conditions for fruit setting. Fruit maturity was delayed by the heavy heading as shown by the smaller proportion of the crop at the first picking, 20% compared with 52% for light pruning.

The heavily headed trees made strong growth in the tops and continued to grow late in the summer. The fruit was poor in color, partly due to excessive shading and partly to the sugars in the leaves going into vegetation rather than into the fruit. Low sugar content of the fruit was associated with a deficiency of red color. Also, these dense trees resulting from heavy heading were slow to dry off after rains and heavy dews, and so were more subject to loss of fruit from brown rot (Upshall 1943).

A combination of heavy pruning, heavy N fertilization, and late cultivation can be very undesirable in peach production. The trees themselves become more subject to winter injury and the fruit is late in maturing, excessively large, coarse, poor in color and quality, and does not carry or hold up well.

When the tree is young the growth is heavy, but as bearing begins the weight of fruit tends to cause the branches to bend down and lean outwards, producing a low, widespread tree. As the trees become older and bear heavy crops, the growth tends to decrease and the fruit to become smaller. The time of development of this condition depends on the soil fertility, moisture supply, cultivar, and pruning.

Trunk diameter may not be a reliable guide to controlled pruning. When trees were pruned to a constant weight of 1-year-old wood removed per unit trunk area gain, the percentage of new wood removed decreased with age of trees 3−6 years old and also decreased with N fertilization. If trees were pruned to a constant percentage of new growth removal, the weight of new wood removed per unit of trunk growth increased with both tree age and N fertilizer. The best method of controlled pruning was based on estimation of percentage of new wood removal by linear measurement, simplified by using a few measured trees as models and pruning by visual comparison.

N fertilizer increased yields largely by increasing the amount of bearing wood and hence number of fruits per tree. The effect of pruning on yield and fruit size was largely by reducing the amount of bearing wood and, hence, number of fruits per tree. Yield of trees whose production was low as a result of heavy pruning was increased by less pruning and higher N fertility with a minimum sacrifice in fruit size.

Until the tree reaches the maximum height desired, a light to moderate thinning out of the surplus branches and heading back to outward-growing laterals continues as a part of the pruning program.

Summer Pruning.—A little summer pruning is sometimes beneficial on young nonbearing trees to relieve the congested growth which may occur in vigorous trees of this age, and in certain areas as a supplementary measure to avoid canker. If the growths are too crowded they may be spindly and light in color, setting few and weak fruit buds. If enough light reaches these individual shoots they are plump, many are filled with strong fruit buds, and the color is a bright red; this characterizes healthy, mature fruiting shoots. Summer pruning should be light and confined to the new growth. It does not take the place of dormant pruning. More summer pruning may be wise in the Deep South where the growing season is longer than in the north.

In California, summer pruning of early cultivars resulted in reduced flower-bud drop and a resultant increase in number of flowers produced per tree, increased yield, and increased shoot size, especially when coupled with a blossom thinning program. July pruning was generally more effective than August pruning (Brown and Harris 1958).

Shaping Young Unpruned Peach Trees.—Trees which have grown for several years without pruning become tall and dense, with weak crotches, surplus scaffold branches, and many suckers and watersprouts.

A first remedial step is to remove the suckers and watersprouts. Eliminate weak crotches as much as possible without "butchering" the trees. Reduce the main scaffold branches to 3−4 and cut back each of these to outward-growing secondary branches. Also, cut back some of these secondaries to outward-growing laterals or buds to force spreading and an open center. Remove growths which grew across the center of the tree.

Pruning Mature Peach Trees

Mature peach trees are pruned heavier than the other fruit trees. Probably no other fruit tree responds to proper pruning and declines with neglect so readily as the peach. Objectives in pruning bearing trees are: promote shoots of the most productive length; keep the height of the tree reasonably low so that picking is convenient; thin out the tree for suitable coverage in spraying; and distribute production over the tree so that the fruit size and color are good. It is a more efficient usage of labor to make a few well-chosen relatively larger cuts than a large number of small pruning cuts.

Bearing Habit.—The peach tree bears its fruit laterally on wood of the past year's growth. New shoots grow from terminal buds and from some of the side buds that did not produce flowers. Hence, the tree tends to produce shoots at the ends of the 1-year-old wood and to extend the fruiting wood farther out each year.

Aim for a terminal growth of 20−46 cm (8−18 in.) each year on most of the outer branches on trees 8−12 years old. Shoots of these lengths carry blossoms from end to end and promote maximum yield. Overvigorous wood 127−152 cm (50−60 in.) long is not fruitful except on its secondary shoots; in fact, shoots more than 61 cm (24 in.) long are often deficient in

FIG. 4.2. PROLONGED DORMANCY IN THE DEEP SOUTH MAY
RESULT IN SOME FRUIT DEVELOPING TO WALNUT-SIZE
WHILE ADJACENT BUDS ARE JUST OPENING

fruit buds. Weak wood of 7.6–10 cm (3–4 in.) is not very productive anywhere in the tree; shoots less than 15 cm (6 in.) long may produce small fruit and carry few fruits per shoot. Comparatively little fruit is found on the short-lived spurs. Pruning should take these facts into account.

It is not always possible to achieve an ideal combination of pruning and other practices which will ensure ample vigor without some undesirable "bull wood" growth. Such growth tends to choke the top areas and produce a dense, shaded condition unfavorable for well-ripened, fruitful growth lower in the tree. It is better to remove such branches entirely in each annual pruning than to head them back. Such removal cuts are also necessary to prevent undesirable increase in tree height.

In mature trees most of the fruit, and the best of it, is produced in the upper 1/3–1/2 of the tree, which is a reversal of the situation with the first few crops borne. The terminal and lateral shoots that have developed at the perimeter of the tree are important in fruit production, as are the shoots and short growths in the center of reasonably open trees.

As the trees become older and produce heavy crops, growth tends to slow down; some heading back is then necessary to encourage the growth of strong, new wood. In making such cuts, first cut back the strongest and longest branches to keep the tree at the desired size. Make cuts to an outward-growing side branch in 2 to 3-year-old wood.

Fruit and Leaf Buds.—The fruit buds of the peach are plump and round. The leaf buds are comparatively small, narrow, and pointed. Using F for fruit bud and L for leaf bud, arrangements such as follows may be found: F, FF, FFF, LFL, LF, LLL, LL, L. The peach tree of bearing age seldom fails to make enough fruit buds for a heavy crop. In fact, it commonly produces many more fruit buds than are desired.

An average mature tree of moderate vigor produces at least 25,000 buds. A good yield for such a tree is 141–176 litres (4–5 bu). Theoretically, less than 5% of the total fruit buds formed on such a tree are enough to produce a crop. In fact, a good crop may be obtained from trees with 90% killed buds, and growers are generally satisfied if 1/2 the buds survive the winter and early spring.

On vigorous shoots, at a given node there may be 1, 2, or 3 fruit buds. Where three buds occur at a node, the center bud tends to be a leaf bud and the two outer buds to be fruit buds. On shorter growths, fruit buds are often borne singly beside a leaf bud. On shoot growths of 76 cm (30 in.) or more long, particularly near the base, the lateral buds may be mostly leaf buds.

Shoots arising from lateral buds in 1-year or older wood are potentially more productive than shoots arising from lateral buds on the current season's growth. Number of buds per shoot, total buds per shoot, and percentage of fruits set are greater in lateral shoots than in secondary shoots. Only leaf buds are suitable for budding nursery trees (Ashley 1940).

When and How Severely to Prune

Prune peach trees every year; otherwise a thick-topped, leggy tree devoid of high fruiting wood develops. Do this in the dormant season, moderately, lightly or heavily, depending on fruit prospects. Moderate pruning may be employed in maintaining the physiological balance necessary for sustained production of well-colored, high-quality peaches. Too severe pruning may result in an unfavorable carbohydrate to N ratio. If too little pruning is done, smaller fruit and higher production costs commonly result. If too much pruning is done, too much N is available per growing point, upsetting the optimal C to N ratio, which in turn results in too much vegetation growth, over-shading, poor fruit color and quality, and reduced yields. Few commercial growers are underpruning their peach trees, but some are overpruning and are obtaining extra vegetation growth at the expense of fruit production and quality (Westwood and Gerber 1958).

The time of year to start the annual pruning is, in part, governed by the hectarage (acreage), available labor, and climate. In small orchards pruning can best be done just before growth starts, but in large orchards it usually must begin earlier. Trees pruned while not fully dormant or too early in the dormant season may suffer from low temperature.

In certain areas, because of the danger of canker, do not prune peach trees before late winter or very early spring. The wounds from winter pruning may become infected before they can form callus tissue and begin to heal over, and winter injury may occur near the cut, which favors the development of canker. Prune the older trees first, finishing with the young trees.

Increase in yield was obtained in Georgia from February, March, and May pruning over November and January pruning. Also, June pruning increased yields when compared with November, December, and January pruning. Further evidence is provided that cold injury is involved at some point in the peach tree decline process (Ferree and McGlohon 1975).

Also on a site in Georgia with a history of short life of peach trees, brown discoloration of trunk cambial area was greater in trees pruned in November or December than in trees pruned in January. Trees pruned in February had less discoloration than those pruned in January.

Corrective Combined Thinning Out and Heading Back

When trees have become 1.8 m (6 ft) or so tall, cut the outward-growing branches to strong laterals, so that it is not too difficult to care for the tree economically. Then moderately head back each year to prevent further increase in tree height.

To maintain desirable growth of terminal shoots 20–30 cm (8–12 in.) long on trees 10–15 years old, it may be necessary to prune heavier than on younger trees. How much cutting back is needed depends on how much new wood the trees are producing. It may be necessary to cut back into older wood, removing entire branches 3–4 years old. As tree growth slows down still more, say at 15–18 years, even more severe pruning may be required to produce good fruiting wood.

It is not wise, as sometimes done, to make the many cuts needed to head back most of the shoots in the top of the tree. The many cuts are laborious and induce a dense growth in the top which shades the lower fruiting wood and interferes with development of good fruit color and quality. Rather, thin out some shoots and cut back some of the 2- or 3-year old wood to desirable laterals.

North.—In Ontario, 10 trees of 6 different cultivars were pruned by 3 methods from the beginning of the third to the twenty-fourth year in the orchard. The treatments, varying primarily in kind and location of wood removed, but also in degree and severity of pruning, were: (1) Basic. Removal of dead, dying, and very weak wood, and low drooping branches. (2) Basic plus heavy heading. As in (1), plus a heavy heading back of the extremities of all branches to a lateral arising from 2-year wood. This is a method used and approved by many growers in the Niagara District. (3) Basic plus lateral thinning. As in (1), plus about 1/2 of the laterals (smaller ones) on 1-year-old wood. This method involves many small cuts and therefore takes considerable time.

Both the basic-plus-lateral-thinned trees outyielded the basic-plus-heavily-headed trees, without adversely affecting the amount of large fruit. Also, the total time required per weight of fruit for pruning, thinning, and picking was greater for heaving heading than for the other two treatments. It would seem that a lighter type of pruning than is practiced in many Niagara orchards would increase yields and profits (Bradt 1955).

South.—On Lakeland sand in North Carolina heavy pruning corresponded with the prevailing local practice and consisted of severe cutting back of upright limbs, considerable thinning out of shoots, and heading back most of the remaining shoots about 1/2 their length. Light pruning consisted of slightly cutting back to outward growing shoots and removal of only crossing, upright, or weak-crotch limbs, and some thinning of shoots. Medium pruning was intermediate between these two extremes (Schneider and Correll 1956).

The heavily pruned trees made the longest terminal growth each year; the lightly pruned trees the least growth, and the medium pruned trees an intermediate amount. New wood on the medium pruned trees was nearly ideal in quality; that on the lightly pruned trees was rather thin and weak; that on the heavily pruned trees was somewhat too succulent and vigorous.

Although differential pruning affected terminal growth, after 4 years it did not have a measurable effect on trunk circumference increase. Hence, trunk circumference is not necessarily a good criterion of tree size and potential production.

Increasing pruning severity tended to result in increased fresh weight of leaves, and increased N and K contents of the leaves. It resulted in decreased percentage of dry matter, and decreased Ca and Mg content of the leaves. The P content was increased the first year and then decreased.

With Dixired, heavier pruning decreased total yields, number of fruits 5.1 cm (2 in.) and up, and amount of red fruit color. Average weight per fruit was not affected. With Redhaven, heavier pruning decreased total yield and red color, and increased average weight per fruit the first and third years of the test. The number of fruits 5.1 cm (2 in.) and up was affected only during the first year of the test.

It is commonly asserted that pruning opens up the tree and lets in more light to improve the color of the fruit, but the data obtained did not support this viewpoint. N content of the tree is known to affect color development. The heavily pruned trees, which had the most foliar N, produced fruit with the least red color. Thus, the N effect did not fully explain all the observed color differences. The heavily pruned trees were very vigorous, and the fruit was produced inside the dense succulent growth. The trees did not spread as the fruit enlarged during maturation; hence, the fruit was heavily shaded, developing only moderate color. Lightly pruned trees made a less dense growth, and when the fruit began to swell and mature, the weight of the crop spread the trees, letting more light reach the fruit and resulting in better color.

Shearing the Top.—The practice of shearing the top is still seen in many orchards. In effect, the fruiting area is greatly reduced, and high yield per tree is sacrificed, often excessively, for size of fruit.

Long Pruning.—Long pruning is essentially a light method, with little or no heading back of the terminal branches. Cutting is mostly confined to the removal of side growth and thinning out unwanted branches. This system may result in removal of only about 70% as much wood as from a moderate treatment.

In British Columbia, higher yields resulted from long pruning during the first 6 years after planting. From the sixth to tenth year, however, yields on the short-pruned and medium-pruned trees increased and became equal to those on long-pruned trees. Due to the initial delay in bearing, however, yields for the full 10-year period were not as great with short and medium pruning as with the long-pruned trees. At the end of the 10-year period the three types of trees were about the same in size, but the vigor of new terminal growth was greater with short-pruned than with medium- or long-pruned trees (Palmer *et al.* 1941). When long pruning is done in the east and midwest, the scaffold branches may have as much as 2.7 to 3 m (9–10 ft) of unfruitful area on the lower portions. Trees pruned in this fashion, even when 7.6 × 7.6 m (25 × 25 ft) apart, practically meet in the centers of the rows by the time they are 9–10 years old. When the trees become older it is difficult to maintain both vigorous renewal growth and desirable fruit size except by greatly increasing the number of thinning-out cuts. Thus, over the lifetime of a planting very little is saved in pruning labor and this method of pruning would call for 9.1 × 9.1 m (30 × 30 ft) spacing and a highly fertile soil (Burkholder and Baker 1957).

Kentucky System.—The Kentucky System is featured by eliminating all but 2–3 selected basic scaffold branches which are formed in the first 2 years. The subsequent framework develops on them. A minimum of pruning is done for several years. Then heading back and corrective pruning is practiced.

Dehorning Peach Trees.—Some years ago a common practice was to "dehorn" or "dehead" peach trees at least once during their bearing age, in a year of crop failure. The practice has fallen into disrepute, and is advocated now only in special cases.

In the first place, it is a mistake to let the bearing area or head of the tree become too high. Secondly, where lowering of the top is necessary, it is done more suitably by moderately heavy pruning than by drastic cutting back.

When dehorning is done, cut all main limbs back to stubs 0.6–1.2 m (2–4 ft) long. The most objectionable upright branches in the center of the tree may be removed entirely. The branches that are to be developed into the main framework will arise from the stubs which are left. Summer pruning often may be practiced to advantage so that the lower portion of the tree will receive adequate sunlight for the production of fruitful types of growth. In the year after dehorning much thinning out is needed.

Drawbacks are: better results are usually obtained by other procedures; weak trees may die as a result of the drastic treatment; a rank, dense growth results which shades the bearing area in the lower part of the tree; the entire crop is lost for 1–2 years; and the main fruiting area may soon become high again in the trees.

Old, weak peach trees can be rejuvenated by moderately heavy cutting back (not dehorning) into 3–4 year-old wood, leaving the main limbs 1.8–2.4 m (6–8 ft) long, with the stubs of lateral branches attached. After such pruning, followed in later years by normal pruning, the trees may bear good crops. Replacing old trees with new cultivars should be considered.

Prevention of Sunscald Following Heavy Heading.—Heavy heading may result in severe sunscald to the bark along the tops of exposed branches. This can be partly prevented by a swabbing the tops of the most exposed branches with a heavy whitewash made of sweet skimmed milk and lime or a latex-thiram mixture.

Effect of Pruning on Root Distribution.—The greater amount of dry matter in the aboveground portion of lightly pruned trees is reflected in the greater total growth and extent of distribution of roots. Less severe pruning with young trees than is often practiced is helpful in obtaining trees of profitable bearing size at an earlier age. Heavy pruning during the early life of the trees may result in lower yields, even some time after the pruning is moderated.

Disposal of Pruning Brush.—Many different kinds of homemade brush burners, carriers, pushers, choppers, and brush rakes have been developed from time to time. Burning brush in or near the orchard creates a fire hazard. Trees and branches may be seriously damaged by a comparatively small amount of fire heat.

Machines that cut or shred the wood and leave it in the orchard as a mulch are now used extensively for brush disposal. Most of these machines incorporate a rotary blade or a rotary hammer. Such equipment is time- and labor-saving.

LOW TEMPERATURE INJURY

How Freezing Kills

Ice formation within the cell is the most common form of killing by freezing. The internal ice formation probably brings about an irreversible change in some colloidal and other properties of the protoplasm.

External ice formation may result in a drying of the cell when the ice is formed from water withdrawn from the cell. The effect is one of dehydration or desiccation. Mechanical strain is a leading factor involved in the killing. Plasmolysis that brings about severe hydration may promote injury as a result of rupture of the surface layer of protoplasm.

A "vital water exotherm" acclimation hypothesis to explain freezing injury or death in plants is, very briefly, as follows: During freezing a point is reached when all readily available water has been frozen extracellularly and only "vital water" remains in the protoplasm. As the temperature continues to decrease, vital water is pulled away from protoplasmic constituents to the extracellular ice. This sets off a chain reaction of denaturation, additional vital water release, and death (Weiser 1970).

Injury to Buds

Winter.—The most common type of winter injury to peaches is killing of the flower buds. There is no definite upper limit of temperatures at which flower buds are killed. Much depends on tree growth during the previous growing season, age and vigor of the tree, stage of bud development, rapidity of temperature drop, its time of occurrence, and duration of the low temperature.

The terms "dormant" and "rest period" are often used interchangeably, but it is pertinent, in some cases, to distinguish between the two. Dormancy is the stage or period from fall to spring when there are no leaves on the tree. During the fall portion of the dormant period in the north, the buds on shoots brought into the greenhouse may fail to bloom within a normal length of time. Such buds are in their rest period and will not grow even though temperature and moisture conditions are favorable. About midwinter the buds complete their rest period, and are

**FIG. 4.3. BLOOM SPRAY FOR BROWN ROT CONTROL IN A
PEACH ORCHARD**

easily forced into bloom. In the late fall and early winter, therefore, the
buds are in both the dormant and rest periods, whereas after midwinter
they are no longer in the rest period but are dormant only. After the rest
period has been broken plant tissues readily become active when exposed
to favorable temperatures.

After the rest has broken, apical buds are likely to start into growth, or
become more physiologically active by warm spells of weather. Thus,

TABLE 4.4
EFFECT OF BUD POSITION ON HARDINESS

Date	Min Temp (°C)	(°F)	Basal (%)	Live Buds Median (%)	Apical (%)
Nov. 22	− 9	16	86.2	77.5	56.4
Dec. 25	−15	5	72.4	48.0	61.2
Jan. 20	−13	8	65.3	58.4	56.9

Source: Knowlton (1937).

low temperatures in midwinter usually injure apical and median buds
more than basal ones. This apparently is true in early winter also, at a
time when buds are still in rest. The greater tenderness of apical and
median buds at this time, however, is probably due to immaturity. Thus,
at all times the dormant basal buds are the most hardy (Table 4.4).

Fruit buds of the peach are less resistant to cold than those of the apple
and pear, and may be killed at −23.3° to −28.9°C (−10° to −20°F) depend-

ing upon the duration of low temperatures and whether they occur suddenly or slowly, relatively early or late in winter, or follow a period of cold or mild weather.

To determine bud injury, cut through the fruit buds. If the centers are brown the buds have been killed. But in attempting to ascertain the extent of the damage at this time and later, examine enough buds from different parts of the tree and orchard to obtain a truly representative sample.

A temperature of −23.3°C (−10°F) is near the minimum for bud survival in most cultivars, although some may tolerate a few degrees colder. Those that do survive may produce only a partial crop, and may be of kinds that are not rated high in quality for the district. For the most part, the differences between commercial cultivars in winter bud hardiness may be relative only, and not a substitute for a good peach site.

Certain cultivars often are said to be hardier than others, e.g.,Veteran over Elberta. Certain cultivars, e.g., Halehaven and Redhaven, may bear a fair to good crop after a high percentage of their buds has been killed because these have large numbers of buds per foot of growth. Cultivars with fewer buds per foot length may have the crop greatly reduced when only a small percentage of their buds is killed.

On Feb. 3, in New York, about 1/2 the buds remained alive when frozen at −25.6°C (−14°F), but on the same date the next year a temperature of −22.2°C (−8°F) killed over 1/2 the buds in all plots. In the second year the mean temperature for the month preceding this date was 2.8°C (5°F) below normal. The prevailing winter temperatures influenced bud hardiness much more than the rate of N fertilization or the cultural treatment (Edgerton and Harris 1952).

The fruit buds in Illinois were hardier than the bark although the leaves were still on the trees in early October and there was little gain in hardiness at leaf fall. Freezing tests showed that the killing point of the fruit buds may rise as much as 6.1°C (11°F) after a cold spell (Chaplin 1948).

The cold hardiness of peach flower buds is subject to marked fluctuation in response to daily temperature changes from midwinter to early spring. Growers commonly worry about damage when a few warm days occur during this period followed by a sudden drop in temperature. A loss in hardiness has been demonstrated even though no visible activity of the bud occurs. The same temperature may damage fruit buds more after the rest period has been broken than before.

The temperature fell to −17.8°C (0°F) or below on Jan. 30−31, 1966, in the Piedmont of South Carolina. Bud injury of 30 cultivars was evaluated on Feb. 3 and 4. Among them, bud mortality ranged from 3−68%. The level of live buds after the damage ranged from 5.4–18.0 buds per 30.5 cm (1 ft) (Stembridge and Sefick 1966).

Fruit buds of several cultivars were compared biochemically in Kentucky. Generally, a high sugar and protein content, and a low total free amino acid were associated with increase in hardiness. Specifically, there was significant correlation between hardiness and a high sugar and protein content when buds were frozen at −19.2°C (−2.5°F), and between two amino acids (arginine and NH_2 butyric) and hardiness at both −19.2°C (−2.5°F) and −20.6°C (−5°F) (Lasheen *et al.* 1970).

Spring.—Usually, the peach blooms so much earlier than the apple that it is more likely to be subjected to damaging spring frosts. In general, the more advanced the bloom, the greater the injury from a given low temperature. Flower buds often are killed while the leaf buds and wood escape.

As flower buds develop in the spring they become increasingly tender. Just before they open they may tolerate about −6.7°C (20°F). Open blossoms show injury at about −3.3°C (26°F). Following petal fall the young fruits are generally killed by minimums of −2.2°C (28°F). Buds and flowers may survive temperatures slightly lower than those just given if the drop is slow and the minimum temperature continues only a short time. If the drop is rapid and the temperatures given are prolonged, considerable damage may occur.

A freeze during bloom in Virginia demonstrated variability in frost-escaping ability of several cultivars. The crop was completely destroyed on Elberta, Sunhigh, Triogem, Redskin, Golden Jubilee, and some others. But Veteran, Vedette, Early Red Free, Sunrise, and several others came through with full crops of fruit. Thus, there may be a variation in the ability of peach cultivars to withstand frost damage to the developing buds, blossoms, and young fruits. The variation in ability to escape frost may be one of a physiological nature and not a result, entirely, of late bloom; nor does it appear to be due to the development of more or less latent blossom buds after the frost damage has occurred to the major wave of buds (Oberle 1957).

The critical stage for peach blossoms in Illinois was −4.4° to −3.3°C (24° to 26°F). The more tender cultivars were injured in the upper limit; Elberta intermediate; and the more hardy cultivars were killed at −4.4°C (24°F). Some years depending on environmental factors, the limits were −5° to −4°C (23° to 25°F). At "shuck fall" the fruit was of about the same hardiness as at bloom, but some cultivars were slightly hardier at the latter stage. Cultivars maintained their relative positions as to hardiness throughout the season and from year to year, but the hardiness differential was not constant (Chaplin 1948).

Unless accurate thermometers are in the orchard, it is easy to be mistaken about the temperatures there. Early spring temperatures, especially on hillsides, may differ at points only 15.2 m (50 ft) apart. The higher part of a sloping site is usually warmer than the lower part, since cold air drains downward. A 15 m (50 ft) difference in elevation on a hillside may mean a difference of as much as 2.8°C (5°F) in minimum temperature on a still night.

One partial protection against spring frost injury is a bare hard surface in the orchard from early pink to ten days after petal fall. Clear, sunny days often precede nights when the frost kill occurs. If the orchard floor is not shaded by a cover crop or natural weed cover, it stores large amounts of heat from the direct rays of the sun. This heat is then radiated into the air at night. Even a low cover of chickweed reduces nature's night heating process. Cultivation just before or during bloom and immediately after may be just as detrimental as a cover crop.

Injury to Twigs, Branches, and Trunks

Peach wood is more tender to cold than is apple wood. But peach wood has a marked capacity to recover from winter injury, and withstands considerably lower temperatures than the fruit buds.

Immaturity of wood often results from new growth in late summer or early fall brought about by high soil fertility, high rainfall, high soil nitrogen content, recent heavy pruning, or excessive crops. Certain processes necessary to the hardening of the tree are carried on only if the trees are exposed to low temperatures gradually.

Trees that have grown poorly are especially subject to winter injury to trunks and branches. Poor soil drainage, nematode injury, low soil fertility, borers, and diseases are common reasons for weak growth. No definite temperature at which wood injury will occur can be given. In general, temperatures that drop to −27.8° to −28.9°C (−18° to −20°F) may injure woody tissues, particularly in the crotches of the large limbs and the base of the tree trunks (the latest parts to mature and harden). Trees that are not fully dormant may be damaged by higher temperatures.

Sapwood of the peach tree may be killed, at least in part, by cold when cambium and bark escape. A typical blackheart condition then arises. In some cases the bark separates from the wood with cambium and sapwood also being injured. Bark injury may be a drying rather than a freezing effect, but the damage is usually associated with cold and immaturity. If the whole trunk is involved the tree may soon die. More often, the injury involves only some of the branches and results in a one-sided tree. Severe blackheart of branches reduces their mechanical strength, resulting in splitting and breaking, and increasing the susceptibility to wood decay. Any such injured wood that is exposed to the air rots rapidly, thus decreasing the productivity and life of the tree.

Even though wood may seem to have been damaged by winter cold, much of it may survive if the trees are not pruned too early or too severely and if quickly available nitrogen is applied somewhat more heavily than usual. If bark on the trunk is loose after a severe drop in temperature, tacking it in place immediately may aid in preventing drying out of the bark and wood and may hasten healing.

Cambium of low chilling Early Amber peach and Sungold nectarine (350 to 550 hr) gained cold hardiness from November through January in Florida even though day and night temperatures were above $-9.4°C$ (15°F). Cultivars attained the greatest hardiness in January surviving $-12°C$ (10.4°F). Early Amber maintained its hardiness until before bud break (February 8) and decreased thereafter. Sungold remained somewhat resistant to cold until February 15 and gradually lost hardiness until March 8. The low chilling cultivars reached an acclimation base higher than that reported for cultivars adapted to higher latitudes (Buchanan et al. 1974).

Over 300,000 peach trees, mostly under 6 years old, died in central Georgia in the late winter and early spring of 1962. In the southeast, minimum temperatures below $-17.8°C$ (0°F) are rarely encountered. In a four-year study, tree damage occurred in three of the years, but only three minimums colder than $-10.6°C$ (13°F) were recorded, all in one winter. During that winter, injury occurred following a January minimum of $-15.6°C$ (4°F) although no injury occurred during two February nights with $-9.4°C$ (15°F) minimums. In two other winters, minimums of $-5.6°C$ (22°F) and $-5°C$ (23°F) in February caused injury. Damage at these temperatures does not fit the classic conception of winter injury (Jensen et al. 1970).

Affected trees either failed to start growth in spring or, more frequently, they started growth normally and then exhibited various degrees of collapse, or the foliage of the entire tree suddenly wilted. Affected trees usually grew well during the previous year. Examination of the cambium layer of the trunks of affected trees revealed discolored tissue extending upward from the groundline. Small sections or the entire circumference of the trunk was involved, and the discoloration sometimes extended into the crotch. Usually the extent of collapse reflected the amount of trunk damage. Root systems of affected trees showed no evidence of injury (Prince 1966).

Weather conducive to physiological movement occurs throughout the winter in the southeast, and soil temperatures are sufficiently high to cause considerably more root activity than occurs in more northern areas. Under these conditions, once the chilling requirement of a cultivar is substantially satisfied, activity proceeds, and the tree becomes progressively more tender and more susceptible to freeze injury (Savage 1970).

High cambium temperatures in late winter as the result of solar radiation impinging on the bark of the trunk causes a loss of hardiness after the rest period has been broken. When these high daytime temperatures are followed by a rapid rate of fall to freezing temperatures at night, severe injury often results which contributes greatly to shortening the life of peach trees. Also, wind may produce local cooling of the cambium tissue and, at critical air temperatures in late winter and early spring, is capable of causing cold damage.

Trunk insulators, e.g., aluminum foil plus fiberglass, modified the micro-climate around the tree to the extent of preventing the extremes that made the trees less hardy and more subject to low temperature injury. Trunk cambium temperature appeared to be a sensitive indicator of the vitality during growth. Trees with low vitality had higher cambium temperature than trees in good condition.

A white latex-thiram mixture painted on the trunk provides significant protection against sunscald, and worthwhile control of canker.

Injury to Roots

Winter injury to roots in the north usually is most severe in light sandy soils, after prolonged cold, and when there is no snow or other covering on the ground. Cover crops may help avoid root injury by directly insulating the roots from cold and by holding protective snow in place. However, as stated earlier, the fact should not be overlooked that a bare hard soil surface may protect against cold damage at bloom. After severe root killing, prune the tops fairly heavily.

Delay Pruning After Severe Winter Injury

In peach areas, cold damage is usually more frequent to the buds than to well-hardened wood. At $-28.9°C$ ($-20°F$) or lower, however, dormant peach wood is often greatly damaged or killed.

Cultural practices strongly affect cold hardiness of xylem, phloem, and cambium of peach trees, and cold hardiness is associated with their longevity in the south. Phloem tissue is more hardy than xylem or cambium. The xylem gains hardiness, then remains at a constant level until the chilling requirement is met, when it loses hardiness rapidly. Cambium gains hardiness rapidly with low temperature but loses hardiness rapidly with the advent of warm temperature. Nitrogen applied alone or in combination with fumigation reduces cold hardiness in early winter but increases vigor and survival. Trees grown in fumigated soil and winter-pruned were hardier than control trees and fall-pruned trees in nonfumigated soil (Nesmith and Dowler 1976).

The following orchard practices may affect tree life. In many cases, calcium may affect tree life. In many cases applications of the surfactant are helpful, as well as pruning in the fall or spring (avoiding winter), subsoiling before planting to break up hardpan, use of nematocides where indicated, mowing ground cover more often to control insect vectors of virus diseases, after-harvest nitrogen applications, painting the southwest side of the tree trunk, and early and adequate thinning. There seems to be no single answer in the southeast (Savage 1970).

When temperatures go below $-23.3°C$ $(-10°F)$ in northern regions, especially when such low temperature periods occur several times during the winter, not only the fruit buds but some of the branches may be killed, along with severe injury to the base of the trunks. Such trees are weak and slow to produce leaves. Early, heavy pruning may result in death of many of these trees. If the trees are not pruned until late spring, heavy N applications and frequent cultivation may aid good recovery by fall.

When, through winter injury or canker, a main limb shows signs of becoming weak it may be possible to find near the trunk an upward-growing watersprout which can be used to replace the original limb. If the whole tree is badly weakened, its condition may call for removal of the tree rather than for remedial pruning.

Delayed Spring Pruning

Overall pruning after petal-fall is wise only when freeze during bloom has caused a crop failure, or when there has been severe winter injury to wood of the tree. Remove the dead limbs, but only lightly prune those with live tissue and growing points.

The bud set varies with different cultivars. For example, some produce 30 or more fruit buds per foot of growth; others 20–30, and still others 15–20 or fewer. In certain orchard areas, some cultivars are injured more frequently than in others. In such areas and on cultivars with the lighter bud set, delay pruning until crop prospects are known more definitely (Christ et al. 1952).

The answer as to how late in spring a peach orchard may be pruned depends, in part, on the total amount of pruning needed. While the trees are dormant, during bloom, and for 10–14 days after normal petal-fall, the influence of moderate pruning on tree vigor is not very pronounced.

But from that time on each week's delay in finishing the pruning means a corresponding reduction in growth response from the pruning cuts. The foliage on trees pruned 2–4 weeks after petal-fall may be pale green for several weeks and with the diameter and length of shoot growth adversely affected. The heavier and later the pruning the greater is the shock to the tree. Vigorous trees with good reserves in the buds and branches show less harmful effect from late pruning than weaker trees. After growth has started some damage may be done to the tree while pruning.

Well-matured Trees

A distinction should be made between (a) winter injury to normal well-matured trees which usually results in killing of fruit buds, or, in severe cases, in the freezing back of terminal branches, and (b) cambial injury affecting the basal parts of the different woody components of the tree, owing to incomplete hardening-off of the trunk and its branches before winter sets in.

When conditions have favored proper maturity of trees, there is less danger of serious cambial injury to the trunk or bases of main limbs as compared with killing back of peripheral branches. When the winter injury consists of fruit-bud killing and freezing back of terminal or weak branches, even though severe, it is safe to prune at the usual time and to the extent desired to remove dead wood and to properly shape the tree, or, in old trees with straggling tops, to renew the top. Trees which freeze back considerably may react somewhat like severely pruned trees, and permit good renewal of the tops after growth has started (Stene 1958).

When no injury takes place and most buds bloom and set, the fruit may be small and the tree checked by the heavy tax on its strength. Pruning, under this condition, helps thin the prospective crop and promotes a more vigorous terminal growth.

When all fruit buds seem to be killed, delay pruning until bloom, especially in small orchards. Then, if no blossoms appear, prune heavily enough to secure a tree of desirable size and height, with proper branch distribution.

In areas where spring frosts often damage the buds, some growers make the necessary thinning-out cuts during the dormant season but do little cutting back until danger of bud killing is past. Then, if the crop is lost, they cut back the tree rather severely in order to lower the fruiting wood.

Prune prolific cultivars heavier than lighter bearing ones. Otherwise, more peaches are left than can be sized properly and much limb breakage may occur.

Imcompletely Matured Trees

Cambial injury to trunk or tip and basal parts of main branches may occur on trees checked by summer droughts and stimulated into late activity by warm weather and rains in fall. Such trees may not harden properly before winter, and so may be damaged by cold. Prune very lightly or not at all until after the leaves are out sufficiently to show the extent of the damage, and apply a fertilizer high in N to help promote good growth.

CHILLING REQUIREMENT

In the north a high chilling requirement keeps peach trees dormant during early winter; otherwise they would bloom too early and lose the crop through freezing. The Deep South requires low chilling for peach cultivars because the growing season is longer and the amount of winter cold is less. When peach buds have not had sufficient exposure to cold to to satisfy their chilling requirement, their leaf and fruit buds are delayed in starting in the spring even though weather is suitable for growth.

Delayed "break" of peaches occurs with northern cultivars when grown in the Deep South, in parts of Italy, South America, South Africa, and in other areas where the winter climate is mild. The amount of winter cold is not enough in these areas for northern cultivars, except, in some cases, at the higher altitudes.

Indications of Prolonged Dormancy

The first noticeable sign of serious dormancy trouble is the failure of the flower buds to swell in January and February in southern regions where such advancement is normal. Leaf buds also may fail to start until considerably later.

Flower buds located near the tips of branches may develop into blossoms in January or early February, while the basal buds are still dormant. Later the basal buds may start a very long bloom period. Some fruits may have developed to walnut size although adjacent flower buds are just opening. The blossoms are small, deformed, and weak; pollen germination is poor. The stigma and style may make no growth after the bud stage, but the ovary and remainder of the flower develop normal size. Such abnormal flowers fail to set fruit. Leaf development is also irregular and scant well into the growing season. Only enough leaf buds may develop to keep the tree alive for a few years. Scaffold branches without the shade protection of foliage may have the temperature of the inner bark raised to an injurious degree.

Cumulative Hours of Cold

Cultivar chilling requirements can be determined by bringing peach twigs into the greenhouse after a certain number of hours' exposure to normal winter cold 7.2°C (45°F). If fewer hours than those indicated for the different cultivars (Table 4.5) occur by Feb. 15 in the Deep South, the trees may suffer from insufficient cold even though additional chilling occurs later.

Chilling effectiveness is not entirely measured by hours below 7.2°C (45°F). Temperatures of 7.2° to 10°C (45° to 50°F) also have considerable effect, and for very low-chilling cultivars such as Okinawa and Flordawon 12.8°C (55°F) is as effective as 7.2°C (45°F) in inducing dormancy break.

Although choice of cultivars requiring more chilling than the average for an area may result in late bloom and reduced spring hazard, fruit set is unsatisfactory with inadequate winter chilling. Cultivars with low chilling requirements, e.g., 300 hr, bloom too early in north Florida, and fruit is generally lost to spring frosts.

The most desirable production is from cultivars that have a chilling requirement slightly higher than the minimum February accumulation of the locality involved. A rest period of this intensity delays bloom following normal or colder than normal winters until the frost hazard is reduced, but does not cause serious prolonged dormancy trouble following milder winters. Cultivars with slightly higher chilling requirements may succeed if they have strong fruit-setting habits.

Warm Period in Winter

In Georgia, 12.1°C (53.8°F) for November through January delayed bloom and foliation slightly, but 13.2° to 14.9°C (55.8° to 58.8°F) was needed for more serious effects. A few days of high temperatures were more serious than −16.1° to −15.6°C (3° to 4°F) continuous elevation through the winter, particularly with the leaf buds. Less chilling was required if leaf buds were not exposed to warm weather. Flower buds, though apparently not retarded so much in development, might suffer abnormalities due to high temperatures that caused delayed bloom and failure to set fruit; 2 days above 32.2°C (90°F) plus 3 days above 21.1°C (70°F) in December could cause these abnormalities (Weinberger 1954).

Temperatures above 21.1°C (70°F) during the chilling period seem to be detrimental in inducing dormancy break. Thus, a winter marked by much cloudy weather usually is followed by better dormancy break than one with an equal accumulation of 7.2°C (45°F) hr that is marked by generally dry, sunny conditions.

TABLE 4.5
FEBRUARY 15TH CHILLING REQUIREMENT OF PEACH CULTIVARS
(FLOWER BUDS)

Cultivars	Hours of Chilling Required at 7.2°C (45°F)
Mayflower	1150
Raritan Rose, Dixired, Fairhaven	950
Sullivan Early Elberta, Triogem, Elberta, Dixired, Georgia Belle, Halehaven, Redhaven, Candor, Rio Oso Gem, Golden Jubilee, Shippers Late Red	850
Afterglow, July Elberta (Burbank),Hiland, Hiley, Redcap, Redskin	750
Maygold, Bonanza (dwarf), Suwannee, June Gold, Springtime	650
Flordaqueen	550
Bonita	500
Rochon	450
Flordahome (double flowers)	400
Early Amber	350
Jewel, White Knight No. 1, Sunred Nectarine, Flordasun	300
White Knight No. 2	250
Flordawon	200
Flordabelle	150
Okinawa	100
Ceylon	50–100

Source: Arranged from Georgia studies (Weinberger 1950A) and Florida studies (Sharpe 1966).

Continuous chilling may break the rest period of more leaf buds of small peach trees than alternating warm (8 hr) and cold (16 hr) periods even when the total hours of chilling are the same. Periods of high temperatures interspersed with periods of low temperatures counteract some of the cumulative influence of low temperature (Overcash and Campbell 1956).

Chilling Differences Between Areas

At Fresno, Calif., in 1959, as few as 440 hr below 7.2°C (45°F) were sufficient for normal development of July Elberta flower buds, 480 hr for Elberta, and 820 hr for Mayflower; in 1965, only 500, 700, and 820 hr were required by the respective cultivars. In contrast 750, 850, and 1150 hr were required at Fort Valley, Ga. In certain other years, cultivars responded differently at the two locations (Weinberger 1967).

The winter weather patterns in Fresno and Fort Valley differ markedly. In the San Joaquin Valley, lower humidity during clear periods permits greater radiation of heat which results in greater diurnal temperature differences. However, foggy periods for as long as two weeks occur frequently during the colder months. These keep daily maximum and minimum temperatures uniform within a narrow range. Also, the tempering influence of the Pacific Ocean less than 160 km (100 miles) away tends to keep daily mean temperatures uniform (Weinberger 1967).

Under special climatic conditions of the Fresno location, where chilling approaches but fails to adequately break rest every year, other temperature relations override chilling in determining rapidity of rest-breaking processes. The chilling-hour requirement of peach cultivars in the southeast was based on certain weather patterns. The 7.2°C (45°F) base selected was an index of chilling weather, only representing definite relations with other naturally occuring temperatures; for the different weather patterns of the San Joaquin Valley, the same index was not applicable (Weinberger 1950A).

Peach flower buds actually may be developing while their rest is being broken. For a normal development, temperatures during this period must be suitable; bud deterioration occurs if minimum temperatures during December and January are too high in the San Joaquin Valley. Even the 7.2°C (45°F) temperature commonly used as an index of chilling weather is not low enough as a minimum temperature to prevent bud drop.

High temperatures in November–December delay bud growth. Besides bud drop and delay in breaking the rest, there is the greater threat of abnormal development of all buds or failure to end dormancy if excessively high temperatures prevail. Breaking of rest of peach flower buds must therefore take place at a critical range of low temperatures for normal bud development (Weinberger 1967).

December–January Mean

In Florida, the chilling hours below 7.2°C (45°F) are normally about 650 for northwestern areas to less than 50 for southern areas. Because of this wide range, a measure other than cumulative hours below 7.2°C (45°F) is desirable with respect to adaptability. The December–January mean can be used satisfactorily; and ranges from 11.1°C (52°F) for northwestern Florida, 14.4°C (58°F) or below for the Gainesville area, to as high as 20° C (68°F) for southern Florida. Converting chilling requirements to December–January temperature permits the predetermination of adaptability of peach cultivars with known chilling requirements to locations where only mean temperature records are available (Sharpe 1966).

In Georgia, December–January temperatures were the chief influence in breaking the rest of peach buds. The average of December–January means; when correlated with dormancy values, gave a coefficient of 0.93. This correlation was equal to that obtained with chilling-hour accumulation and was reliable. Data on mean temperatures and chilling-hour accumulation covering 145 winter locations were summarized, and comparable values for the two indices were calculated. Where 750 hours of chilling occurred to February 15 on the average, December—January means averaged 9.8°C (49.6°F); where 950 hours occurred, the means averaged 9.1°C (48.3°F) (Weinberger 1950A).

In Florida, the cambium of low chilling Early Amber peach and Sungold nectarine (350 to 550 hr) gained cold hardiness from November through January, even though day and night temperatures were above 15°C (59°F). The cultivars attained the greatest hardiness in January, surviving −12°C (10.4°F). Early Amber maintained its hardiness until just before bud break (February 8) and decreased thereafter. Sungold remained somewhat resistant to cold until February 15 and gradually lost hardiness until March 8. The low chilling cultivars reached an acclimation base higher than that reported for cultivars adapted to higher latitudes (Buchanan et al. 1974).

Shortening the Chilling Requirement

Breeding for cultivars tolerant of mild winters, plus budding on a nematode-resistant rootstock, is the most promising approach to the problem. The quality of northern peaches is being combined with the low cold requirements of the South China peach, and is serving to extend commercial peach production into semitropical areas. Nematode-resistant rootstocks have been discovered and this is serving to remove another serious problem in the Deep South.

SOIL MANAGEMENT

Tillage practices on any orchard site may affect conservation of the soil itself, conservation of moisture, organic matter, and the fertility of the soil.

There seem to be three major points of general agreement of soil-management systems: (a) permanent cover crops and no tillage may be best on sites where erosion may be serious; (b) permanent cover crops reduce production and growth as compared with cultivated checks; (c) this reduction in growth and productivity has been caused by competition for water or for nutrients, chiefly N, or both.

Cultivation

For many years it was customary to cultivate thoroughly during much of the year. Stress was placed on clean appearance of the orchard, on orchard sanitation, which presumably accompanied the clean orchard, and on making available more N and other plant nutrients to stimulate growth. However, besides encouraging erosion, cultivation speeds up the

**FIG. 4.4.   SOD AND MULCH INCREASE ABSORPTION OF WATER
AND PREVENT EROSION**

oxidation and disintegration of organic matter and breaks down the structure of the soil. Yields may be reduced under such conditions and, where erosion is serious, the soil may eventually become too depleted for profitable management. Today, more importance is attached to the conservation of organic matter in the soil (which is not easily replenished), to preventing soil erosion (especially on sloping sites), and to saving of soil moisture and nutrients.

Cultivation has been resorted to as a method of preventing competition for moisture by other vegetation. Although the value of cultivation in moisture conservation has, in many cases, been publicized, even greater moisture, soil, and nutrient losses were overlooked. Even though cultivation has been advocated for the maintenance of a dust mulch as a protection against moisture losses, this operation usually liberates moisture and may make the soil susceptible to crusting after heavy rains, resulting in less penetration and more runoff. A less uniform seasonal supply of moisture may result.

The release of available N by cultivation may result in some stimulation of the trees. But it is usually wise to add N and to retain the organic matter in the soil for controlling erosion, maintaining soil porosity, and improving the general physical structure of the soil. Peach orchards are cultivated less now than they were years ago, and that change is justified.

Where irrigation is not used, cultivation of young trees is usually wise in late spring and early summer. Even in mature orchards restrict cultivation to spring and early summer, but it may be continued slightly later with young trees. If the slope is slight, it may be helpful in controlling erosion to end each cultivation by going across the slope.

Make the first cultivation as soon as the soil is in good workable condition, since early cultivation encourages the new growth to develop earlier in the season. Then check the growth by ending cultivation in midsummer and sow a cover crop. Do not work the soil more than 7.6–10.2 cm (3–4 in.) deep beneath the trees; deeper tillage may cut many shallow feeding roots.

In areas where land is high priced, some growers practice intensive culture and plant vegetables or other annual crops between the rows of young trees.

Stubble Mulch or Trash Cultivation

In this system, varying amounts of cover crop residue, either a seeded crop or weeds, are left on the soil surface. The litter, trash, and roughness of the soil surface reduce soil erosion, especially on steep slopes and when heavy rains occur in short periods of time. Stubble mulch also prepares the soil for absorption of rain and in building up a soil moisture reserve for drought periods.

A disk is generally used to stir the soil without covering the stubble, but other implements that break up, disarrange, and only partly cover the material with soil may be satisfactory.

Cover Crops

In many good peach orchards the conventional system of culture is cultivation with cover crops and an annual application of a fertilizer.

Some merits of cover crops in peach orchards are: retarding erosion, runoff, and leaching; helping maintain or increase the level of organic matter originally present in the soil; aiding in water penetration; serving as an indicator of moisture or nutrient deficiencies; checking growth in the fall and thereby encouraging maturing of the tree tissues.

After midsummer, a crop such as millet, soybeans, or weeds may be needed to take up surplus nitrates so that tree growth will be checked at the proper time, thus permitting the sugars manufactured by the leaves to go into the fruit instead of into new growth.

At Vineland, Ontario, soil nitrates were determined at 2-week intervals over a 5-year period on plots seeded May 15 and July 15 (Table 4.6).

At the 0–15 cm (0–6 in.) depth, the growth of a green manure crop seeded May 15 kept the nitrate supply at the low level of 1 ppm from late July to early September. In the plots seeded July 15, the nitrates had accumulated to such an extent that they still were 4 ppm in early September. An excess of nitrates in late summer and early fall is detrimental to the color of peaches (Upshall 1946).

TABLE 4.6
FIVE-YEAR AVERAGE OF EFFECT OF EARLY SEEDING ON SOIL NITRATES

Time of Seeding	Late July (ppm)	Early Aug. (ppm)	Late Aug. (ppm)	Early Sept. (ppm)
May 15	1	1	1	1
July 15	12	8	5	4

Source: Upshall (1946).

In California, the few permanent roots in the surface 30.5 cm (1 ft) of soil in a clean cultivated orchard was due to excessive soil temperature during the growing season. Absorption of fertilizers may be negligible in this zone, even where a high concentration of an element such as K had been built up over a period of years. A summer cover crop helped avoid this problem (Proebsting 1943).

In Delaware, all cover crops slightly retarded tree growth for 1–2 years after the orchard was planted. By the fourth year this detrimental effect of cover crop was overcome by a double cover crop system of soybeans in summer and rye-vetch in winter (Hitz 1954).

Sometimes crops that winterkill, such as spring oats and millet, are grown during late summer and into the fall, or weeds are left to develop during that period. If danger of soil erosion is fairly serious, it may be wise to grow a summer soil-protective crop even though it competes with the trees. Some summer cover crops are soybeans, 9–13 litres per ha (1–1 1/2 bu per acre); sudan grass, 16.5–27.5 kg (15–25 lb); millet, 11–16.5 kg (10–15 lb); and domestic ryegrass, 13–16.5 kg (12–15 lb). If an overwintering crop is not sown in late summer or early fall, leave the summer crop on the surface over the winter and disk it in the spring.

Frequently used fall and winter cover crops in northern peach orchards are rye and mixtures of rye and vetch. Rye starts to grow very early in spring, so disk it in before it begins to seriously compete with the trees. The disking need not be thorough. Merely break down the crop, working the litter into the soil in later diskings, but leaving the soil in trashy condition. A relatively heavy seeding of rye helps control weeds, gives a finer stem to the rye, and lessens the likelihood of its being excessively tall in young orchards before it is turned under. Yield may be increased substantially with N fertilizer.

Purkof wheat is used as a winter cover by some growers. It is hardy, makes an excellent ground cover, and does not head out as quickly as rye in spring. Fairfield winter wheat is also used for this purpose (Burkholder and Baker 1957). In Washington, performance of trees under

TABLE 4.7
COVER CROP MIXTURES GIVING FAVORABLE RESULTS IN MARYLAND

Mixture	Amount of Seed to Sow (kg per ha)	(lb per Acre)	Seeding Date
No. 1			Aug. 20–Sept. 1
Crimson clover	9	8	
Winter vetch	9	8	
Winter rye	7	6	
Golden millet	2	2	
No. 2			Aug. 20–Sept. 10
Crimson clover	9	8	
Winter vetch	9	8	
Winter rye	9	8	
Wong barley	9	8	
No.3			Aug. 20–Sept. 1
Winter rye	26 (litres)	3 (pecks)	
Winter vetch	22	20	
No. 4			Sept. 10–Sept. 20
Winter rye	53 (litres)	1 1/2 (bu)	

Source: Vierheller (1949).

clean cultivation with vetch cover (no supplementary N) was similar to that in rye, but yields were slightly greater and maturity later (Proebsting 1958).

To obtain a good stand of cover crops, cultipack or roll the seedbed after seeding. This practice is a must with any cover crop combination containing small seeds such as clovers. If grasses or ladino clover is used, cultipack both before and after seeding.

Improper preparation or condition of the soil for the sowing of a winter cover crop is often responsible for serious fall and winter erosion, especially in the south where the cover crop may be sown too late under such dry conditions that little or no growth is made until late the following spring. Thus, a good natural cover of weeds may be destroyed only to be replaced by a poor growth of the sown crop.

Deep-rooted cover crops, e.g., alfalfa, sericea lespedeza, and coastal bermuda, used in a rotation between orchard plantings have been the most successful in increasing tree longevity in Georgia. Crops such as these help to maintain good physical condition of the soil so that O_2 levels are adequate for root growth and nutrient absorption. No new absorbing roots were produced at O_2 levels below 15%. Many peach orchards in the Georgia Piedmont have O_2 levels of 15% or less during a great part of the growing season. Under these conditions preplant subsoiling to 56 cm (22 in.) has increased growth, yield, and tree longevity (Savage 1970).

Sod Cover

The best sites for peaches are often located on elevated land to reduce the hazard from spring frosts. Often erosion will deplete the soil and water losses from runoff will occur because infiltration is reduced. Thus, a deficiency of water and nutrients may result and a general decline in yield occur. A sod system may be desirable to reduce erosion in orchards on hillsides.

If the orchard is not irrigated the sod's use of soil nutrients is less serious than its use of water, since fertilizer can be added. A sod area requires more N than one under cultivation, for the first 4–6 years. A mulch helps conserve moisture in a sod orchard, but peach trees root extensively and the sod may compete for moisture with the tree roots even 3.7–4.6 m (12–15 ft) away from the trunk.

Some growers use a sod cover but disk the soil every few years. Others use sod strips, across the slope, between cultivated tree rows. The sod strips control erosion, and the cultivation reduces plant competition with the trees. The strips are useful also for moving sprayers and harvesting equipment, especially during wet weather. Orchard sod should be mowed occasionally.

In Washington, orchard grass competed strongly for N. With orchard grass sod, under equal N and irrigation applications, growth and yield were lowest, maturity earliest, and fruit was the most highly colored of the cover-crop systems. Yield in alfalfa sod was low, and maturity was delayed excessively. Mulching trees whose vigor had been seriously depressed by three years in orchard grass promoted good results. Yields were highest in this treatment for the 5-year period, owing primarily to very high yields in 1 year. Land kept bare of vegetation with 2–3 sprays of an oil-dinitro general contact weed spray, using no tillage implements except for reditching once every 2–3 years, gave yields that averaged between vetch and rye; fruit maturity was early in relation to yield (Proebsting 1958).

On a deep silt loam in Ohio, moisture was more important than N in the growth of 1-year-old trees. Their growth in sod was retarded during the first year when rainfall was below average. In years when soil moisture was not limiting, the trees made as much growth under a cultivation-sod system during their first 3 years in the orchard. During this time the growth of the trees was not increased by N fertilizer in excess of 45 g (0.1 lb) per year of tree age (Judkins 1949).

In Ohio, six soil management systems were used: (a) bluegrass sod; (b) blue-grass sod plus straw mulch under the branches of the trees; (c) spring cultivation followed by a summer cover crop of soybeans and a winter cover of rye; (d) same as treatment (c) except 36 tonnes per ha (16 tons per acre) of manure were applied each spring; (e) same as (c) except 31.5 tonnes per ha (14 tons per acre) of chopped corn stover were applied each spring and disked into the soil; (f) spring cultivation followed by a summer cover crop of sudan grass and a winter cover of rye. Each spring all plots received 440 kg per ha (400 lb per acre) of ammonium sulfate. During the first 10 years the different systems had no effect on bloom or ripening date, quality or color of fruit, or winter injury of wood or buds. Yield was proportional to the size of trees (trunk circumference). Tree growth and yield were not proportional to total soil organic matter, undecomposed organic matter, or soil moisture in the silt loam soil. The chief factor limiting the growth of the trees in the sod plots was a lack of available N (Judkins and Wander 1945).

In Kentucky, trees growing in lespedeza sod made as good growth as those in cultivated plots after the trees reached maturity (Olney *et al.* 1950).

Organic Matter and Farm Manure

Organic matter in the soil promotes the development of soil organisms which aid in supplying nutrients. It also improves soil water-absorbing and -holding capacities, and soil aeration. With suitable supplies of organic matter present in the soil, peach trees utilize N and P more efficiently. The deep incorporation of organic matter in the soil may not be as important, or as easily done, as is often thought. Many growers use large heavy disks too much and too often for the best development of the trees; deep tillage injures many roots which feed in the fertile topsoil.

An average application of farm manure, e.g., 22.5 tonnes per ha (10 tons per acre), does not provide adequate N or other elements for maximum

FIG. 4.5. A MULCH OF CORN COBS PROVED MOST BENEFI-
CIAL WHEN COMPARED WITH OTHER ORGANIC MULCHES

production; it may supply some minor elements besides the major ones. In general, supplement it with N fertilizer, probably at 1/2 the usual rate. Poultry manure may cause damage if applied in excess of 11.2 tonnes per ha (5 tons per acre), particularly on sandy soils.

Herbicides

In South Carolina, the residual herbicides Sinbar, Casaron, Lorox, Karmex, and Simazine at 4.4 kg active ingredient per ha (4 lb per acre) gave excellent weed and grass control for the growing season with no injury to young peach trees. Sinbar was also effective in controlling nutgrass when applied premerge or postmerge to grass. Paraquat at 0.55 kg (1/2 lb) and Sinbar at 4.4 kg (4 lb) per acre controlled existing large grasses and kept the area weed-free around young trees for the remainder of the season. Bermuda grass could be effectively controlled by Dalapon, using repeat applications at a low rate. In considering herbicides use only those which have been cleared for young and/or bearing peach trees and then follow manufacturers directions (Gambrell and Rhodes 1967).

FERTILIZER

It is usually necessary to fertilize peach trees annually, except in years of successive crop failures. Amount and kind of fertilizer to use depend on the soil type and its fertility, structure, humus content, climate, and other variables. Base fertilizer applications largely on length of the previous season's terminal growth, and crop prospects. Trees that carried heavy crops or those with sparse or light-colored foliage should receive increased amounts.

In planning a fertilizer program, aim to achieve and maintain a proper balance between the various nutrient elements. This does not mean a similar amount of each fertilizer element. The present trend, e.g., with N fertilizer, is to make recommendations per tree or per hectare (acre) on an actual N basis and this is related to the N percentage in different fertilizers. Excessive amounts of one nutrient element can induce a deficiency of another.

As might be expected, fertilizer recommendations will vary among regions. Soil and leaf analyses have become increasingly efficient and available to growers. Growers should avail themselves of these services.

Nitrogen

N is often the limiting factor in peach tree growth. The peach is a Ndemanding tree. N can be applied in several forms. The most noticeable effect of available soil N is a stimulation of vegetative growth. Where other factors are optimum, trees well supplied with N grow luxuriantly. In general, the shoot growth is longer and thicker, the leaves are darker green and more numerous, larger, and retained later than those lacking sufficient N.

In Pennsylvania, during the prebearing period all levels of N increased the effectiveness of the trees in absorbing and utilizing the available K and Mn. Application of N increased the average yield per tree during the first fruiting year. It also resulted in greater fresh and dry leaf weights, larger leaf areas, darker green leaf color, and thicker trunks (Ritter 1957).

Fruit from trees low in N is small, generally highly colored, and earlier in maturity than from trees well supplied with N. Most fruit from trees low in N may be harvested in 1−2 pickings. Where N is abundant the first picking may be light, the second heavier, with a third and sometimes a light fourth picking. Peach maturity is delayed under conditions of high N supply. When the N supply is excessive, the fruit may begin to soften and ripen with little yellow or red color developing. Yields are usually heavy on trees well supplied with N.

Time of Application.—N fertilizer is usually applied in early spring. Split applications require more labor but let the grower adjust the amount to crop needs. For example, if the bloom or the fruit set is very heavy, the amount of N can be increased or, in the case of a serious late spring freeze, the application can be reduced or even omitted. In the south, part of the N fertilizer is often applied after harvest.

The depth of heaviest root concentration in cultivated orchards is usually at 25.4−45.7 cm (10−18 in.); in noncultivated orchards roots may be nearer the surface. If N fertilizer is applied in late winter in the milder areas, it should reach the heaviest concentration of roots before active spring growth starts.

Tree roots are active at temperatures just above freezing. For conduction of N from roots to the aboveground portions of the tree a temperature of 7.2°C (45°F) or higher is necessary. Daytime temperature often exceeds the 7.2°C (45°F) minimum over a considerable period from fall to bloom, thereby providing ample opportunity for translocation of N nearer the growing points where it is used.

Fall fertilization and late disking resulted in deepest red skin color of Elberta, but did not affect that of Halehaven and Triogem. Fruits ripened earliest under fall fertilization, but time of ripening was not affected by disking time (Havis 1957).

Rate of Application.—The rate of application of N fertilizer should be such as to give proper amount of tree growth and maximum annual yields of fruit of good size, color, quality, and grade. Tree spacing, depth, type and management of soil, age, size, cultivar, and condition of trees, length of shoot growth, use of crop, pruning method, and available water supplies are factors to consider in rate of application.

A general rate of application of a 20% N fertilizer is about 59–89 g per cm (1/3–1/2 lb per in.) of trunk diameter near the base. Another guide is to apply 136 g (0.3 lb) of nitrate of soda for each year of the tree's age up to a maximum of 1.8–2.3 kg (4–5 lb) per mature tree. Apply ammonium nitrate at about 1/2 the rate of nitrate of soda because of the double percentage of N.

In deciding on the amount of N to apply, bear in mind that 17.6 kl (500 bu) of peaches contain about 12.2 kg (27 lb) of actual N or the amount in 61.2 kg (135 lb) of ammonium sulfate. Some of the N in the fruit may come from N already in the soil and tree. The rate should be high enough to produce moderately vigorous shoots as well as a full crop each year. The use of N can be overdone. Symptoms of that are excessive growth, which continues late in the season; larger and darker green leaves than normal; poor to fair color development; lack of normal sweetness and aroma.

Adding large amounts of N, particularly late in the growing season, reduces fruit color and delays ripening. With highly colored cultivars, a reduction in color may not be serious. New cultivars have been developed that color well even under fairly high N conditions. Some growers take advantage of the delay in fruit maturity caused by high N to extend the ripening season. Color reduction, however, can be serious in a cultivar like Elberta.

Method of Application.—Most N fertilizers broadcast on the soil surface soon penetrate to the root zone, and are translocated to the top of the tree within two weeks, if there is ample moisture. Broadcast or spread N fertilizer mostly in a 61 cm (2 ft) band under the outer edge or perimeter of the tree where the feeding roots are most numerous. It is not necessary to work it into the soil.

Band placement of fertilizer into the soil near irrigation furrows often gives better results than broadcasting, especially where elements other than N are included in the fertilizer.

Foliar Sprays.—The peach can absorb and utilize some of the N applied to the foliage as urea. However, the peach is less efficient than the apple in this respect (Eckert and Childers 1954).

Recommendations for Different Areas.—Recommendations on the best fertilizer formula and rate and time of application vary with different climatic conditions, soils, and many other factors. Local government authorities can supply the most suitable information for a particular peach site. In many cases, the recommendations can be made on a basis of soil and tissue tests in relation to known responses to peach trees in the area.

Peach growers should avail themselves of the local information from universities, agricultural experiment stations, departments of horticulture, departments of soils, extension services, county agents, agricultural representatives (in Canada), fieldmen, and other useful services.

Phosphorus

Yields may be reduced where P is deficient in the soil, but many cover crops have high needs for it and P fertilizers often benefit orchards more indirectly by improving cover crop growth than through direct nutrition of the trees.

Some symptoms of P deficiency are: older leaves mottled with light green areas between dark green veins; progressive defoliation of mottled leaves from base to tip of twigs as the season advances; and purplish pigment development in leaf petioles in cool summer weather.

At one time the standard recommendation for peach orchards in South Carolina was a 6-9-12 fertilizer. This has changed to P applications only when soil tests indicate need. When soil tests indicate adequate P, and/or when complete fertilizers have been added to cover crops, only equivalent rates of N and K are added. Many growers are using compounds to supply these elements, without adding P, and are producing quality fruit.

Tissue analyses indicate that peach trees remove little P from the soil. Average leaf samples in South Carolina are: 0.15–0.20% P (dry weight). At Clemson, S.C., trees receiving only N and K from K nitrate (23.75% nitrate N, 44.5% potash) in 4 years did not decrease in foliar P content.

Low temperatures accentuate P deficiencies by a reduced root growth and reduced rate of absorption. As a nutrient ion, P absorption is strikingly dependent on temperature, light (energy), and oxygen and is depressed by metabolic inhibitors. Total absorption proceeds less rapidly than for the other major nutrients, and mobility is more restricted than for K. N, particularly as ammonium, applied with P accelerates its absorption and utilization. High or adequate rates of P fertilization may depress the uptake and utilization of Fe and Zn. The site of action appears to be not in soil fixation of Zn by phosphate but in a restriction of absorption at the root surface (Wittwer 1969).

Time and Rate of Application.—Time of application of P to peach trees does not seem so important as with N. The total amount of available P already present in the soil and whether cover crops are used influence the optimum rate of application. Probably P fertilizer, equal in available phosphate content to the quality of N applied in fertilizers annually, should be applied in bearing orchards at least every 2–3 years.

Method of Application.—Many tests have shown no response of deciduous fruit trees to applications of P; lack of response often is the result of surface rather than deeper placement. Place P fertilizer at least 10.2 cm (4 in.) deep and deeper if suitable equipment is available for this purpose.

In arid soils of the west the water-soluble P in a fertilizer reacts with the lime in the soil to become insoluble in water, but if it is near the roots it can again be made available to some extent by root action. If it is on the surface of the soil and becomes insoluble in water it is of little value to the trees because it does not move downward quickly to the roots.

Combined Effects.—Acid soils inherently low in nutrients, as in the coastal plains of New Jersey, can fix considerable P as relatively insoluble Fe and Al compounds. When such soils become deficient in Ca and Mg, they may be strongly acid and low in P. Thus, acid peach soils that are deficient in K may also be deficient in P.

P deficiency in peach trees may develop independently of, and may coexist with, a deficiency of K, Ca, and Mg. Omission of P from a nutrient treatment may result in increased pigmentation, and in narrow, dark green leaves, regardless of whether or not the treatment was also deficient in a base. In all combinations involving a deficiency of P as well as of a base, however, symptoms of the base deficiency are more prominent than those of the P deficiency.

Potassium

K aids in formation of carbohydrates, tends to improve quality, and may increase the resistance to disease and cold.

Potassium Deficiency.—Symptoms of K deficiency are as follows: The trees are smaller than usual. The leaves near the ends of the terminals and on lateral shoots are small, narrow, and the ends tend to roll upward. Early in the season the leaves are light green, but become reddish-yellow, with marginal scorch and other necrotic areas late in the summer, giving the tree a bronzed appearance. Shoots and branches are smaller than usual and with fewer fruit buds (Baker 1949).

Some K values (dry weight) listed for August samples are as follows: range in field, 0.3–2.9%; deficiency symptoms, 0.8%; optimum, 1.5–2.2%. K should be in balance with the other elements (Forshey 1969).

Deficiency symptoms may be brought about on soils of low K availability by some particular soil treatment that temporarily fixed the K. K fertilizer is not necessarily essential throughout an orchard where K deficiency is observed. Soil areas of low K are often localized in peach orchards. It is likely, however, that where K deficiency symptoms have been recognized or determined by leaf analysis, the trees will respond to K.

Potassium Fixation.—Fixation of K in the soil is sometimes due to the combined effects of two actions: (a) the effect of free alumina in forming muscovite or some closely related mineral which contains K in an unavailable form; and (b) the result of alternate wetting and drying of the exchange material containing K, thus causing the latter to be held in an unavailable form. The difficulty of fixation was obviated in orchards by using a heavy mulch of straw, hay, or similar plant material. The K which leaches from unweathered mulch material remained more mobile and penetrated deeper in an available form than did K supplied as fertilizer on top of a sod. This was due to the conditions which occurred under a mulch, i.e., more constant moisture content of the soil, insulation effect allowing for a more constant soil temperature, increased organic matter supply, and increased porosity.

K fixation at or near the soil surface when applied as a K fertilizer can be largely eliminated in mulched orchards. When 4.5 kg (10 lb) of potash as 60% KCl were added to a heavy mulch the soil to a depth of 0.3 m (1 ft) was greatly increased in exchange K within 3 years. Within 4 months of the start of such treatment, soil to a depth of 20 cm (8 in.) was increased in available K. In no other known way, except by mechanical placement, can K be increased to such a degree (Wander 1945).

Results in Different Areas.—In New York, recovery from scorch and rolling symptoms in 7-years-old Elberta trees and the coincident increase in leaf size and color and shoot growth were striking as a result of applying 1.4 kg (3 lb) of 60% muriate of potash per tree in both fall and spring. The recovery was accompanied by 100–400% increase in K in the leaf samples and marked increases in the proportion of ash in leaf dry weight (Boynton 1944).

Peach leaves with acute K deficiency symptoms analyzed less than 1% K content (dry weight). The range frequently was 0.5–0.9% for the more seriously affected leaves. Leaves with 1.1–1.25% K showed no marked deficiency, but trees from which they were taken responded to K fertilizer. Where the leaf content was 1.5–2.0% it was possible to increase the amount of K salts to the soil, but the fertilizer did not seem to produce a beneficial tree response. When the K content of the leaf was above 2%, a beneficial response did not occur from added K.

Cores or holes, bored 46 cm (18 in.) deep, into which K and P had been placed, were left to remain under field conditions for 3 years. By digging a trench and exposing a vertical section of the core, soil samples were taken at 2.5 cm (1 in.) intervals laterally away from the core. Determinations revealed that 100–200 ppm of available K might be expected 12.7–15.2 cm (5–6 in.) from the core; the content of the native soil was 30–50 ppm. Therefore, over 6% of the soil in the region of greatest root concentration had been changed from a low to a high content of available K in 3 years. P remained immobile under the same conditions for the same length of time (Wander 1940).

In Ontario, P, K, and farm manure were applied alone and in combinations. N was applied to the whole orchard each spring. Each fall, manure was applied on the surface and minerals at a depth of 30.5 cm (12 in.) at the branch tips. In the sixth year, all treatments except P alone and K alone gave yield increases over the checks. The PK combination was a beneficial treatment. Manure alone gave good response but at a higher cost. Some of the effect was due to tree size differences, but most of it to fruit size. Red color of fruit was increased by manure treatments (Upshall *et al.* 1955).

In South Carolina, peach tree abnormalities resulted from N alone. For good development of cover crops, and the foliage, twigs, fruit buds, and fruit of trees, it was necessary to apply P, K, and limestone, as well as N. Although a good cover crop resulted from N, P, and limestone, abnormal peach trees were not restored to normal condition until adequate K was applied (Rawls 1940).

In North Carolina, on Lakeland sand, application of K or Mg was associated with an increase in yield of Elberta peaches greater than 5.7 cm (2 1/4 in.) in diameter. Treatments resulted in a delay in maturity of about one day for each increment of fertilizer added. Red coloration of the fruit was improved by K fertilizer; Mg decreased the redness. Results indicated that the level of Mg and especially of K required for optimal production was much greater than that needed to alleviate visual deficiency symptoms. Applications of either Mg alone or K alone had no positive effects on yield of Elberta. But when these two nutrients were applied together, yields increased markedly. The balance between cations may be just as important as the level of individual cations (Cummings 1965).

In California, very young leaves were lower in K than older leaves in mid-summer. The smallest change in K content occured in June and July. K content of the leaf was not influenced by tree age, but was affected by amount of fruit on the tree in late July and thereafter. Variations in soil moisture above the wilting point did not change the leaf K content, but severe drought reduced it. No difference in leaf K occurred between cultivars. The K in the dry matter of the leaf in June averaged 2% ranging from 0.6−3.4% in 130 orchards (Lilleland and Brown 1941).

Boron

Although peach trees can absorb and retain appreciable amounts of B, they will not tolerate as much B as apple trees. Deficiency or shortage symptoms include failure of buds to break in spring, death of wood, and firm brown spots in the flesh.

Excess boron symptoms consist of leaf distortion, withering and drying back of terminal shoots and suckers in mid and late summer; small cankers along the shoots accompanied by gum exudation, particularly in crotch areas; rough bark; prominent lenticels; swelling near the buds; excessive development of lateral shoots, resulting in a bushy type growth; and reduced yield (Kamali 1970).

B-deficient leaves show compactness of the palisade and spongy mesophyll layers, the cells of which assume a palisade-like appearance. Necrosis of the phloem of the leaf and stem, bark splitting and distorted bark lenticels also are evident. Boron toxicity does not affect the phloem tissue of the leaf, but there is a sloughing off of the necrotic ground parenchyma cells along the abaxial side of the midrib. External appearance of the fruit is similar for deficient and toxic treatments. However, parenchyma cells of the inner mesocarp of B-deficient fruit is corky with pronounced suberization of the cell walls (Kamali and Childers 1967).

Based on growth response, symptoms expression, and fruiting, the B content of peach leaves can be correlated as follows: severe deficiency, less than 10 ppm; deficient, 11−17 ppm; low, 18−30 ppm; optimum, 31−59 ppm; high, 60−80 ppm; excess, 81−155 ppm. Classes of B status in the fruit were given as follows: deficient, less than 10 ppm; normal, 11−29; and excess, 44−124 ppm. An application of fritted trace elements (FTE), 220−286 kg per ha (200−260 lb per acre) to sand gave good vegetative growth and fruiting the second year. An additional application of 44 kg (40 lb) or more per ha (acre) near the end of the second growing season afforded normal growth, fruiting, and fruit quality for the third year.

Zinc

In Zn deficiency, crinkling, waving, and chlorosis appear in leaves at the terminals of twigs and shoots. In severe cases where considerable defoliation occurs, rosettes of small leaves may form at the terminals, and leaves in these rosettes may be without leafstalks, very small, and rigid.

In North Carolina, Zn deficiency was found in all peach orchards where the Zn content of midshoot leaves in August was below 6 ppm. The mean Zn content for all samples from orchards with chlorosis was 10.3 ppm compared with 19.2 ppm for orchards without chlorosis. A survey showed a decline in Zn content of leaves and an increase in chlorosis over several previous years. The almost complete change from a lead arsenate-zinc sulfate program for insect control to a schedule containing organic compounds without Zn may account for these changes in Zn status. Foliar Zn sprays were more efficient in correcting the chlorosis than were soil applications or dormant sprays (McClung 1954).

In Florida, all fertilizer for peaches should contain 1–2% Zn oxide equivalent when used on young trees. On older trees, Zn may be applied as part of the regular spray program by including 2 kg (2 lb) of neutral Zn per kl (per 100 gal.) of water in 1–2 cover sprays each year, or it may continue to be applied in the regular spray program. Sprays of Zn-sulfate-lime, 4-4-100, for control of Zn deficiency also helps control rust.

Magnesium

Mg is the only mineral found in chlorophyll. Mg helps plants absorb P, an element which is rather difficult to absorb. Plants fertilized with Mg usually contain more P than plants growing under the same conditions but without additional Mg.

The incidence of two abnormalities of peach leaves, one a marginal chlorosis and the other an interveinal necrotic spotting, occurring on 5 to 7-year-old trees on a sand in North Carolina, was greatly reduced by soil applications of Mg sulfate. Midshoot leaves from trees not fertilized with Mg had 0.13–0.19% Mg; those from fertilized trees had 0.31–0.39%. Two rates of fertilizer were used at each level of Mg; increased K increased incidence of the chlorosis in both years (McClung 1954).

If liming is needed, use a dolomitic limestone. It contains Mg.

Manganese

Interveinal leaf chlorosis due to Mn deficiency occasionally occurs. Then, apply an early summer spray of $MnSO_4$. Since there is little carry-over or storage of Mn applied in sprays, this treatment may need to be made annually if there is a deficiency (Boynton *et al.* 1951).

In New Jersey, Mn deficiency was more pronounced where soil values were 6.5 or higher at all 3 soil depths tested, viz., 0–25, 25–51, and 51–76 cm (0–10, 10–20, and 20–30 in.). Deficiency symptoms were mor intense during dry seasons than wet (Bell and Childers 1956).

Iron

In lime-induced Fe chlorosis there is yellowing between the veins of the younger leaves although the veins remain green. The entire leaf ma\) become yellow or whitish and may then die. Frequently only one branch of a tree is affected or perhaps only a few scattered trees in the orchard are chlorotic. In severe cases the entire orchard may be affected.

In some acid soils the amount of Fe is too low and an actual Fe deficiency exists. In other acid soils, excessive Mn or Cu may interfere with the plant's use of Fe, even when adequate Fe is in the soil.

In Utah soils, bicarbonate seemed to be the main cause of Fe chlorosis. It reduced movement of Fe to the leaves, and reduced respiration in roots of species susceptible to Fe chlorosis by inhibiting the Fe-cytochrome enzymes (Wiebe 1956).

Certain Fe chelates, applied on the soil in early spring, 0.23–0.45 kg (1/2–1 lb) per mature tree, may overcome chlorosis. Moisture helps bring the chelates in contact with the roots. The chelates are broken down in the soil by too much water and the Fe can then be tied up.

Calcium accumulates to a higher level at a given treatment in prune leaves than in peach leaves (Abdalla and Childers 1973).

Interrelations of Fertilizer Elements, Pruning, and Irrigation

Earlier fruit ripening occurred on moderately pruned trees and with low K, in contrast to later ripening on heavily pruned trees and high K. Heavy pruning usually resulted in increased K in the leaves. Increased N applications resulted in reduced K content where no application of K was made. Where the rates were low the percentage in leaves was reduced during the season under heavy crops; but with no crop and high rates of K, the leaf K increased. The differences were especially large where the soil K application rates were low and the N application was high (Havis and Gilkerson 1951).

There was a positive relation between N, P, and K in the leaves until a certain nutrition level was reached, beyond which there was a reduced rate of K absorption although higher P levels exerted an antagonistic action on the N absorption and on new growth. Higher K absorption may also depress N absorption, and other elements may also have an influence (Kenworthy 1948).

Increasing pruning severity decreased yield, red color on fruit, and de-layed fruit maturity. It increased fruit size, terminal growth, and foliar N percentage. Within limits, increasing N fertilizer from 0.14 to 0.54 kg (0.3 to 1.2 lb) per tree per year increased terminal growth, foliar N per-centage, and delayed fruit maturity. It decreased fruit color and had no measurable effect on fruit size. Increasing N application to 0.41 kg (0.9 lb) resulted in increased yield and trunk cross-sectional area with no further response from increasing N to 0.54 kg (1.2 lb) per tree. Time of application did not greatly alter any of the factors measured. It is neces-sary to know the level of pruning to be followed before accurately pre-dicting yield response to fertility level changes (Schneider and McClung 1957).

Interactions with cultivar, N rates, and pruning indicated that Redskin responded to irrigation more than did Elberta. Droughts usually oc-curred during Redskin ripening. Rainfall in July, prior to Elberta ripening, was relatively heavy thus eliminating the need for additional irrigation during the Elberta ripening season. Irrigated Redskin trees produced (a) more harvested fruits in the greater than 5.1 cm (2 in.) size class than did those not irrigated, but this became less as the trees were more heavily pruned; (b) more harvested fruits on both numbers and kilograms (pounds) basis than those not irrigated, and this effect became more significant with each succeeding harvest season; (c) fewer har-vested fruits in the small, less than 5.1 cm (2 in.) size, than did nonirrigated trees when the N rate was increased from N_1 to N_2, but a further increase in N rate to N_3 negated this response (Ballinger et al. 1963).

In the above experiment, responses of peach trees to irrigation seemed to be influenced by season, pruning, N rates, and the cultivars involved. These should be considered in the development of an irrigation system. Rainfall during the period of the experiment (above) was similar to the average for 28 years, with most rain in July and August. Supplemental watering generally was required whenever the number of days during a given month with rainfall greater than 2.5 mm (0.1 in.) was 7 or less. There was a trend toward larger fruit on the irrigated trees.

IRRIGATION

In the eastern peach areas, 2–3 weeks of dry, hot weather during pit hardening and the final-swell period may adversely affect fruit size. A deficiency of water during the growing season may also result in less terminal growth and reduced winter hardiness.

In Maryland, 90% of the roots were in the first 0.6 m (2 ft). Moisture in the third 0.3 m (foot) was near the wilting percentage on the dry plot during most of August. Evidently there was sufficient moisture available to some of the roots in this plot, since the trees did not show permanent wilting (Cullinan and Weinberger 1933).

In Pennsylvania, irrigation resulted in larger yields and a higher percentage of fruits of larger sizes. Fruit from the irrigated trees ripened uniformly and was harvested in 1 picking; fruit from nonirrigated trees required 2 pickings. Also, irrigated fruit ripened 2–3 days earlier than did nonirrigated fruit (Feldstein and Childers 1957).

In Washington, the first irrigation is usually applied about April 15, and thereafter every 7–20 days during the growing season. If a cover crop is growing, irrigate as early as water is available. At harvest time, irrigate between pickings, if only for overnight, as this tends to keep the fruit firm, especially in hot, dry weather. Adequate soil moisture during attainment of fruit maturity is essential because of the rapid increase in fruit size (Overholser et al. 1941).

At every irrigation, completely wet the soil where most of the roots are developing. On deep, well-drained, and aerated soils, a maximum depth of wetting (1.8 m [6 ft] at each irrigation) may be sufficient in irrigated areas. On light, shallow soil apply enough water at each irrigation to penetrate only to the soil depth where the roots are found; otherwise leaching of nutrients in the soil may result. When water and nutrients are carried down to gravel where no roots are growing, the trees can not recover them and both are lost. Hence on shallow soils, apply less water at each irrigation, but irrigate more frequently.

Excessive irrigation may result in a rise of the water table and thus cause injury to peach roots. Overwatering usually causes early yellowing and leaf fall. These, however, may also be signs of nutrient deficiency or of disease. A high water table not only seriously limits the root development, but may also affect the accumulation of alkali. Keep the water table 1.8–2.4 m (6–8 ft) or more below the surface. Irrigation water should be free from an excess of alkali salts.

SEED AND FRUIT DEVELOPMENT STAGES

There are three stages of peach fruit development.

(1) Embryo development is arrested. This is a period of rapid development, chiefly in the seed area: in Elberta, mostly within 30–40 days following bloom. The stone reaches nearly full size.

(2) There is retarded pericarp increase during midseason. Embryo development is rapid. In Elberta, this period starts about 40 days after bloom and continues for about 4 weeks. It is featured mainly by pit hardening with little increase in size.

(3) There is limited pericarp to fruit maturity. With early cultivars this stage is initiated during rapid development of the embryo, and such cultivars have abortive embryos. With late cultivars it does not begin until the embryo is morphologically maximum size, and such cultivars have a high proportion of viable seed. In Elberta, it starts 65–70 days after bloom and lasts about 6 weeks. During this period, especially near maturity, the greatest increase in volume of the flesh and about 66% of the weight increase of the peach takes place. This late growth explains why there is somewhat more benefit from late peach thinning than from late apple thinning.

Destruction of the embryo early in Stage 2, or earlier, abruptly checks fruit development, causing shriveling and abscission. Embryo destruction in the transition between Stages 2 and 3 results in growth of the fruit at the normal rate but for a limited period, ending with earlier ripening and failure to reach full size. In Stage 3, embryo abortion results in increased growth rate and earlier ripening, and occasionally increased size over that attained normally. Wounding of the flesh and stony pericarp either alone or together without injuring the seed does not alter the growth rate of the fruit. The nearer the date of normal fruit maturity at which the embryo is destroyed, the more rapid the growth increase of pericarp. Early ripening of some cultivars is due, at least in part, to embryo abortion, whether a genetic relationship or not.

Increase in fruit volume after stone formation is due to increase in amount of flesh (over 60% as dry weight). Fruit growth rate is slow during stone hardening, regardless of thinning time, because stone and kernel are dominant over flesh. Increase in size from late thinning is due to removal of competition of developing flesh, largely for carbohydrates, during the final swell. Stone and kernel are also made up predominantly of complex carbohydrates. Since removal of carbohydrate competition gives larger size at maturity, thinning soon after the June drop is more effective. Early thinning, particularly with large-stoned cultivars, saves carbohydrates that otherwise go to stones and kernels. Even though early thinning does not always promote largest fruit, it conserves carbohydrates for fruit growth, formation of fruit buds, vegetative extension, or for storage.

The slow fruit growth period is due to slow growth in flesh only; both stone and kernels make their fastest dry weight increase at this time. Also, the stone may become physiologically mature during this period, since it reaches its maximum concentration of all materials except sugar, which decreases. The final swell shows a marked increase in size, dry weight, and sugar content of the flesh; a decrease in dry weight in the stone, accompanied by a decrease of N, carbohydrates, and ash; a decrease in dry weight of the kernel, but increasing N, sugar, hemicellulose, ash, and ether extract, thus indicating that the kernel has not reached physiological maturity at harvest. Materials were being translocated from stone to kernel during this period, since all stone constituents are decreasing while those of the kernel, except starch, are increasing. There may also be some translocation from stone to flesh, since all constituents of the flesh, except hemicellulose, are increasing.

Size, green weight, and dry weight of the stone, and N and other extract content of the kernel increase with the time required for the cultivar to mature. During Stage 2 in Elberta, the stone and kernel are the dominant parts of the fruit, but in Mayflower (early) flesh is dominant over stone and kernel to the extent that the stone never becomes completely hard and the cotyledons develop only enough to fill the integuments of ripe fruit. Dry weight, amount of N, and ether extract in Mayflower are so small as to preclude the ability to reproduce itself without artificial culture.

There may be an overlapping of the processes associated with the different developmental periods rather than an abrupt cessation of one and the beginning of another. Thus, the linear growth may be resolved into two parts: the growth represented by the coincidental development of pit and flesh, and the growth of the flesh in the final swell. What has previously been termed Stage 2 may be formed by components of the two parts mentioned above. Increase in size during the final swell may consist of increase in size of cells of the flesh attendant upon the rapid rise in soluble solids and water. Development during the first period may be quite complex. Increase in cell number and cell size of both pit and flesh all contribute to the total increase in size.

MORPHOLOGICAL CHARACTERISTICS OF
FLOWERS AND YOUNG FRUITS

Lott and Simons (1966) hav. described ten specific morphological stages, beginning at full bloom (Stage I) to style abscission (Stage X). These are readily observable reference points for timing of some production practices, e.g., fruit thinning.

Since Stages IV and X each occupy only a few days they are probably the most useful as developmental reference points.

Stage IV: Abscission zone evident as a pale greenish-yellow line around the outside of the floral cup about 25% of the distance from the bottom of the cup to its rim; no visible separation of the floral tube from the floral-cup base; purplish-red the dominant external color in the floral tube; floral-cup base predominantly green externally with red absent or nearly so; anthers dry, filaments erect and dry 6.4 mm (1/4 in.) below tip, remainder purplish-red; stigma dry, style erect and dry 3.2

mm (1/8 in.) below stigma, remainder purplish-red down to pubescence.

Stage X: Style abscised, leaving suberized tip of fruit; if not abscised it falls off when touched; top of floral-cup base less in diameter than base of fruit immediately distal to it, showing rapid enlargement in fruit base.

Stage X is characterized by the abscission of the style. Stage IV occurs before fertilization, and coincides approximately with the beginning of embryo development. Since a larger percentage of the fruits reach Stage X at the same time than reach Stage IV, it has preference as a physiological reference point.

THINNING THE FRUIT

Thinning improves the size, grade, and finish of the fruit, reduces limb breakage, and maintains tree vigor. The actual cost of labor of thinning may be considered as an expense early in the season rather than at harvest time.

A frequent error is failure to remove enough peaches, thus requiring a second thinning. If thinning is done during or shortly after bloom, the hazard of a further reduction in fruits due to freeze must be kept in mind.

TABLE 4.8
RELATION BETWEEN YIELD AND FRUIT SIZE IN 7-YEAR-OLD ELBERTA PEACH THINNED BY PRUNING

Fruits per Tree	Yield (kg)	(lb)	Diameter (cm)	(in.)
100	16.8	37	7.1	2.8
300	47.2	104	6.9	2.7
500	72.1	159	6.6	2.6
700	91.6	202	6.1	2.4
900	106.6	235	5.6	2.2

Source: Cain (1956).

It is cheaper to reduce the crop with the pruning shears in the dormant season than to try to do so entirely by thinning (except possibly with chemicals) after the fruit is set. But to attempt to thin by pruning so that no fruit thinning will be needed to secure the desired size and quality may reduce the crop too much.

Peach trees are not, like some apple trees, biennial bearers. Even with an excessive crop, peach trees can produce a crop the next year. But overloaded trees may suffer subsequently in winter hardiness and in other ways.

Many large commercial growers thin by the chemical method (discussed later). However, many growers still thin manually. Both should consider certain basic factors.

When to Thin Early Cultivars

Early cultivars have a much shorter fruit development period than Elberta. For many early cultivars the final size of fruit may not be greatly increased by thinning delayed until after the fruits have begun to show signs of pit hardening, especially if dry weather prevails early in the growing season and at harvest. With early cultivars that tend to set heavily, reduction in number of fruits well before the June drop increases fruit size, shoot growth, and leaf size.

When to Thin Later Cultivars

With midseason and later cultivars, one practice is to wait until after the first drop and then remove many of the small and imperfect fruits. If the set is only moderate and the June drop is fairly heavy, thinning after that usually promotes fruits of good size and color. If thinning is delayed until well along in the pit-hardening period, it still benefits the final size of fruit. Even if done during the final swell, thinning may increase fruit size. If thinning is done quickly near the end of the first growth period (before pit hardening), the result is almost as good with Elberta as from somewhat earlier thinning; and the cost may be much less.

High correlation exists between the diameter at harvest of cling peaches early in the season ("reference date") and their diameter at harvest in California. Reference date is a time, arbitrarily selected, that is ten days after the extreme tip has begun to change from a white to a creamy yellow and has hardened enough that the blade of a sharp knife hesitates when successive thin sections are cut through the distal end of the fruit. This date is near the start of the second growth period of the canning cultivars. It is also at the beginning of the normal thinning season and marks a time when the fruit has made about 1/2 its suture diameter growth (David 1950).

In Washington, a reference date 14 days after the beginning of pit hardening in Elberta permitted a 10-day period during which thinning could be done at a time when the fruit was growing at its slowest rate. The reference date was 70 days after full bloom. Peach fruits at this time had made an average growth of 20% of their harvest size. A reference size for Elberta of 3.3–3.5 cm (1.31–1.37 in.) was necessary for a harvest size of 6 cm (2 3/8 in.). Orchards with fruits averaging 6.7 cm (2.63 in.) or less had more than 10% smaller than 6 cm (2 3/8 in.) (Batjer and Westwood 1958).

With Elberta, the stages of development of fruit, nucellus and integuments, and embryo offer more exact reference points for comparison of such operations as thinning in different years than do calendar dates. Thinning before the increase in number and size of cells of the stony pericarp had been completed resulted in an increase in size of the pit. It also resulted in the largest and best color, and less reduction in total yield in any one year (Tukey 1945). Over a 3-year period, early thinning resulted in more blossoms in a light crop year, and in size and number of leaves per tree, and so tended to promote more regular bearing (Shoemaker 1934).

Under favorable growing conditions, when there is a heavy crop, removing the small or injured fruits in late thinning tends to even up the size of fruit by taking advantage of the final swell of the fruits remaining on the tree.

Degree of Thinning

The proper number of fruits to remove in thinning depends, in part, on the size of the tree and its bearing capacity. Leave only enough of the peaches that can develop to the desired size. When a tree has a uniformly heavy set of fruit, it is possible to thin by hand to a fixed spacing such as 15 to 20 cm (6 to 8 in.) along the twig. When thinning, consider leaf area, tree vigor, and bearing capacity.

In the total load system of thinning, if the aim is 91 kg (200 lb) with fruits averaging 6 cm (2 3/8 in.) 1–2 per kg (3–4 per lb), leave about 800 peaches per tree (Upshall 1946).

Because No. 2 peaches are of less value than No. 1 grade, thin sufficiently to guard against failure to reach No. 1 size even in a dry season. The amount of thinning year after year also pays dividends indirectly by helping to avoid exhaustion of the tree through the production of an overabundance of pits, the development of which draws heavily on the tree's mineral supply.

Leaf to Fruit Ratio

Elberta peaches increased in size with increase in leaf area up to 75 leaves per fruit, with no indication that larger peaches might not have been obtained with more leaves. Five leaves produced fruit having a volume of 69.3 cc (5.1 cm [2 in.] in diameter); 10 leaves, 95.8 cc (5.7 cm [2 1/4 in.]); and 20 leaves, 117.4 cc (6 cm [2 3/8 in.]). However, 10 leaves did not produce twice as much increment in size as 5 leaves, and the fruit with 20 leaves was less than double the size with 10 leaves. Fruit on branches having more leaves per fruit than 20 were larger with increase in leaf area, but the increments per unit leaf area were smaller. The efficiency of the individual leaf to food production decreased greatly when the opportunity for utilization of the products was lessened (Weinberger 1932).

The peaches grown with 5–10 leaves per fruit were of poor quality, tasting bitter, flat, and disagreeable (5.3% sugar); with 20–30 leaves the fruit was sweeter, with a good flavor (7.6% sugar); with 40 or more leaves the fruit was of very good flavor and quality. Further increase in sugar was not obtained with increase in leaf area up to 40 leaves per fruit; but fruit grown with 50 leaves had 8.8% sugar and with 75 leaves 9.1%.

Thinning, or increasing the number of leaves per fruit on ringed branches, caused differentiation of more fruit buds. Branches with 20 leaves per fruit had almost 6 times as many fruit buds per 100 nodes as branches with 10 leaves per fruit. With ratios above 20, the increase was only about 10% for each 1-leaf increment. With the same leaf to fruit ratios, increase in moisture supply promoted larger fruit. The minimum leaf area which favored quality fruit varied under different soil conditions (Jones 1933).

Fruit had more colored area where the leaf area was low than where it was high, and the colored area decreased as leaf area per fruit and size of fruit decreased (Table 4.9). Fruit in the interior of trees was about as

TABLE 4.9
ELBERTA COLOR IN RELATION TO LEAF AREA PER FRUIT
AND TO SIZE OF FRUIT ON RINGED BRANCHES

Leaves per Fruit	No. of Fruits	Leaf Area per Fruit (cm^2)	Avg Wt. of Fruits (g)	Solid Red on Exposed Half of Fruit (%)
10	245	1.1	71.6	44.7
20	248	61.8	115.7	32.7
30	149	87.5	157.9	29.5
40	143	120.1	185.3	22.5
60	112	170.4	217.2	16.5

Source: Kinman (1941).

well colored as outside fruit, but fruit enclosed in black cloth bags did not develop any red (Kinman 1941).

The heavier the thinning the greater is the reduction of the crop, but the size of fruit is increased. The number of leaves per fruit to be left after thinning varies with cultivars and with trees of a given cultivar. Vigor of tree, with which leaf size is correlated, and the moisture supply also are significant. Usually, however, 30–40 leaves per fruit are desirable, considering size and quality of fruit and future tree performance.

Thinning Methods

Hand Thinning.—Rub off the surplus fruit by hand to the desired spacing and break up clusters, e.g., 15–25 cm (6–10 in.) spacing according to crop and cultivar. Early thinning is slower and more costly than late thinning, but thinning of Elberta can be done to advantage up to the time of final swell. Hand thinning is slow and costly work.

Courtesy Hort. Inst. Ontario

FIG. 4.6. REDHAVEN THINNED

Pole Method.—The time required to thin a peach crop by hand may be cut in half by using a hose-tipped pole without serious adverse effect on tree growth, yield, or grade of fruit. The work is done with a rubber hose 38–46 cm (15–18 in.) long, on the end of a 2.5–3.8 cm (1–1 1/2 in.) pole.

With the use of two poles, one 0.61 m (2 ft) long and the second 1.2–2.4 m (4–8 ft) long, no ladder is needed. However, some growers prefer to use only a short stick, or the hose without a stick, and a ladder. Striking into clusters along the branches toward the outside of the tree or limb, 0.6–0.9 m (2–3 ft) from the end of the branch where it is 3.8–5.1 cm (1–2 in.) in diameter, generally gives good results. After thinning from the inside of the tree, make a round from the outside, breaking up clusters that were missed. When the set is very heavy there may be some value in preliminary use of a shaker or in supplementary hand thinning of peaches from the weaker growth and clusters that were missed with the hose-tipped pole.

The fruits are most easily knocked off when they are of walnut size. If thinning is started too soon, the peaches will not come off readily. Some cultivars are easier to thin effectively than others.

Tree and Branch Shakers.—Machine shakers may be useful to thin peaches in some instances. Do the thinning before the pits harden. The shakers can do a fast job (2–3 min per tree). Yields often are reduced because of a tendency to overshake the top portion of the tree where the fruits separate most readily. The claw causes some damage unless precaution is taken for protection. Certain pests, e.g., lesser peach borer and canker, may cause increased damage. Supplementary hand and/or pole thinning is desirable since shakers only reduce the number of fruits on the tree and do not promote even spacing of the developing fruits.

Chemical Method.—Chemical sprays have become widely used commercially for thinning young developing peach fruits. They greatly reduce the labor required, and may be supplemented later by hand and pole methods.

Three main factors have contributed to the widespread adoption of chemical thinning: (a) less labor required; (b) information which permits a precise timing of the spray; and (c) the development of improved spray materials for greater effectiveness without adversely affecting the tree and fruit.

Timing the Thinner Spray.—A number of guides or indices are useful to determine the proper time to apply fruit thinning sprays, e.g., ovule, seed, or pericarp length, seed diameter, and fruit volume. These may be used singly, or in combination for check purposes. These indices are much more precise and consistent than such guides as days to full bloom, days after petal fall, accumulated heat units, and calendar dates.

Chemical thinning is probably best done when about 1/3 of the young fruits on the tree have reached the completely cellular stage and are resistant to chemical thinning.

In general, an ovule length of 7–10 mm is the optimum stage for most peach cultivars in northern regions. However, different cultivars respond differently and some, e.g., Redskin, seem more difficult to gage efficiently than others. There is a need for more information on the way different cultivars respond to spray thinning at a given index in the same and different regions.

The low-chilling, short-cycle cultivars in Florida, e.g., Maygold, require 24 mm diameter and/or an ovule length of 13–14 mm or more for the best chemical thinning results. This stage occurred 29 days after full bloom when cytokinesis was completed in 35% of the fruit. During cytokinesis (endosperm becoming cellular from the free nuclear stage), the fruit and seed sizes are larger in cultivars with a shorter development cycle than in the north (Sherman *et al.* 1970).

To determine the proper ovule length for chemical thinning, cut through enough of the fruits along the suture for a representative sample.

The water-displacement index is also simple and easy to determine. Select 100 young peaches at random. Place them in a 1000 ml graduated cylinder holding enough water so that the pubescent fruits can be submerged before the volume reading is taken. The desired index usually is 280 ml per 100 fruits; of course, start testing before the fruits become too large.

Materials.—Caustic dinitros (DN), naphthaleneacetic acid (NAA), 3-chlorophenoxy propionic acid (CPA), and naphthyl phthalamic acid (NPA) are all useful peach thinners; their degree of success varies among areas and from one season to the next.

The DNs have given the most satisfactory thinning results over a period of years in some areas, e.g., Niagara. Failures have more often been due to insufficient than to overthinning. Occasional overthinning has usually been attributed to slow drying conditions. Sprays should be applied when about 90–95% of the blossoms are open. Concentrations vary from low (235 ml [1/2 pt]) for easy-to-thin cultivars, e.g., Golden Jubilee and Elberta, to high (705 ml [1 1/2 pt]) for hard-to-thin cultivars, e.g., Veteran and Redskin.

NAA has been generally erratic and inconsistent as a thinning agent for peach but some satisfactory results have been attained. Timing appears to be critical and although 30–45 days after full bloom has frequently been satisfactory, the early stages of cytokinesis, i.e., when fruit length is 27–31 mm, has been considered most likely to be sensitive to NAA.

NPA has given a good degree of thinning at petal-fall stage where thorough coverage has been obtained and weather conditions were favorable for absorption. Elberta may be thinned by 100—150 ppm while heavy setting cultivars such as Redhaven and Halehaven require higher concentrations, e.g., 200—300 ppm.

CPA applied 150 ppm moves slowly, accumulating in the margins and veins of leaves, and in the epidermal and subepidermal layers on the micropylar end of the ovule. Equal degrees of fruit thinning result when fruit or leaves are treated, but treating both leaves and fruit has an additive thinning effect (Martin and Nelson 1969).

Uniform application throughout the tree is necessary to avoid gaps in coverage. But more thinning by CPA occurs in shady areas of the tree, which indicates that light plays a role in its effectiveness. Overthinning of fruit in shaded areas, therefore, may prove a problem in orchards with trees densely planted in hedgerows, or insufficiently open-pruned (Bauscher *et al.* 1970).

Physiology of Chemical Thinning.—Caustic materials are applied after the first (and usually the most strongly set) blossoms have been pollinized and growth of the pollen tube down the style is complete or well advanced. The spray destroys the anthers as well as the stigmas and the adjacent stylar tissue of the newly opened blossoms, thus rendering them sterile.

It has been proposed that during the critical growth, when the auxin content of seeds is at a maximum and the auxin destructive factors like indolebutyric oxidases are low, the application of an exogenous auxin like CPA or NAA may result in auxin changes which bring about the onset of fruit abscission. The endosperm is closely associated with sensitivity to chemical thinning agents. Exogenous "hormones" may critically damage the endosperm, preventing its development, and result in embryo abortion and fruit thinning (Leuty and Bukovac 1968).

NAA and related hormone-like substances act to control the formation of the corky cells of the abscission zone. The site of their action appears to be within the embryo but the actual mode of action is not understood.

Spray Additives.—Certain spray additives, e.g., sodium sulfates of mixed long chain alcohol-fatty acid esters and diethylene glycol abietate, e.g., Surfactant F, may increase the thinning response from NAA and CPA by increasing CPA absorption. The use of suitable spray additives may allow some reduction in concentration of the thinner and perhaps reduce some of the variability.

Effect on Hardiness.—Fruit thinning may increase the cold hardiness of the flowers and buds through the following bloom period. Yield increase associated with this apparent effect seems to be related to the degree of thinning in previous years rather than to use of a specific thinning agent per se. The increase in hardiness is probably related to carbohydrate concentration obtained in thinning the fruit (Stembridge and Gambrell 1969).

SPLIT-PIT AND GUMMING

Years in which peaches are large are also years in which there are many split pits. Occurrence of split-pit is closely associated with those conditions which cause an increased growth rate of the fruit. Split pits of Elberta are most numerous where the downward translocation of food is interfered with so that the top and twigs are high in carbohydrates. The cross diameter of the mature fruit is relatively greater in proportion to the length than normally.

Embryos of split-pit fruits are aborted in a high percentage of cases, but the flesh develops to maturity as in normal fruits. Fracture of the pit takes place mainly along lines corresponding to the dorsal and ventral sides. The split along the ventral suture leaves a smooth, unfractured surface, except for the inner portion from the point of attachment of the ovule to the base of the pit. This region and the dorsal side break down with a fractured surface as if interlacing cells had been torn apart. Split-pit occurs mostly during pit hardening, the percentage being high under conditions that favor large fruits.

Gumming on Phillips Cling occurs during pit hardening, either on the dorsal side opposite the ventral suture and about 1/3 the distance from the distal end, or at the distal end. Abortion of the embryos is associated with gumming (Ragland 1934).

Gumming of Phillips Cling falls into at least three types:

(1) Due to mechanical injury, e.g., a limb rub or insect puncture. It may appear any place on the periphery of the fruit and has no pocket in the flesh beneath the gum.

(2) When the pit begins to harden at the tip or a little earlier. It always occurs on the ventral suture, but the point of appearance may be any place on the suture from the distal to the stem end. The gummy mass is large and globular; the flesh and pit seem to be split; and a continuous mass of gum extends from the endocarp and entirely fills the cavity of the pit. The fruit falls within 7–10 days after the gum appears.

(3) Gumming begins to appear about two weeks after the pit begins to harden at the tip. The number of these gummy fruits increases rapidly; 80–90% of the total has appeared 3 weeks after the first gumming became evident. The gum is twisting in nature, and small compared with the second type. It occurs on the distal end or dorsal side near the distal end and is accompanied by a pocket in the flesh. This type of gum causes heavy losses at harvest time. No organism has been associated with gumming.

Split-pit is a serious quality defect affecting both cling and freestone peaches. Any factor that enhances fruit growth may contribute to split-pit. Some factors appear to be more effective if they influence growth prior to the time of thinning. But factors that enhance fruit growth after thinning also seem to contribute, where fruit size is such as to make pits susceptible to splitting. Split-pit varies from year to year even under the same cultural regime; therefore, climatic factors that affect fruit growth also appear to contribute to potential split-pit. One may speculate that mechanical forces causing split-pit are associated with cell rigidity, particularly when pits are in the most susceptible stage for splitting. Favorable temperature and soil moisture are the critical conditions in this regard. Since pit size and fruit growth are genetic in nature, a likely solution is through the genetic approach. But an understanding of the factors that affect severity of split-pit indicates means of partial control (Claypool *et al.* 1972).

UNEQUALLY SIDED FRUITS

Peaches, cherries, and plums often have fruits which are unequally sided; one side is more fully developed than the other. Also, the larger side may be more highly colored, more highly flavored, and may soften and ripen earlier.

The pit also shows development similar to the flesh; the overdeveloped half of the pit is on the same side as the overdeveloped half of the flesh. This unequalness of halves is associated with development of the seed. A peach fruit consists of a single carpel containing two ovules. Just before full bloom and prior to fertilization, one ovule usually aborts. The remaining ovule develops into a seed.

Examination of the pit at the point of attachment of the seed in unequally sided fruit shows that the seed is attached to the larger and better developed side. It is thus possible to ascertain to which side the seed is attached by observing the larger and better developed side, either of the flesh or of the pit (Tukey 1945).

LENGTH OF FRUIT DEVELOPMENT PERIOD

Interest is expressed each year by growers, shippers, and others in the probable harvest date. The time of full bloom marks a definite stage in the development of fruit. The interval between full bloom and fruit maturity is fairly specific for each cultivar, with variations from year to year due to various factors. An early bloom or late bloom does not necessarily result in a respective early or late time of maturity, but an early bloom tends to result in a longer development period and a late bloom in a shorter development period.

The elapsed interval between full bloom and fruit maturity for Elberta in New Jersey and New York was 128 days. High temperatures following bloom shortened the interval between bloom and harvest (Blake 1948).

In Colorado, the industry has used a guide for Elberta of 126 days from bloom to the first carload for shipping. In an unusual year, with an interval as few as 119 days or as much as 131 days, picking could miss the maturity date by 6–7 days. A more exact method developed by U.S. Dept. of Agr. is based on daily maximum temperatures prior to July 1, together with an additional adjustment for earliness of bloom. For cool days, a day or part of a day is added toward maturity date. An early or late prediction can throw the entire marketing operation off, because, long before the maturity date, growers and shippers plan for the pack-out. This involves the availability of government inspectors, truckers, brokers, and buyers. If the prediction is too early, pickers may stand idly by. If too late, they may move to another job, and growers find themselves with ripe fruit and a shortage of labor.

Many other factors aside from temperature, such as size of crop, age of tree, growth of tree, and N supply, affect the length of the fruit development period. However, by correcting the average length of the fruit development development period for early season effects, more than half the variability is eliminated between the expected and the actual date of harvest.

At Fort Valley, Ga., a southern limit for Elberta and where prolonged dormancy may occur, the elapsed time from full bloom to first picking in a 10-year period was 104–124 days, with a mean of 113 days. An estimate of maturity date based on date of full bloom and the average of 113-day fruit development could be off as much as 11 days. The date of full bloom varied in the same year from as early as March 1 to as late as April 4. Where the rest period is not broken until relatively late in the season, the buds are not greatly influenced by relatively high temperatures occurring during midwinter. Often, Elberta trees at Fort Valley bloom only a few days earlier than in areas 250 miles to the north, but relatively warm weather following bloom advances Fort Valley peaches to a much earlier harvest (Weinberger 1932).

Stages of Maturity

"Mature" emphasizes fullness of growth and completeness of development, and "ripe" suggests readiness for use. Accumulation of carbohydrates and other materials and enlargement of the cells during growth and development of the fruit are maturity changes that in-the ultimate quality when peaches become ripe. Softening of the flesh by hydrolysis of the pectic materials and the development of aromatic constituents are ripening changes. Maturation occurs only while the fruit is attached to the trees, but ripening may occur either before or after harvest.

In the hard green peach, the middle lamella or cementing material between the individual cells holding them tightly together is probably an insoluble calcium salt of protopectin. As the peach ripens the proto-pectins are converted by the enzyme protopectinase into pectins and pectinic acids. The individual cells are no longer held firmly in place; they slip and slide and lack rigidity; therefore the fruit becomes soft. As the peach becomes softer and overripe the pectins and pectinic acids are further hydrolyzed by the enzyme pectinase to pectic acids or pectates and methyl alcohol, and the fruit becomes mushy. These changes are retarded by low temperatures (Overholser *et al.* 1941).

The term "shipping ripe" is often used to describe peaches picked while still hard. This is an incorrect usage of "ripe"; fruit in this condition is not suitable for consumption. It might better be described as "shipping mature." But if the fruit is allowed to remain on the tree until soft and suitable for eating, it is correctly described as "tree ripe."

Five commercial stages of harvesting maturity are recognized. Normally, 3—4 days are required for peaches to pass from one maturity stage to another at 21.1°C (70°F).

Hard.—The peach does not yield to moderate pressure. This is the stage at which peaches are often picked when long keeping or carrying qualities are the prime considerations. However, hard peaches commonly are not sufficiently mature to be picked and under some industry regulations are regarded as culls.

Firm.—The peach yields very slightly to moderate pressure. With some packinghouses this is the minimum maturity allowed.

Firm-ripe.—The fruit has a slight "give" when pressed in the cupped hand, and has a yellowish-green ground color in yellow-fleshed cultivars. It is fairly palatable, but is not in prime eating condition. It will store for a reasonable period.

Ripe (Tree Ripe).—The fruit yields readily to moderate pressure and is in prime eating condition with a useful life of 1—2 days. The storage life, even at 0°C (32°F), is very limited because the fruit tends to soften rapidly when removed from storage. The fruit has little resistance to slight pressure and should move directly to the consumer instead of being stored.

Soft (Soft Ripe).—This stage is not suitable for trade channels, even though the fruits are still of high dessert quality.

Dessert quality may be affected by cultivar, size of crop, soil fertility, soil moisture, weather, picking maturity, temperature during storage, and ripening. Good dessert quality in Elberta is associated with at least 7% total sugars, 5.0–5.5% sucrose, and less total acidity than the equivalent of 15 ml N per 10 Normal NaOH in 10 ml of juice.

Size Increase of Fruit

Peach growers are tempted to start picking at too early a stage of fruit development, often because of price advantage at the time or because of the risk of unfavorable weather when a large hectarage (acreage) is involved. Fruit picked before the ground color becomes appreciably yellow never develops good quality. And when picked too early the maximum volume of the crop is not obtained, since peaches increase in size immediately before picking time.

In Utah, the increase in volume of Elberta was 25% in the last week before optimum maturity (at or near the time of disappearance of the green in the ground color (Coe 1933). In British Columbia, the increase averaged 29% for Rochester, Vedette, and Veteran (Fisher and Britton 1940). In Washington, pickings spaced six days apart resulted in an increase in fruit weight up to 20% between pickings (Snyder 1952). In Illinois, pickings began on August 15. For every 3.5 kl (100 bu) picked on that date, 3.8 (108) could have been picked on August 17, 4.1 (117) on August 20, and 4.4 (125) on August 22 (McMunn and Dorsey 1934). In Ontario, Golden Jubilee, South Haven, Veteran, and Elberta increased in tonnage 41, 23, 20, and 18%, respectively, during the week before optimum maturity. When there was a high proportion of No. 2 peaches, the gain by delaying picking was considerable. Growers profited both through increased tonnage and higher unit prices for the fruit (a larger proportion of the crop coming into the higher priced No. 1 grade) (Upshall 1943).

After being picked, peaches no longer increase in size, and may actually shrink due to loss of moisture. Hiley fruits shrink 0.8 mm (1/32 in.) in least diameter during transit. When peaches are closely graded this shrinkage may cause many of the fruits to become smaller than the minimum allowed for the grade. Inspectors may take this into account, and allow the fruit to be 3 mm (1/8 in.) undersize at terminal markets without reversing shipping-point certificates.

Effect of Growth Regulators on Fruit Maturation

Succinic acid-2,2-methyl hydrazide (Alar) had no effect on fruit size or yield, but when applied at 2000 ppm near pit-hardening it accelerated ripening, caused uniform ripening, and improved fruit color (Gambrell *et al.* 1967).

When canned, peach cultivars which are high in red pigments or which tend to discolor after canning have a greater tendency toward discoloration when Alar is used. But if the cultivar has no tendency to discolor, the use of Alar enhances the development of a desirable yellow-orange color. Limited observations showed that those peaches treated with Alar had smaller pits, none of which split (Van Blaricom *et al.* 1970).

Alar may hasten ripening, hasten the occurrence of the climacteric, increase internal flesh color and yellow skin color, slightly decrease percentage of soluble solids and decrease flesh firmness. Increases in red and yellow skin pigmentation and internal flesh color are increased in fruits of the same firmness. Terminal growth is not altered greatly by Alar, even at high concentration, unless applied near the beginning of the pit-hardening stage (Byers and Emerson 1969).

Alar applied as post-bloom sprays to nine cultivars for 6 years accelerated maturation and reduced the number of pickings required for most cultivars. Although Alar did not affect the number of fruits per tree, yield, or fruit size, it advanced the maturity date of Ranger one week and Blake four days. The early Cardinal was not noticeably affected by applications at different stages of development. Alar caused fruit to abscise more readily from the stem and left less fruit remaining on the trees when machine harvested. Alar had no detrimental influence on Redglobe stored at 10°C (50°F) for three weeks. These effects support the feasibility of using Alar as an aid in machine harvesting freestone peaches intended for fresh market (Gambrell *et al.* 1967).

Ethephon (30 and 50 ppm) and GA (50 ppm) applied as a whole spray to Early Amber (seed length 12 mm, 2 weeks after petal fall) in central Florida consistently prevented browning of puree and sliced peaches for 24 hr and failed to darken whether frozen, freshly harvested, or held at about 25°C (77°F) for seven days after harvest (Buchanan *et al.* 1970).

When to Pick

Proper timing of picking is very important. Because peaches mature unevenly it is usually wise to make two or more pickings, depending on the cultivar, crop, weather, and market. Each grower must decide when there are enough sufficiently mature fruits to warrant the labor of picking.

The grower's decision on when to pick is influenced by how he intends to market the crop. If peaches are to be trucked to nearby markets and will be sold within 1−2 days, let them become nearly ripe on the trees. For long shipment, pick when the fruits are still firm (mature ripe). Fruit that is to be cooled before shipment or shipped in fan cars can be left on the trees until somewhat more mature than if it were to be shipped under standard refrigeration in ordinary cars.

The processing outlet, particularly in certain regions, takes much of the crop. Relations with the canners often would be greatly improved if fruit of uniform maturity is supplied. All fruits in a container chould be of the same maturity, and containers of the same maturity should be grouped. For example, in a container of peaches the bulk of which require 6−8 days to ripen at 4.4°C (40°F), the canner does not want some peaches that are almost eating ripe. While he waits for most of the peaches to ripen, the peaches that were nearly ripe when he received them may rot and cause the others to rot.

Peaches in transit about 4 days, from Utah to market, for example, advance about 1 stage in maturity. A peach picked at the hard stage in Utah is usually of firm maturity when it reaches the retail store in Kansas City. Most customers consider a firm peach to be inedible. A peach picked at the firm stage arrives firm-ripe at the Kansas City store, i.e., edible but still not of best eating quality. Peaches picked while firm-ripe are mostly ripe upon arrival at the midwestern store, and are preferred by customers. Growers may need to pick more often and take more care in picking only those peaches that are of the proper stage of maturity in order to deliver more peaches of the most desirable maturity to the consumer. The extra pickings increase costs. Also, it may be more costly to market ripe fruit because of the greater care needed in handling. The increased yield, larger size, and higher price may more than compensate for the increased cost (Lamborn 1955).

Ground Color.—As peaches approach maturity, the leaf-green ground color changes to a lighter shade of green. Simultaneously, a gradual yellowing, starting on the sides exposed to the light, takes place. As the peaches become riper, a deep orange-yellow develops on yellow-fleshed cultivars and a cream color on white types. The rate of change is proportional to the rate of ripening and is more rapid at the higher temperatures.

There is no better practical index for orchard use than the disappearance of the green from the ground color of the peach. Pickers should note the ground color of each fruit.

Red color is not a reliable index of maturity; a red-cheeked peach may actually not be sufficiently ripe. A definite break in color toward yellow usually occurs first in a small area on the exposed side of the peach. At this time, the rest of the peach may show no appreciable change. The yellow area gradually deepens in color and enlarges until the entire fruit becomes yellowish. In certain cultivars, e.g., Elberta, particularly early in the harvest period, all fruits showing this definite break in ground color in only a small area may be picked.

The following ranges of ground color are commonly found in the varying degrees of firmness of Elberta that have reached maturity: hard, turning from dark green to light green; firm, light green to turning yellow; firm-ripe, light green to yellow; ripe, turning yellow to yellow; soft, yellow.

Pressure Test

The pressure test can be used to determine the stage of maturity, and a ground color range that corresponds to this can be established for the pickers.

Differences in pressure have been found in five different fruit positions. The order of firmness was: left cheek (when facing suture with fruit hanging on the tree); right cheek; suture; opposite suture; apex. Use a definite set of positions to determine the relative firmness of different fruits. Pressure tests should be made on both pared cheeks and averaged.

Elberta peaches testing 5.4 to 6.4 kg (12 to 14 lb) with an 8 mm tester on the pared cheeks hold up in refrigerated shipment for several days and ripen with good dessert quality. Somewhat greater firmness may be required of peaches intended for long shipment.

In Ontario, out of 1000 peaches of each cultivar, 636 Elberta, 992 Fisher, and 948 Golden Jubilee fruits adhered to the tree to a pressure of 2.3 kg (5 lb) or less (8 mm [5/16 in.] tester through the skin of both cheeks). The Elberta drop began 4 to 10 days before the optimum picking time for shipping, but this early drop was more than offset by the normal swell of the remaining peaches. Hence, for economy as well as quality, picking before the optimum time was not justified. A pressure range of 4.5 to 8.2 kg (10 to 18 lb) was a compromise between quality and safety from loss through dropping. The drop period averaged 25 days for Elberta and 17 days for Golden Jubilee (Upshall 1943).

In Michigan, peaches testing more than 5.4 kg (12 lb) with an 8 mm (5/16 in.) plunger were too immature and did not ripen to make a good processed product. Therefore, pick peaches for processing only when the maximum firmness is 5.4 kg (12 lb) or less. Peaches may be picked at lower pressure tests if they are firm enough to withstand the necessary handling and transportation without bruising. Peaches on the softer side that tested 2.3 kg (5 lb) or less with an 11 mm (7/16 in.) plunger on the pared surface generally bruised in normal handling (Bedford 1956).

Light absorbance and random vibration techniques for estimating firmness of fruit of four peach cultivars nondestructively were evaluated. Correlation between the Magness probe pressure and the difference in light absorbance at 690 and 540 nm was significant for Early Redhaven, Garner Beauty, and Rio Oso Gem. The linear regression equations differed with cultivar for both systems, but either system could be used to estimate firmness (Watada *et al.* 1976).

Sense of Touch.—Firmness of a peach can be judged to some extent by touch. A ripe peach placed in the palm of the hand yields to press as the hand closes. The "give," which is considerable compared with that of a hard object such as a stone, increases with maturity.

TABLE 4.10
INCREASE IN QUANTITY, GRADE, APPEARANCE AND QUALITY OF
VETERAN CULTIVAR IN FINAL WEEK BEFORE
OPTIMUM MATURITY (OM)

	Yield per Tree		Above No. 2 Grade (%)	Surface with Red Color (%)	At Ripe Stage		
	(kg)	(lb)			Total Sugars	Total Acids	Taste
Picked 7 days before OM	71	156	73	14	5.7	1.07	Poor
Picked at OM	91	200	94	44	7.0	0.91	Good

Source: Upshall (1946).

Windfalls.—Peaches increase in weight 3–5% per day. A possible loss from wind of 2% per day, for instance, would not in itself justify picking. Besides size increase, there is also a rapid improvement in grade, appearance, and quality.

In most peach cultivars there is little danger of loss from windfalls up to the time of disappearance of the green from the ground color. With Elberta, however, which tends to drop, pick while there is still a trace of green in the ground color. There is more need of picking to color in Elberta than in many other cultivars. Picked too soon they are of poor quality, and left too long they are likely to drop. Greater than usual tendency to drop is sometimes associated with the prevalence of the Oriental fruit moth; wormy fruits dropping before the good fruit.

Overmaturity.—Growers often fear that with limited labor the fruit will mature faster than they can have it picked, particularly during a heat wave. They feel disposed to "more than keep up" with the picking as a measure of insurance against loss from overmature fruit. Daily weather forecasts take away some of this fear.

Pit Discoloration.—Pit browning is a fairly consistent index of maturity in certain cultivars. Amount of browning increases as the fruit nears maturity. Pit browning is a disadvantage in processing.

In cultivars ripening before Elberta in the south, the red color "bleaches out" and disappears a few months after canning.

Freeness of Pit.—In freestone cultivars, the pit does not become free until it nears maturity. This condition is often used as a maturity index, but is less reliable than the ground color or pressure test. Certain cultivars may be distinctly freestone in most years but only partly so in years when there is excessive rain.

Taste.—Tasting an occasional peach helps in determining maturity, perhaps more so for local trade than for shipping. Much of the bitter and astringent taste of immature fruit may be lost as it becomes mature.

Sugar to Acid Ratio.—In a 3-year test in Ontario with peaches picked at optimum maturity and tested at the best eating stage, Veteran, Elberta, and Golden Jubilee averaged 0.9, 1.2, and 1.3 acid, respectively; the differences in sugar content were insignificant. In most fruits, a decrease in acid content occurs in the week before optimum maturity for picking, and after picking (Upshall 1946).

In Ohio, Halehaven peaches picked at 3, 5, and 7 days from full ripeness were analyzed at picking time and after various periods of storage at 0°C (32°F). They were postripened 4 days at 23.9°C (75°F) before analysis. All but one of the 3-day samples gave a high sugar to acid ratio at a low percentage of total acid. Also, all but 1 of the 5-day samples gave lower or medium sugar to acid ratios at a higher total acid content. The 7-day samples had a wider range in their sugar to acid ratios and total acid contents, but in general were lower than the 5-day samples. This wider range in sugar to acid ratios was due to the difficulty of selecting fruits just seven days from full ripeness. Selection of proper maturity was more accurate with the 3-day and 5-day samples. The riper the fruit at harvest, the higher the sugars and sugar to acid ratio and the lower the acid (Comin 1956).

Avoid Picking Before Optimum Maturity

Many reasons are given by growers for picking before optimum maturity. All have to do with the fear of direct and immediate financial loss. In thinking of personal loss, the grower may forget that the consumer is also subject to loss and can "make or break" him by accepting or rejecting his product. Quality is increased only while the fruit is on the tree, not in the basket.

Growers who pick fruit before it reaches optimum maturity consistent with good carrying qualities fear loss from one or several of the following factors:

Price Drop.—This is primarily an economic problem that growers cannot ignore. When a glut is likely to occur on a certain market, make provision for wider distribution, cold storage facilities, or an advertising campaign in order to prevent undue price depression.

Temperature and Ripening

At high outdoor temperatures that prevail during the marketing season, picked fruit ripens rapidly and also is subject to rapid decay. Peaches harvested at the shipping stage of maturity become eating ripe in 4 to 8 days at 21.1° to 26.7°C (70° to 80°F). Ripening proceeds about one-half as fast at 15.6°C (60°F) as at 21.1° to 26.7°C (70° to 80°F), and about half as fast at 10°C (50°F) as at 15.6°C (60°). At 15.6°C (60°F) and above peaches ripen with good aroma and flavor; if held at 10°C (50°F) until ripe they have undesirable flavor. At 4.4°C (40°F) ripening proceeds about half as fast as at 10°C (50°F) and the peaches usually break down internally before they become fully ripe.

Low temperatures retard the heat of respiration of the fruit. At 21.1°C (70°F) a carload of peaches gives off enough heat to melt 454 kg (1000 lb) of ice per day; at 0°C (32°F) it generates only enough heat to melt 45 kg (100 lb) per day.

Because of the decay hazard and of the influence of temperature on ripening, peaches should be marketed rapidly and under cool conditions. If they can be delivered to market the morning after they are picked, an unrefrigerated truck usually serves the purpose.

In refrigerator cars 2–3 days may be required to cool the fruit to 10°C (50°F) on the top part of a load and to 4.4°C (40°F) in the bottom part. In cars equipped with air circulating fans, the desired low temperatures are reached more quickly and the top part of a load is kept nearly as cool as the bottom part. Thus, in fan cars the fruit ripens less and more uniformly than in ordinary cars.

Rot.—In most peach regions decay of the fruit during marketing is a serious problem, particularly when humidity is high during or shortly before the harvest period. A small percentage of rot in the orchard will mean a loss of several times this amount to the consumer. For example, when only 5% of peaches on the tree show rot spots and only nonspotted fruits are packed, by the time the package reaches the purchaser the wastage may be 50%.

An idea of the potential inoculum from one partly rotted peach can be seen in Table 4.11. At least 90% of these spores may infect additional peaches in the orchard and in transit.

TABLE 4.11
NUMBER OF SPORES PRODUCED BY 3 BROWN ROT-INFECTED PEACHES
OVER A 3-DAY PERIOD

Peach No.	Date	No. of Spores per Microscopic Field[1]	Total Spores Produced[2]
1	Aug. 30	20	100,000,000
	Aug. 31	15	75,000,000
	Sept. 30	6	30,000,000
2	Aug. 30	19	95,000,000
	Aug. 31	10	50,000,000
	Sept. 30	8	40,000,000
3	Aug. 30	39	195,000,000
	Aug. 31	12	60,000,000
	Sept. 1	3	15,000,000
Avg during 3-day period			220,000,000

Source: Burkholder and Sharvelle (1951).
[1]Each count of 10 Petroff-Hauesser field of 1.0 × 10 × 0.02 mm.
[2]Computed from volume of 100 cc of water used in washing each peach.

At high temperatures both brown rot and *Rhizopus* fungi develop and spread rapidly. They can be greatly checked by cooling the fruit soon after harvest and keeping it cool throughout the marketing period. Ripening also is retarded by the cool temperatures, and at some point during marketing or after the peaches reach the consumer they may be exposed to temperatures high enough to ripen them. Under such temperatures decay develops. Thus, although cool temperatures during transit and marketing may make it possible to market the fruit before decay becomes serious, they do not generally reduce the amount of decay that develops in the fruit by the time it is ripe enough to use.

Mechanized Harvesting

Mechanized harvesting of cling peaches in California has become economically feasible, even though canker diseases may invade the bark tissue injured by the shakers. Hand pickers recover more salable fruit and also cause less mechanical damage than do machines. However, fruit damage losses related to mechanized harvesting can be reduced to some extent by modification of tree training and pruning. The size of an economic unit for mechanized harvesting cannot be accurately given at this time, but even with improvements in tree training only the relatively larger operator can expect to compete with hand harvesting (Claypool *et al.* 1965).

The total system must be well designed and properly managed for economic soundness: (A) The shaker must remove at least 95% of the fruits. (B) The frame must be of size and design to catch essentially all fruits. (C) The catching surface must have all hard portions effectively padded, and all portions over and near where density of fruit is high must have decelerators. (D) The conveying and bin filling system must cause essentially no bruising. (E) Trees must be pruned to minimize interference with equipment. (F) Fruit-bearing hangers must be kept short so that the vibration imparted by the shaker will be transmitted efficiently to the fruits. (G) Major limbs should be pruned to form a modified vase with no major limbs located one beneath another, minimizing impact injury from fruits hitting large branches. (H) Maturity must be relatively uniform (Fridley 1969).

Under the above conditions, machine-harvested fruit was not significantly different in quality from hand-picked fruit 24 hr after harvest. When some of the above essentials were absent, however, fruit losses approached or exceeded 10% (about the maximum loss which can be justified economically). Thus, not only must the machine be of good design but, of equal importance, the trees must be adapted to machine harvest. Further, losses during storage must be minimized by processing the fruit within 24 hr. Minor bruises and cuts which do not downgrade the fruit during the first 24 hr lead to rot and fruit breakdown if processing is delayed.

Tests of mechanized harvesting of freestone peaches are limited. They present a combination of the problems of both cling peaches and apricots. Some additional factors are a fungicidal treatment during postharvest ripening, and ability to harvest a high percentage of the crop without excessive trips over the orchard by the machine.

Harvesting Aids

Man-positioners have potential with crops especially subject to damage, e.g., fruits intended for the fresh market. This approach is practical only as long as sufficient labor is available, since the productivity associated with such aids is small. Probably the average advantage with most man-positioners is 20 to 25%, being greatest with average or slow pickers. It is difficult to improve the productivity and efficiency of good pickers given an incentive wage. Economic justification of such machines is difficult since investment per man is largely dictated by the slowest worker when one machine carries several workers.

Man-positioners may require a well-shaped, uniform hedge. A picker can harvest faster when he does not have to reach far, but if the machine crowds the tree row to reduce the reach required, limbs interfere and slow the pickers.

Before this approach to harvesting comes into extensive use, questions must be answered about yield, harvest costs, tree training, and other factors determining total profit. What tree spacing, row spacing, and hedgerow thickness are optimum? How can trees best be trained, and how can limbs best be kept from leaning out into the aisle (Fridley 1969).

Precooling

Precooling refers to the process of cooling the fruit soon after harvesting—generally before it is loaded into refrigerator cars—or before it is started on its trip to distant points. The nearer that peaches are picked to the proper stage of maturity for best development of eating quality, the greater the need to rapidly lower the fruit temperature. The temperature of the peaches should be lowered quickly to retard ripening, deterioration of quality, and the growth of rot organisms.

When individual peaches are wrapped and tightly packed, the rate of precooling is retarded, as contrasted with peaches unwrapped in the field containers. Even when placed in cold storage at 0°C (32°F), it may require 40–45 hr to cool the interior of wrapped packed peaches in the center of the box to 0.6° to 1.7°C (33° to 35°F). Each 5.6°C (10°F) drop in temperatures of peaches from field temperatures 23.9° to 29.4°C (75° to 85°F) down to 0.6° to 1.7°C (33° to 35°F) increases the period of marketability and reduces the loss from rot.

In precooling, air may be circulated by means of fans permanently installed in cars or by propeller type fans installed temporarily in the top bunker openings. The space around a fan installed in a bunker opening is closed by means of baffles. In either method, the fans reverse the normal direction of air movement and cool the top part of the load more than the bottom part. By these methods the average temperature of a load of peaches can be lowered about 11° to 14°C (20° to 25°F) in 5–6 hr. Further precooling requires replenishing the ice in the bunkers and continued operation of the fans.

Cold storage precooling is adapted to concentrated fruit areas, or shipping centers. Portable car precoolers may be used in small shipping centers that do not have cold storage facilities.

Hydrocooling

In hydrocooling, peaches are subject to an ice-water drench or spray, usually including a germicidal agent, in order to reduce bruising and decay during transit and to enable growers to ship riper peaches. Field heat is thus quickly removed from the fruit. When the water is kept at −1.1° to 0.6°C (30° to 33°F) the produce loses about 22.4°C (40°F) of field heat in 10–20 min, as it is conveyed through the machine.

Hydrocooled fruit held at 15.6°C (60°F) or at room temperature developed a greater amount of more attractive bright yellow ground color than uncooled fruit from the same orchards. Hydrocooled fruit held at lower temperatures showed this same response, but to a lesser extent (Schneider and Correll 1956).

Chlorine (100 to 200 ppm) has been the most commonly used fungicide in water of the hydrocooler. It has been fairly effective in keeping the water relatively clean, presents no residue problem, is economical to use, and is easily tested for concentration by packinghouse personnel. Chlorine probably has little or no effect on peaches already infected with brown rot or *Rhizopus* organisms.

With Redskin peaches at shipping stage of maturity, 2,6-dichloro-4-nitroaniline (Botran or DCNA), 300 ppm in the precooler, reduced decay to 2% compared with 18% for chlorine in one test. A second test showed that peaches treated with Botran had no decay after three days, whereas peaches with no treatment had 32% decay, and those treated with both Botran and Captan had 5% decay. In the third test, the decay in Botran-treated peaches was 6%, chlorine treatment showed 38%, and water with no fungicide treatment showed 42% (Ridley and Sims 1964).

When spores of brown rot and *Rhizopus* were added to the water in the hydrocooler, 0.25% 2-aminobutane carbonate (Tutane) was totally effective in its control of brown rot, but not *Rhizopus*. Tutane-Botran was totally effective in control of brown rot and *Rhizopus*; no brown rot or *Rhizopus* was found after seven days of storage (Ridley and Sims 1964).

Slightly more than a ton of ice is required to cool the water in a 7.6-m (25-ft) cooler before any fruit can be cooled each day, depending on prevailing temperatures. Cooling each 35 litres (bushel) of peaches 19.4°C (35°F) from 29.4° to 10°C (85° to 50°F) requires about 9.5 kg (21 lb) of ice.

Peaches are hydrocooled in many peach-producing areas. Growers without cooling equipment may need to decide if they should invest in this equipment. An important determining factor here is whether the grower can sell his fruit on central markets if it is not cooled. Volume greatly affects the cost of cooling a bushel of fruit; hence, the small

grower may wish to consider alternative arrangements, such as a cooperative arrangement with other growers for packing and cooling.

Hydraircooling

Hydraircooling, simply stated, is the circulation of cold air through a
mist of chilled water that is sprayed into the airstream as it impinges on
the fruit. Hydraircooling combines the advantages of hydrocooling and
air cooling while minimizing their respective disadvantages (Bennett
and Wells 1976B).

Defuzzed Peaches

At many packinghouses, peaches are defuzzed by a brush machine and
water spray in the hydrocooler line at the time of grading and before
they are packed. Defuzzed peaches may be more susceptible to rot than
those which have not been treated, since the process involves removal of
the protective hairs. But using chlorine or other efficient fungicides in
the hydrocooler solution aids in rot control.

PACKINGHOUSES AND PACKING

About 0.1 m² (1 ft²) of packinghouse floor space is required for each 113
kg (250 lb) of peaches handled. When more than 180 tonnes (200 tons) is
to be handled, at least 250 to 270 m² (2700 to 3000 ft²) of packing area
plus 108 m² (1200 ft²) of cold storage space should be available.

Packinghouses require good lighting. Two-lamp industrial fluorescent
fixtures with egg-crate louvres and 40-watt Delux warm lamps placed
over the sorting and grading areas are useful.

Theoretically, the fruit-farm packinghouse should be situated in the
center of the orchard area. However, other considerations may make this
inadvisable. A well-drained area is important, permitting access by
truck at any time during the year. Where natural drainage is questionable, install tile underdrainage. If sufficient space is available, a
roadway (preferably hard-surface) circling the building lends itself to
speedy unloading and to dust-free conditions. Trucks should not be
cramped for room. Availability of electricity and water may largely
determine the location of the building but the grower should compare
costs of building and operation in various locations (Upshall 1955).

Peaches are subjected to several operations in peach packing sheds,
including dumping onto conveyer rolls, belts, or into water tanks. Sometimes the tanks of water contain sodium carbonate in order to better

float the peaches. Peaches are conveyed on various types of conveying equipment with steel rolls, rubber rolls, and flat rubber rolls. Defuzzing has been mentioned earlier. Peaches are hydrocooled either in boxes or baskets or in bulk hydrocoolers where the fruits are several layers deep. Somewhere during these operations a number of factors or combination of factors may cause discoloration of peaches.

The peach industry continues to employ new and more scientific equipment for handling and packing fruit, including numerous chemicals. Growers, packers, shippers, buyers, and consumers have complained about a discoloration or abnormal color of peaches in some instances. The discoloration is referred to as "brown," "black," "spotted," "streaked," and "purple." Some of the injury has been apparent as brown discoloration and depressed areas in the skin.

Any operation which causes peaches to rub, roll, or scrub against any substance or against themselves may cause discoloration. The treatment which contributed the most injury was the movement of peaches on new rubber drying rolls. Addition of 10 ppm of Fe to the water caused purple discoloration after brushing and drying. Adopt procedures which will subject peaches to less physical handling during packing, and avoid soluble Fe compounds in the water which is used for conveying, hydrocooling, or wet washing (Van Blaricom and Webb 1965).

The use of polyurethane drying rolls in place of latex rolls resulted in a striking reduction in discoloration (Van Blaricom and Ridley 1976).

Hail

Fruits of Redhaven and Redskin were lightly to severely injured by hail 18 days after bloom; four days later the wounds were dry and apparently suberized. After 18 days, wound surfaces were moderate brown, russeted, and sometimes included small islands of pubescence. Dark red pigment frequently was present in the uninjured skin adjacent to the wound. The extension of the wound to a diameter many times that when the hail occurred was due to the lack of anticlinal division in the phellogen.

The few layers of hypodermal cells in Redhaven in comparison with Redskin is associated with the greater firmness of Redskin fruits at maturity, and is a factor that should be considered in all handling operations (Simons and Lott 1969).

STORAGE

Peaches may be stored for short periods to extend the marketing and canning seasons, if cooled promptly to 0°C (32°F) after picking. Even under the best conditions, freestone peaches are not stored longer than two weeks. The proper time to bring the fruit out of storage to market must be determined by experience with the cultivar.

Peaches that are stored under unsuitable conditions or are stored for too long lose their flavor and natural bright color, become dry and mealy (or wet and mushy), and show marked browning of the flesh, especially around the stone. Loss in flavor is more rapid at 2.2° to 4.4°C (36° to 40°F) than at 0°C (32°F) and breakdown develops sooner at 2.2° to 4.4°C (36° to 40°F) than at either lower or higher temperatures. Clingstone peaches are often stored for three to four weeks before canning. The lower the temperature at which peaches are held, the less moisture and food reserves are used up in respiration.

Firm-ripe Halehaven and Elberta were held at 16.7°C (62°F) successfully for only five or six days, at 11.1°C (52°F) for seven to eight days, at 5.5°C (42°F) for nine to ten days, and at 0°C (32°F) for two to three weeks. Since changes in the fruit were slowest at 0°C (32°F) and since at this temperature the fruit retained the ability to ripen normally for a longer period than at higher temperatures after removal to room temperature, 0°C (32°F) was preferred for storage (Comin 1955).

Temperatures of 0°C to 2.2°C (32° to 36°F) hold ripening and decay almost to a standstill. After being exposed for more than one week to 10°C (50°F) or lower, the fruit usually breaks down internally or becomes mealy or off-flavored if ripened at room temperatures. Peaches can be held somewhat longer at 0°C (32°F) than at 2.2° to 10°C (36° to 50°F) without losing the capacity to ripen normally.

Three peach and one nectarine cultivar were stored in CA atmosphere or in air at 18.3°C (65°F) for two days and then returned to CA or air. Fruits stored in CA retained better quality and had lower respiration rates than those stored in air. Skin browning frequently developed during ripening at 18.3°C (65°F). This disorder and decay remain serious problems for successful long-term storage (Anderson and Penny 1975).

Freezing temperature of peaches in storage is about −1.1°C (30°F). To monitor for cold spots in the storage room and otherwise provide a margin of safety, keep peaches in a room where a recording thermometer registers any air temperature lower than −1.1° to 0°C (30° to 32°F).

Early studies with ozone to extend the life of peaches both in transit and in storage gave variable results. Most of the studies were conducted before the newer, more efficient ozone generators were available. More recent tests indicated that the use of ozone during storage of peaches acted favorably (Ridley and Sims 1964).

HIGH DENSITY PEACH ORCHARDS

The purpose of high density orchards is to provide more and smaller

Courtesy of Prof. G.E. Stembridge, Clemson Univ.

FIG. 4.7. A HIGH DENSITY PEACH ORCHARD IN SOUTH
CAROLINA

trees per ha (acre) than in the older standard practices. The advantages
are much the same as given for pome fruits.

Fruiting Habits

Peaches are borne laterally on the terminal growth produced the
previous year (see Index). This fact, and also the fact that summer
pruning has a greater dwarfing effect than dormant pruning is basic to
success in developing smaller peach trees for high density plantings.

Role of Dormant Pruning

Some dormant pruning may be needed for correction and thinning-out
in high density trees, but to a much more limited extent than with
standard trees. In general, the winter pruning is invigorating to the
trees, especially around the pruning wounds.

Role of Summer Pruning

Summer pruning is devitalizing. Research in Indiana over an eight year
period has drawn attention to the dominant role of summer pruning in
the development of high density peach trees on peach rootstock, and to
maintain the trees in good balance between fruit-bud formation and
vegetative growth (Hayden and Emerson 1975).

Summer pruning may remove about 50% of the current season's growth, if done when 20 to 25 cm (8 to 10 in.) of growth has occurred. It may be necessary to repeat the pruning or shearing a month later. This type of pruning tends to cause breaking of lateral buds when accompanied by dormant season pruning. As one proceeds past bloom in pruning a point is reached at which vegetative growth is discouraged. This provides the extra dwarfing characteristic of the summer-pruned trees.

Tree-row Systems

Flat-fan System.—The actual training in this system seems to be relatively unimportant except for first and second fruiting years, during which time more severe pruning must be done to obtain a desired shape.

The individual tree may be conceived as an elongated box with all trees in the row sheared to the same dimensions. Each year after planting the dimensions of the box may be increased until the entire row becomes a solid hedgerow with a width of 1.2 to 1.5 m (4 to 5 ft) and a height of 3 to 3.6 m (10 to 12 ft). Probably the height of the tree wall can be any height that best suits the grower's harvesting procedure (Hayden and Emerson 1975).

Tilted-tree System.—"Tilting" peach trees to a 45-degree angle to form hedgerows has produced high yields in a young South Carolina orchard. One upright branch is allowed to develop near the base of the trunk.

The design (as of 1974) utilizes a hedge-row concept, but with a different twist: short hedgerows are spaced close together, only 2.4 m (8 ft) apart, spanning 7.6 m (25 ft) wide beds. Individual hedgerows are oriented diagonally across the bed, so that each row is 12.2 m (40 ft) rather than 7.6 m (25 ft) long and consists of 5 trees spaced 2.4 m (8 ft) apart.

"Instant" hedgerows are formed by planting dormant-budded trees at a 45-degree slant in the row and heading them only slightly before the first growing season. Ten-foot middles for equipment separate the beds. The planting results in a density of 196 trees per ha (484 trees per acre).

Summer pruning is essential, and is done at least twice each summer in climates with moderately long growing seasons, to remove upright shoot growth and to force new growth in a more horizontal direction.

Palmate, Hedgerow, and Standard Systems

Peach at 190 and 142 trees per ha (470 and 350 per acre) and trained as either palmate or hedgerow produced nearly twice as much fruit in the first five years as did trees trained to a modified central leader at 65 trees per ha (160 per acre) in Ontario, Canada (Phillips and Weaver 1975).

In addition to some dormant pruning to shape the trees and restrict their size, the palmate and hedgerow plantings were pruned each year in midsummer when about half of the current year's growth was removed. Insects and mites were no great problem, and brown rot was less in the high than in the low density planting.

The palmate-trained trees were pruned either flat or gable roof-shaped at the top of either 2.1 m (7 ft) or 2.4 m (8 ft) high, with the sides sloped or uncut. The sides of the hedgerow were shaped with a base width of either 1.7 m (5.6 ft) or 2.0 m (6.6 ft) and either cut flat in top at either 2.4 m (8 ft) or 2.7 m (9 ft) high, or left uncut.

Two-arm Oblique Palmette System

This system involves a special planting and training procedure, including 1.8-m (6-ft) high posts between each tree in the row, and a single No. 9 gage wire running down the row from the top of the posts to deadmen set in each tree hole, then up to the next post and so on.

After the two main arms reach a height of 2.1 or 2.4 m (7 or 8 ft) by the middle of the second summer, they are gradually tied down to the wires to keep tree height low.

Because the two branches to be selected as arms must arise around 38 cm (15 in.) from the ground, they must be carefully selected so that none of the lower branches have been trimmed away (Fisher 1977).

The above information from the several sources provides an insight into the present status of High Density Peach Orchards. It is evident that the matter is controversial at the moment, and that certain features still need to be resolved.

INSECTS AND DISEASES

Pest control schedules vary in different regions, and change frequently. Follow up-to-date schedules which are usually available in each region.

Insects Attacking the Fruit

Plum Curculio.—The adult is a rough-looking grayish-to-dark brown snout beetle with a hump on the middle of each wing cover. The chewing mouth parts are on the end of a long snout which projects from the head somewhat like an elephant's trunk. Soon after the petals have fallen, the beetles attack the fruit, which they injure both by feeding and by egg-laying. Injury shows on the fruit as crescent-shaped cuts, which enlarge as the fruit grows and form D-shaped russeted scars. Except in early cultivars, the peaches usually drop when the larvae are nearly full grown, so the larvae are seldom found in peaches at harvest. The beetles may introduce spores of brown rot during feeding and thus indirectly cause destruction of the fruit. From Kentucky south, summer brood adults may cause damage to the fruits.

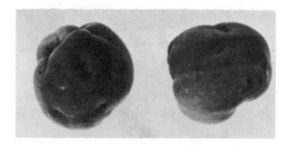

FIG. 4.8. "CATFACING" DUE TO CURCULIO AND PLANT BUG
INJURY

Oriental Fruit Moth.—The larvae of the first brood attack the tips of growing shoots and burrow downward for a short distance. Terminal leaves wilt, turn brown, and die. Larvae of later broods attack both shoots and fruits, but generally do not attack the fruit until it is nearly 2/3 grown. Wormy fruits result. The larvae are pinkish when mature and crawl rapidly (curculio larvae are yellowish-white when full grown and are sluggish and grublike in appearance).

FIG. 4.9. CROWN GALL ON PEACH ROOTS

Japanese Beetle.—Adults are about 1.3 cm (1/2 in.) long, and have a shiny green body with reddish-bronze wing covers and a row of white dots around the hind parts. They skeletonize the foliage and eat holes in ripening fruit.

Tarnished Plant Bug.—Adults are small, dark brown, and have dark markings about 6 mm (1/4 in.) long. The most conspicuous injury is deformed fruit.

Stink Bugs.—Several species of these sucking bugs introduce an enzyme which breaks down the cellular tissue at the point where the beak is inserted. If the scar is formed early in the season the result is a "cat-faced," dimpled, or otherwise deformed fruit. Stink bugs may also cause fruit gummosis and dark green depressed areas having a water-soaked appearance.

Insects and Mites in Foliage Attack

Red Mites.—These arachnids rasp the surface of the leaves and withdraw the liquid contents of the cells. Feeding by large populations results in a bronzed or grayish leaf color and impairs the quality and size of the fruit. When red mites are abundant on peaches the hands of pickers may become covered with small reddish spots which result from crushing the mites.

Two-spotted Spider Mite.—Unlike the red mite, the two-spotted spider mite tends to congregate on the lower sides of the leaves near the midvein. Heavily infested leaves usually have whitish specks concentrated in irregular bands near the midvein and visible from the upper surface. These mites also produce a webbing which is seldom found on leaves infested by red mites.

Peach Silver Mite.—This mite sometimes attacks nursery stock, causing the lower leaves to become curled and spotted. Feeding by large populations on mature trees may result in "silver leaf" injury.

Green Peach Aphid.—Aphids cause curling of the leaves in spring. The leaves usually retain normal color but may be distorted and contain either whitish aphid skins or small green aphids.

Leafhoppers.—Besides the actual injury, leafhoppers may transmit peach yellows and other virus diseases. The damage, which usually occurs late in the season, resembles that caused by the two-spotted mite except that the whitish specks on the leaves are larger and more scattered.

Insects That Attack Woody Parts

San Jose Scale.—In heavy infestation, San Jose scale causes the entire surface of the bark to be covered by a grayish layer of overlapping scale insects.

Terrapin Scales.—These are small and oval, concentrate on the lower sides of twigs and branches and, when numerous, may cause death of these parts. The scales secrete a honeydew and sometimes foliage and fruit are almost covered with this material. A sooty, black fungus develops in the honeydew and renders the fruit unmarketable.

White Peach Scale.—In the Deep South serious damage can be caused by this scale if it is allowed to become established. The woody parts become covered with the white (male) crawlers; the female scales, under the crawlers, do most of the damage. Check carefully and spray with 3% oil twice in early winter. In problem orchards use a suitable material in late summer.

Shot-hole Borer.—Weakened trees are often the victims of the shot-hole borer which causes small spots of gummy exudate (without frass) to occur on the twigs, smaller branches, and sometimes on the trunk. When the gum is removed, beneath it is found a small circular hole, caused by the adult. The very small, blackish beetles are not usually found in these holes because the exudation of gum forces them from the burrows before the breeding chambers are constructed.

Fungus Diseases

Peach Leaf Curl.—The infection first shows as a reddening and arching of the young leaves as they begin to appear. Later these become thickened, curled, or crinkled, and yellowish with a silvery sheen. Diseased leaves fall in early summer, thus weakening the tree.

Brown Rot.—First appearing as a small, brown spot, the blemish increases in size until the whole fruit may become soft, brown, and rotten, and show the fruiting pustules. Besides control by fungicides, it is helpful to control curculio and other insects that puncture the fruit, and to remove mummified fruits from the ground in the north; mummies which hang on the tree in the south rarely carry the disease through winter.

Peach Scab (Freckles or Black Spot).—Symptoms are small, round, olive-black spots on the fruits, usually on the upper part, which appear about six weeks after petal-fall. When the spots run together, the fruit

FIG. 4.10. PEACH LEAF CURL INFECTION

is stunted and misshapen; the fruit may crack and thus permit brown-rot infection. Twigs and leaves may be killed.

Peach Rust.—The disease is prevalent in the south, and may cause premature defoliation. The spots are yellow on the upper surface and bear masses of powdery brown spores on the lower surface of the leaf.

Mushroom Root Rot.—See Peach Replants.

FIG. 4.11. THIS TWELVE-YEAR-OLD ORCHARD IS IN RAPID
DECLINE DUE TO CANKER INFECTION

Cercospora Canker.—A conspicuous feature is a large open wound on the trunk, on scaffold branches, or in the crotch, with sunken bark covered with gum. Dead branches may be covered with an enormous number of tiny pimples erupting through the bark. There are the fruiting bodies of the disease-causing fungus.

Peach canker is a fungus disease caused by one or both of two fungus organisms. Bacterial canker (*Pseudomonas*) is a major cause of gummosis.

Virus Diseases

Phony Virus.—Leafhoppers or certain other insects that penetrate to the xylem tissue carry the disease from infected to healthy trees. Infected trees have abnormally dark green leaves, stunted terminal twig growth, and fruit greatly reduced in size and quantity. This is often present but symptomless in wild plums.

Peach Yellows.—This virus causes the tree to develop bunches of slender, willowy shoots which branch to give a "broom" effect, with leaves that are narrow, yellowish, and red-spotted. Fruits ripen prematurely, have reddish streaks in the flesh, and are insipid. The disease is transmitted by leafhoppers.

X-Disease.—See under Virus Diseases on Cherries in Chap. 5.

REFERENCES

ABDALLA, D.A., and CHILDERS, N.F. 1973. Calcium nutrition of peach and prune relative to growth, fruiting, and fruit quality. J. Am. Soc. Hort. Sci. *98* (5): 517-532.

ALDERFER, R.B., and SHAULIS, N.J. 1953. Effects of cover crops in peach orchards on runoff and erosion. Proc. Am. Soc. Hort. Sci. *42,* 21-29.

ANDERSON, R.E., and PENNY, R.W. 1975. Intermittent warming of peaches and nectarines in a controlled atmosphere. J. Am. Soc. Hort. Sci. *100* (2): 151-153.

ANON. 1965. Dare report. Inst. Food. Agr. Univ. Florida.

ARCHIBALD, J.A. 1966. Orchard soil management. Ontario Dept. Agr. Food *457.*

ARMSTRONG, W.D. 1976. Topworking, a valuable tool for fast test and mistakes correction. J. Am. Soc. Hort. Sci. *11* (3): 226.

ASHLEY, T.E. 1940. Productiveness of secondary and lateral peach shoots. Proc. Am. Soc. Hort. Sci. *37,* 208.

ATKINSON, F.E. 1956. Maturity Manual. British Columbia Tree Fruits, Vancouver, B.C.

BAKER, C.E. 1949. Effectiveness of organic mulches in correcting potassium deficiency of peach trees on a sandy soil. Proc. Am. Soc. Hort. Sci. *45*, 21-22.

BAKER, G.A., and DAVIS, L.D. 1951. Growth of the cheek diameter of peaches. Proc. Am. Soc. Hort. Sci. *57*, 104-110.

BALLINGER, W.E. *et al.* 1963. Interrelationships of irrigation, nitrogen fertilization, and pruning of Redhaven and Elberta peaches in the sandhills of N. Carolina. Proc. Am. Soc. Hort. Sci. *83*, 248-258.

BATJER, L.P., and MOON, G.C. 1965. Influence of night temperature on growth and development of Early Redhaven peaches. Proc. Am. Soc. Hort. Sci. *87*, 139-144.

BATJER, L.P., and WESTWOOD, M.N. 1958. Size of Elberta and J. H. Hale peaches during the thinning period as related to size at harvest. Proc. Am. Soc. Hort. Sci. *72*, 101-105.

BAUSCHER, M.G. *et al.* 1970. Peach cultivar responses to fruit thinning with CPA. J. Am. Soc. Hort. Sci. *95*, 500-503.

BEDFORD, C.L. 1956. Maturity and quality in freestone peaches. Natl. Peach Council Ann. Rept., 9-10.

BELL, H.H., and CHILDERS, N.F. 1956. Effect of manganese and soil culture on growth and yield of the peach. Proc. Am. Soc. Hort. Sci. *67*, 130-138.

BENNETT, A.H., and WELLS, J.H. 1976A. Peach precooling aims at perfection. Am. Fruit Grower. May, 13-33.

BENNETT, A.H., and WELLS, J.H. 1976B. Hydrair cooling. A new precooling method with special application for waxed peaches. J. Am. Soc. Hort. Sci. *101*(4): 428-431.

BLAKE, M.A. 1948. Length of fruit development period of Elberta and some other varieties of peaches. Proc. Am. Soc. Hort. Sci. *51*, 175-179.

BONAMY, P.A., and DENNIS, F.G. 1977A. Abscissic acid levels in seeds of peach. I. Changes during maturation and storage. J. Am. Soc. Hort. Sci. *102* (1): 23-26.

BONAMY, P.A., and DENNIS, F.G. 1977B. Abscissic acid levels in seeds of peach. II. Effects of stratification. J. Am. Hort. Soc. *102*(1): 26-28.

BOYNTON, D. 1944. Responses of young Elberta peach and Montmorency cherry trees to potassium fertilization in New York. Proc. Am. Soc. Hort. Sci. *57*, 175-179.

BOYNTON, D. *et al.* 1951. Leaf and soil analysis for magnesium in relation to interveinal leaf chlorosis in sour cherry, peach, and apple trees in New York. Proc. Am. Soc. Hort. Sci. *57*, 175-179.

BRADT, O.A. 1955. Pruning methods for fruit trees. Rept. Hort. Res. Inst. Ontario Dept. Agr. Food Bull. *430*.

BRASE, K.D. 1948. Field nursery tests with newly harvested and one- or two-year stored Lovell peach seed. Proc. Am. Soc. Hort. Sci. *51*, 258-262.

BROWN, D.S., and HARRIS, R.W. 1958. Summer pruning trees of early maturing peach varieties. Proc. Am. Soc. Hort. Sci. *72*, 79-84.

BROWN, D.S., and KOTOB, F. 1957. Growth of flower buds of apricot, peach, and pear during the rest period. Proc. Am. Soc. Hort. Sci. *69*, 158-164.

BUCHANAN, D.W. 1969. Peach fruit abscission and pollen germination as influenced by ethylene and 2-chloroethane-phosphonic acid. J. Am. Soc. Hort. Sci. *94*, 327-329.

BUCHANAN, D.W. *et al.* 1970. Influence of Alar, Ethrel, and gibberellic acid on browning of peaches. HortScience *4*, 302-303.

BUCHANAN, D.W. *et al.* 1974. Cold hardiness of peach and nectarine trees growing at 29−30° latitude. J. Am. Soc. Hort. Sci. *99* (3): 256-269.

BURKHOLDER, C.L., and BAKER, C.E. 1957. Peach production. Indiana Agr. Expt. Sta. Bull. *330.*

BURKHOLDER, C.L., and SHARVELLE, E.G. 1951. Peach brown-rot control in orchard, storage, and transit. Indiana Agr. Expt. Sta. Bull. *364.*

BYERS, E.E., and EMERSON, F.F. 1969. Effect of succinic acids 2,2-dimethyl hydrazide (Alar) on peach fruit maturation and tree growth. J. Am. Soc. Hort. Sci. *94,* 641-645.

CAIN, J.C. 1956. New look at old facts about peach production. N.Y. Farm Res. *22,* 16.

CAIN, J.C., and MEHLENBACHER, R.J. 1956. Effects of nitrogen and pruning on trunk growth in peaches. Proc. Am. Soc. Hort. Sci. *67,* 139-143.

CARLSON, R.F. 1943. Treatment of peach seed with fungicides for increased germination and improved stand of seedlings in the nursery. Proc. Am. Soc. Hort. Sci. *48,* 105-113.

CHAPLIN, C.F. 1948. Artificial freezing of peach fruit buds. Proc. Am. Soc. Hort. Sci. *52,* 121-129.

CHAPLIN, C.E., and SCHNEIDER, G.W. 1974. Transmission of hardiness in peach rootstocks. J. Am. Soc. Hort. Sci. *98* (3): 231-234.

CHRIST, E.G. *et al.* 1952. Pruning fruit trees. New Jersey Agr. Expt. Sta. Bull. *259.*

CIBES, H.R. *et al.* 1956. Boron toxicity induced in a New Jersey peach orchard. Proc. Am. Soc. Hort. Sci. *66,* 13-20.

CLAYPOOL, L.L. 1972. Split-pit of Dixon cling peaches in relation to cultural factors. J. Am. Soc. Hort. Sci. *97* (2): 181-185.

CLAYPOOL, L.L. *et al.* 1965. Horticultural aspects in mechanization of cling peach harvesting. Proc. Am. Soc. Hort. Sci. *86,* 152-165.

CLAYPOOL, L.L. *et al.* 1966. Pomological aspects of mechanizing tree fruit harvesting. HortScience *1,* 47-50.

CLINE, R.A. 1967. Soil drainage and compaction effects on growth, yield, and leaf composition of peaches and cherries. Rept. Hort. Res. Inst. Ontario, 28-34.

COE, F.M. 1933. Peach harvesting studies. Utah Agr. Expt. Sta. Bull. *141.*

COMIN, D. 1955. Success storage of peaches depends on maturity and post-ripening. Ohio Farm Res. *40,* 63-65.

COMIN, D. 1956. Relation of sugars and acids in peaches to dessert quality. Ohio Agr. Expt. Sta. Spec. Circ. *93.*

COVILLON, G.A., and HENDERSHOTT, C. 1974. Characterization of the "after rest" period of flower buds of two peach cultivars of different chilling requirements. J. Am. Soc. Hort. Sci. *96* (3): 320-322.

CULLINAN, F.P., and WEINBERGER, J.H. 1933. Influence of soil moisture on growth of fruit and stomatal behavior of Elberta peaches. Proc. Am. Soc. Hort. Sci. *29,* 28-33.

CULPEPPER, C.W. *et al.* 1955. Effect of picking maturity and ripening temperature on the quality of canned and frozen eastern grown peaches. U.S. Dept. Agr. Tech. Bull. *1114.*

CUMMINGS, G.A. 1965. Effect of potassium and magnesium fertilization on yield, size, maturity, and color of Elberta peaches. Proc. Am. Soc. Hort. Sci. *86,* 133-140.

CUMMINGS, G.A., and REEVES, J. 1971. Factors affecting chemical characteristics of peaches. J. Am. Soc. Hort. Sci. *96* (3): 320-322.

DANIELL, J.W. 1973. Effects of time of pruning on growth and longevity of peach trees. J. Am. Soc. Hort. Sci. *984:* 383-386.

DANIELL, J.W., and WILKINSON, R.E. 1972. Effect of ethephon-induced ethylene on abscission of leaves and fruits of peaches. J. Am. Soc. Hort. Sci. *97* (5): 682-685.

DAVID, L.D. 1950. Size in canning peaches. Relation between diameter at harvest and growth after reference date. Proc. Am. Soc. Hort. Sci. *55,* 216.

ECKERT, J.W., and CHILDERS, N.F. 1954. Effect of urea sprays on leaf nitrogen and growth of Elberta peach. Proc. Am. Soc. Hort. Sci. *63,* 19-22.

EDGERTON, L.J. 1953. Peach growing. Cornell Univ. Agr. Expt. Sta. Bull. *869.*

EDGERTON, L.J., and HARRIS, R.W. 1952. Effect of thinning peaches and cultural treatment on Elberta fruit-bud hardiness. Proc. Am. Soc. Hort. Sci. *55,* 51-55.

EDGERTON, L.J., and HOFFMAN, M.B. 1952. Effect of thinning peaches with bloom and postbloom sprays on cold hardiness of peach buds. Proc. Am. Soc. Hort. Sci. *60,* 155-159.

FELDSTEIN, J., and CHILDERS, N.F. 1957. Effect of irrigation on fruit size and yield of peaches. Proc. Am. Soc. Hort. Sci. *69,* 126-130.

FERREE, R.J. 1957. Nitrogen-potash balance for peaches. Potash News Letter, 101.

FERREE, M.E., and MCGLOHON, N.E. 1975. You can prevent peach tree short life. Am. Fruit. Grower. (Aug.), 20, 27.

FISHER, D.V. 1977. Palmette proves practical. Am. Fruit Grower. (Aug.), 15-18.

FISHER, D.V., and BRITTON, J.E. 1940. Maturity and storage studies with peaches. Sci. Agr. (Canada) *21,* 103.

FORSHEY, C.G. 1969. Potassium nutrition of deciduous fruits. HortScience *4,* 7-9.

FRIDLEY, R.B. 1969. Tree fruit and grape harvest mechanization and progress. Am. Soc. Hort. Sci. *4,* 7-9.

FULEKI, T., and COOK, F. 1976. Relationship of maturity as indicated by flesh color to quality of canned clingstone peaches. J. Inst. Canada Sci. Technol. Aliment. *9* (1).

GAMBRELL, C.E., and RHODES, W.H. 1967. Influence of several residual and contact herbicides on weed control and growth of young peach trees. S. Carolina Agr. Expt. Sta. Res. Serv. *102,* 1-11.

GAMBRELL, C.E. *et al.* 1967. Effects of four growth regulators on thinning, yield, size, and maturity of Ranger peaches when applied for one year and two years in succession. S. Carolina Agr. Expt. Sta. Res. Serv. *101,* 1-21.

GAMMON, N., and SHOEMAKER, J.S. 1963. Effects of fertilizers on mineral levels in the leaves of Flordawon peach. Proc. Florida State Hort. Soc. *76,* 380-384.

GAMMON, N., and SHOEMAKER, J.S. 1964. Effect of nursery stock size on peach yields. Proc. Florida State Hort. Soc. *77,* 384-387.

GUARDIAN, R.J., and BIGGS, R.H. 1964. Effect of low temperature on terminating bud dormancy of Okinawa, Flordasun, Flordahome, and Nemaguard peaches. Proc. Florida State Hort. Soc. *77,* 370-379.

HAUT, I.C., and GARDNER, F.E. 1935. Influence of pulp disintegration upon the viability of peach seeds. Proc. Am. Soc. Hort. Sci. *32,* 323-327.

HAVIS, L. 1957. Effect of times of applying nitrogen and of disking on bearing peach trees. Proc. Am. Soc. Hort. Sci. *68,* 70-76.

HAVIS, L., and GILKERSON, A.L. 1951. Interrelationships of cultural systems to soil organic matter. Proc. Am. Soc. Hort. Sci. *57,* 24-30.

HAYDEN, R.A., and EMERSON, F.H.   1975.   The case for summer pruning.   Am. Fruit Grower (June), 27.

HERNANDEZ, E., and CHILDERS, N.F.   1956.   Boron toxicity in a New Jersey peach orchard.   Proc. Am. Soc. Hort. Sci. *67*, 121-129.

HIGGINS, B.B. *et al.*   1950.   Effect of nitrogen fertilization on cold injury of peach trees.   Georgia Agr. Expt. Sta. Bull. *226.*

HITZ, C.W.   1954.   Effect of soil management practices on growth and fruitfulness of peach trees.   Delaware Agr. Expt. Sta. Bull. *300.*

JENSEN, R.E. *et al.*   1970.   Effect of certain environmental factors on cambium temperatures of fruit trees.   J. Am. Soc. Hort. Sci. *95*, 286-292.

JONES, I.D.   1933.   Influence of leaf area on fruit growth and quality in the peach.   Proc. Am. Soc. Hort. Sci. *29*, 34-38.

JUDKINS, W.P.   1949.   Relationship of leaf color, nitrogen, and rainfall to the growth of young peach trees.   Proc. Am. Soc. Hort. Sci. *53*, 29-36.

JUDKINS, W.P., and WANDER, I.W.   1945.   Effect of cultivation, sod, and sod plus straw mulch on growth and yield of peach trees.   Proc. Am. Soc. Hort. Sci. *46*, 183-186.

KAMALI, A.R.   1970.   Growth and fruiting of peach in sand culture as affected by boron and a fritted form of trace elements.   J. Am. Soc. Hort. Sci. *95*, 652-656.

KAMALI, A.R., and CHILDERS, N.F.   1967.   Effect of boron nutrition on peach anatomy.   Proc. Am. Soc. Hort. Sci. *90*, 33-38.

KAMINSKI, W., and ROM, R.   1974.   A possible role of catalase in the rest of peach flower buds.   J. Am. Soc. Hort. Sci. *99* (1): 84-86.

KENWORTHY, A.L.   1948.   Interrelationship between the nutrient content of soil, leaves and trunk circumference of peach trees.   Proc. Am. Soc. Hort. Sci. *51*, 209-215.

KINMAN, C.F.   1941.   Extent of colored area on Elberta peaches in relation to leaf area per fruit and to fruit size.   Proc. Am. Soc. Hort. Sci. *38*, 191-199.

KNOWLTON, H.L.   1937.   Hardiness of peach and apple buds.   Proc. Am. Soc. Hort. Sci. *34*, 238-241.

LAMBORN, E.W.   1955.   How ripe should we pick peaches?   Utah Farm Home Sci. *16*, 54-55.

LARSEN, F.E.   1973.   Stimulation of leaf abscission of tree fruit nursery stock with ethephon-surfactant mixtures.   J. Am. Soc. Hort. Sci. *98* (1): 34-36.

LASHEEN, A.M. *et al.*   1970.   Biochemical comparison of fruit buds in five peach cultivars of varying degrees of cold hardiness.   J. Am. Soc. Hort. Sci. *95*, 177-181.

LAVEE, S., and MARTIN, G.C.   1974.   Ethephon in peach fruits. 1. Penetration and persistance.   J. Am. Soc. Hort. Sci. *99* (2): 97-99.

LAYNE, R.E.C.   1975.   New developments in peach varieties and rootstocks. The Compact Fruit Tree *8* (4): 69-77.

LAYNE, R.E.C. *et al.*   1977.   Influence of peach seedling rootstocks on defoliation and cold resistance of peach cultivars.   J. Am. Soc. Hort. Sci. *102* (1): 80-92.

LEUTY, S.J., and BUKOVAC, M.J.   1968.   Effect of napthaleneacetic acid on abscission of peach fruits in relation to endosperm development.   Proc. Am. Soc. Hort. Sci. *92*, 124-134.

LILLELAND, O., and BROWN, J.G.   1941.   Potassium nutrition of fruit trees. Proc. Am. Soc. Hort. Sci. *32*, 37-48.

LOMBARD, P.B., and MITCHELL, A.E. 1962. Anatomical and hormonal development in Red Haven peach seeds as related to the timing of naphthaleneacetic acid for fruit thinning. Proc. Am. Soc. Hort. Sci. *80,* 163-171.

LOTT, R.V., and SIMONS, R.K. 1966. Floral tube and style abscission in the peach and their use as physiological reference points. Proc. Am. Soc. Hort. Sci. *85,* 141-153.

LOWE, R.H. 1970. Cited by H.B. Tukey in Am. Fruit Grower. October, 18-19.

MALO, S.E. 1967. Nature of resistance of Okinawa and Nemaguard peaches to root-knot nematode. Proc. Am. Soc. Hort. Sci. *90,* 39-46.

MARTIN, G.C., and NELSON, M. 1969. Thinning effect of 3-chlorophenoxy-a-propionamide (3-CPA) in Palora peach. HortScience *49,* 206-208.

MARTSOLF, J.D. *et al.* 1975. Effect of white latex paint on temperature of stone fruit tree trunks in winter. J. Am. Soc. Hort. Sci. *100* (2): 122-129.

MCCLUNG, A.C. 1953. Magnesium deficiency in N. Carolina peach orchards. Proc. Am. Soc. Hort. Sci. *62,* 123-130.

MCCLUNG, A.C. 1954. Occurrence and correction of zinc deficiency in N. Carolina peach orchards. Proc. Am. Soc. Hort. Sci. *64,* 75-80.

MCMUNN, R.L., and DORSEY, M.J. 1934. Delayed harvesting of Elberta peaches. Illinois State Hort. Soc. *68,* 491-501.

MOORE, C.S. 1968. The course of variation in fruit tree experiments. J. Hort. Soc. (England) *43,* 121-136.

NESMITH, W.C., and DOWLER, W.M. 1976. Cultural practices affect cold hardiness and peach tree short life. J. Am. Soc. Hort. Sci. *101* (2): 116-119.

OBERLE, G.D. 1957. Breeding peaches and nectarines resistant to spring frosts. Proc. Am. Soc. Hort. Sci. *70,* 85-92.

OLNEY, A.J. *et al.* 1950. Effect of cover crops and tile drainage on growth and yield of peaches. Kentucky Agr. Expt. Sta. Bull. *547.*

OVERCASH, J.P., and CAMPBELL, J.A. 1956. Effects of intermittent warm and cold periods in breaking the rest period of peach leaf buds. Proc. Am. Soc. Hort. Sci. *66,* 12-31.

OVERHOLSER, E.L. *et al.* 1941. Peach growing. Wash. Agr. Expt. Sta. Bull. *162.*

PALMER, R.C. *et al.* 1941. Soil management and pruning methods for peaches and apricots. Canada Dept. Agr. Bull. *34.*

PHILLIPS, J.H., and WEAVER, G.M. 1975. A high density peach orchard. HortScience *10* (6): 16-17.

PRINCE, V.E. 1966. Winter injury to peach trees in central Georgia. Proc. Am. Soc. Hort. Sci. *88,* 190-196.

PRINCE, V.E. 1972. Influence of pruning at various dates on peach tree mortality. J. Am. Soc. Hort. Sci. *97* (3): 303-305.

PROEBSTING, E.L. 1943. Root distribution of some deciduous fruit trees in California orchards. Proc. Am. Soc. Hort. Sci. *46,* 1-4.

PROEBSTING, E.L. 1958. Yield, growth, and date of maturity of peaches as influenced by soil management systems. Proc. Am. Soc. Hort. Sci. *72,* 92-100.

PROEBSTING, E.L. 1963. Role of air temperatures and bud development in determining hardiness of dormant Elberta peach buds. Proc. Am. Soc. Hort. Sci. *83*, 259-260.

PROEBSTING, E.L. *et al.* 1957. Relationship between leaf nitrogen and canning quality of Elberta peaches. Proc. Am. Soc. Hort. Sci. *69*, 131-140.

RAGLAND, C.H. 1934. Development of peach fruit, with special reference to split-pit and gumming. Proc. Am. Soc. Hort. Sci. *32*, 1-21.

RAWLS, E.H. 1940. Peach tree fertilizer demonstration results. Proc. Am. Soc. Hort. Sci. *37*, 85-86.

REEVES, J., and CUMMINGS, G. 1970. Influence of some nutritional and management factors on certain physical attributes of peach quality. J. Am. Soc. Hort. Sci. *95*, 338-241.

RIDLEY, J.D., and SIMS, E.T. 1964. Preliminary investigations on use of ozone to extend the shelf-life and maintain the market quality of peaches and strawberries. S. Carolina Agr. Expt. Sta. Res. Serv. *90*, 1-13.

RITTER, C.M. 1957. Effects of varying rates of nitrogen fertilization on young Elberta peach trees. Proc. Am. Soc. Hort. Sci. *68*, 48-55.

ROWE, R.H., and CATLIN, P.B. 1971. Differential sensitivity to waterlogging and cyanogenesis by peach, apricot, and plum roots. J. Am. Soc. Hort. Sci. *99* (3): 305-308.

SAVAGE, E.F. 1954. Peach tree longevity in Georgia. Proc. Am. Soc. Hort. Sci. *64*, 81-86.

SAVAGE, E.F. 1955. Effect of size of nursery stock on subsequent growth and production of peach trees. Proc. Am. Soc. Hort. Sci. *65*, 149-154.

SAVAGE, E.F. 1970. Cold injury as related to cultural management and possible protective devices for dormant peach tress. HortScience *5*, 425-428.

SCHNEIDER, G.W., and CORRELL, F.E. 1956. Hydrocooling peaches. Am. Fruit Grower. (August), 6, 13.

SCHNEIDER, G.W., and MCCLUNG, A.C. 1957. Interrelationships of pruning, nitrogen rate, and time of nitrogen application in Halehaven peach. Proc. Am. Soc. Hort. Sci. *69*, 141-147.

SHARPE, R.H. 1966. Peaches and nectarines in Florida. Florida Agr. Expt. Sta. Circ. *299*.

SHARPE, R.H. 1969. Breeding peaches for root-knot nematode resistance. J. Am. Soc. Hort. Sci. *94*, 209-212.

SHERMAN, W.B. 1970. Endosperm development in Maygold peach. HortScience *5*, 41.

SHERMAN, W.B. *et al.* 1970. Comparative endosperm cytokinesis in two low chilling short cycle peaches and a nectarine. HortScience *5*, 109.

SHOEMAKER, J.S. 1934. Certain advantages of early thinning of Elberta. Proc. Am. Soc. Hort. Sci. *30*, 223-224.

SHOEMAKER, J.S., and GAMMON, N. 1963. Effects of fertilizers on yield, size, color, and tip disorder of fruit of Flordawon peach. Proc. Florida State Hort. Soc. *76*, 384-387.

SIMONS, R.K., and LOTT, R.V. 1969. Effect of early season hail injury on the morphological and anatomical development of peaches. Proc. Am. Soc. Hort. Sci. *87*, 154-162.

SIMS, E.T. *et al.* 1974. Influence of (2-chloroethyl) phosphonic acid on peach quality and maturation. J. Am. Soc. Hort. Sci. *99* (2): 152-154.

SKELTON, B.J., and SENN, T.L. 1966. Effect of seaweed on quality and shelf-life of Harvest Gold and Jerseyland peaches. S. Carolina Agr. Expt. Sta. Res. Serv. *86*, 1-9.

SMITH, W.L., and ANDERSON, R.E. 1975. Decay control after controlled atmosphere and air storage. J. Am. Soc. Hort. Sci. *100* (1): 84-86.

SNYDER, J.C. *et al.* 1952. Growing peaches. Wash. Agr. Expt. Sta. Bull. *462.*

SPENCER, S., and COVILLON, G.A. 1975. Relationship of node position to bloom date, fruit size, and endosperm of the peach. J. Am. Soc. Hort. Sci. *100* (3): 242-244.

STEMBRIDGE, G.E. 1965. Effect of various growth regulators on fruit set in Redskin peach. S. Carolina Agr. Expt. Sta. Res. Serv. *71*, 1-10.

STEMBRIDGE, G.E. 1977. Summer pruning: a century old "new" idea. Am. Fruit Grower. (June), 23-24.

STEMBRIDGE, G.E., and GAMBRELL, C.E. 1969. Thinning peaches with chlorophenoxy-a-propionamide. J. Am. Soc. Hort. Sci. *94*, 570-573.

STEMBRIDGE, G.E., and GAMBRELL, C.E. 1976. High density peaches go "full tilt." Am. Fruit Grower. (January), *96* (12): 8-11.

STEMBRIDGE, G.E., and SEFICK, H.J. 1966. Relative bud hardiness and bud set of thirty peach varieties. S. Carolina Agr. Expt. Sta. Res. Serv. *80*, 1-5.

STENE, A.E. 1938. Pruning of winter injured peach trees. Proc. Am. Soc. Hort. Sci. *35*, 147-150.

TUKEY, H.B. 1945. Breaking the dormancy of peach seeds by treatment with thiourea. Proc. Am. Soc. Hort. Sci. *36*, 181-186.

UPSHALL, W.H. 1940A. Transplanting shock in peach seedling rootstocks and its effect on the nursery trees in the nursey row. Proc. Am. Soc. Hort. Sci. *37*, 340-342.

UPSHALL, W.H. 1940B. Summer pruning of peach trees in the nursery row. Proc. Am. Soc. Hort. Sci. *37*, 343-344.

UPSHALL, W.H. 1942. Methods of handling Elberta peach seeds in relation to nursery germination. Proc. Am. Soc. Hort. Sci. *40*, 279-282.

UPSHALL, W.H. 1943. Increase in quantity, grade, and returns from peaches as they approach maturity. Sci. Agr. (Canada) *23*, 12.

UPSHALL, W.H. 1946. Fruit maturity and quality. Ontario Dept. Agr. Food Bull. *450.*

UPSHALL, W.H. 1955. The packing house on the fruit farm. Ontario (Vineland) Ann. Rept., 7-19.

UPSHALL, W.H. *et al.* 1955. Phosphate, potash, and farm manure for peach trees. Ontaro (Vineland) Rept., 7-19, 1555.

VAN BLARICOM, L.O. 1968. Evaluation of peach varieties for processing in the southeast. S. Carolina Agr. Expt. Sta. Bull. *539.*

VAN BLARICOM, L.O., and RIDLEY, J.D. 1976. Skin discoloration of fresh market peaches. HortScience *11* (3): 225.

VAN BLARICOM, L.O., and SEFICK, H.J. 1967. Anthocyanin pigments in freestone peaches grown in the Southeast. Proc. Am. Soc. Hort. Sci. *90,* 541-545.

VAN BLARICOM, L.O., and WEBB, H.J. 1965. Discoloration of fresh peaches. S. Carolina Agr. Expt. Sta. Res. Serv. *74,* 1-21.

VAN BLARICOM, L.O. *et al.* 1970. Effect of Alar on processing peaches. S. Carolina Agr. Expt. Sta. Res. Serv. *130,* 1-6.

VIERHELLER, A.F. 1949. Considerations in peach culture. Maryland Agr. Expt. Sta. Mimeo *15.*

WANDER, I.W. 1940. Lateral movement of potassium and phosphorus in an orchard soil. Proc. Am. Soc. Hort. Sci. *37,* 27-31.

WANDER, I.W. 1945. Increasing available potassium to greater depths in an orchard soil by adding potash fertilizer on a mulch. Proc. Am. Soc. Hort. Sci. *46,* 21-24.

WATADA, A.E. *et al.* 1976. Firmness of peaches measured nondestructively. J. Am. Soc. Hort. Sci. *101* (4): 404-406.

WEINBERGER, J.H. 1932. Relation of leaf area to size and quality of peaches. Proc. Am. Soc. Hort. Sci. *71,* 11-19.

WEINBERGER, J.H. 1950A. Chilling requirements of peach varieties. Proc. Am. Soc. Hort. Sci. *56,* 122-128.

WEINBERGER, J.H. 1950B. Prolonged dormancy of peaches. Proc. Am. Soc. Hort. Sci. *56,* 129-133.

WEINBERGER, J.H. 1954. Effects of high temperatures during the breaking of the rest of Sullivan Elberta peach buds. Proc. Am. Soc. Hort. Sci. *63,* 157-162.

WEINBERGER, J.H. 1962. Studies on flower drop in peaches. J. Am. Soc. Hort. Sci. *91,* 78-83.

WEINBERGER, J.H. 1967. Some temperature relations in natural breaking of the rest of flower buds in the San Joaquin Valley, California. Proc. Am. Soc. Hort. Sci. *91,* 84-89.

WEISER, C.J. 1970. Cold resistance and acclimation in woody plants. HortScience *5,* 403-409.

WESTWOOD, M.N., and GERBER, R.K. 1958. Seasonal light intensity and fruit quality factors as related to method of pruning peach trees. Proc. Am. Soc. Hort. Sci. *72,* 85-90.

WIEBE, H. 1956. Chlorosis. Utah Farm Home Sci. *17,* 607.

WITTWER, S.H. 1969. Regulation of phosphorus nutrition of horticultural crops. HortScience *4,* 320-321.

5

Cherries

Botanically, cherries (like peaches) are drupe fruits. Horticulturally, the two main types of cherries are sweet (*Prunus avium*) and pie, tart or sour (*P. cerasus*). There are also two main groups of both sweet and sour cherries, as well as a separate Duke group (*P. avium* × *P. cerasus*, and vice versa). The terms red, pitted red, red pitted, tart and pie cherry are preferred for *P. cerasus* because the term sour tends to be somewhat of a handicap to the industry.

Commercial cherry production is concentrated largely in certain northern regions (Table 5.1). Processing (canning and freezing) is a dominant factor in large-scale cherry marketing.

The most concentrated plantings of sweet cherries are in the Bay Areas of California, western Oregon, central Washington, and parts of Utah, Idaho, Colorado, and Montana. In the east, concentrated plantings occur in the Great Lakes region and Hudson River Valley. In Canada, most of the sweet cherry plantings are in British Columbia and Ontario.

The most concentrated plantings of tart cherries are in the eastern shore region of Lake Michigan, in the Door County peninsula of Wisconsin, in western New York, northern Ohio, western Pennsylvania, and in Oregon, Washington, Colorado, and Utah. Ontario leads in tart cherry production in Canada.

CULTIVARS (VARIETIES)

Sweet Cherries

There are two horticultural groups of sweet cherry cultivars.

The Heart Group.—Also called the Guigne group by French pomologists, and the Gean cherries in England. This group is characterized by soft, tender flesh, and heart-shaped fruit. It includes dark-colored cultivars with reddish juice, and light-colored cultivars with nearly colorless juice.

The Bigarreau Group.—Characterized by firm, crisp, breaking flesh. The fruit is mostly round, but some cultivars have somewhat heart-shaped fruit. The group includes dark red or black cultivars, and light-colored or yellowish ones.

Sweet Cherry Cultivars.—Presented in order of maturity (H for heart; B for Bigarreau):

Seneca (H).—Purplish-black, medium size, soft, juicy. Extreme earliness is its chief merit.

Vista (B).—Larger, firmer, and more attractive than Black Tartarian. Planting of Vista has greatly increased in Ontario.

Black Tartarian (H).—Medium purplish-black; soft, juicy, rich; the leader of its group in the east. It is the chief pollinator for Bing and

TABLE 5.1
COMMERCIAL CHERRY PRODUCTION

	Tart		Sweet		Total	
	(Tonnes)	(Tons)	(Tonnes)	(Tons)	(Tonnes)	(Tons)
Michigan	65,070	72,300	7,740	8,600	72,810	80,900
New York	20,880	23,200	3,690	4,100	24,570	27,300
Wisconsin	13,500	15,000	—	—	13,500	15,000
Ontario	9,450	10,500	4,500	5,000	13,950	15,500
Pennsylvania	9,000	10,000	—	—	9,000	10,000
Oregon	2,880	3,200	19,800	22,000	22,680	25,200
Washington	2,070	2,300	17,550	19,500	19,620	21,800
Utah	2,070	2,300	3,420	3,800	5,490	6,100
California	—	—	28,800	32,000	28,800	32,000
Idaho	—	—	2,880	3,200	2,800	3,200
British Columbia	—	—	11,790	13,100	11,790	13,100

Lambert in the west, but is being replaced by other cultivars for this purpose.

Venus (B).—Good size and quality; shiny black; heavy cropper.

Bing (B).—Dark red, black when ripe; large, firm, juicy, aromatic; ships well; cracks in rainy seasons; excellent canner. A leading cultivar, both east and west.

Deacon (B).—Slightly smaller than Bing; highly resistant to cracking; used as a pollinator in the west; its fruit has better canning and shipping value than Black Tartarian.

FIG. 5.1. DIFFERENT SHAPES OF CHERRY TREES: (LEFT TO RIGHT) MORELLO TYPE OF TART CHERRY; EMPRESS EUGENIE (DUKE CLASS); MONTMORENCY TYPE OF TART CHERRY; SWEET CHERRY

Napoleon (Royal Ann) (B).—Leading firm, light-colored cultivar; white flesh. Well adapted to canning, brining, and for maraschino.

Sue (B).—Not quite as large as Napoleon, but sweeter in the west.

Republican (B).—Dark red shipper and pollinator in the west. Its production has decreased. Seldom grown in the east.

Hedelfingen (B).—Dark purplish-black, large, firm. Lambert type but cracks less.

Van (B).—Good pollinator for Bing, Lambert, and Napoleon, and vice versa. Fruit black, blockier than Bing; medium size more resistant than Bing to cracking; short stem; good canner.

Sam (B).—Black; approaches Lambert shape; good pollinator for Bing, Lambert, and Napoleon, and vice versa; good canner.

Lambert (B).—Purplish-red; somewhat conical; cracks very badly in rainy weather; sound fruit ships well.

Others.—Early Rivers (H), Lyons (H), Yellow Spanish (B), Emperor Francis (B), Schmidt (B), Sodus (B), Giant (B), Victor (B), Gil Peck (B).

Tart Cherries

As in sweet cherries, there are two horticultural groups.

The Amarelle Group.—Characterized by light red flesh and juice.

Early Richmond.—Early, but has only fair size and quality. The tree is slightly hardier than that of Montmorency.

Montmorency.—This is the only tart cherry of commercial importance. It responds to good care and feeding, and at the present time this is the only way of "improving" the cultivar. Except for virus disease infection which is common with the cultivar, separate strains have not been conclusively demonstrated. Plant only trees propagated from buds of virus-tested Montmorency. Eliminate young trees showing virus symptoms before they come into bearing.

The Morello Group.—Comparatively dwarf trees, they are grown to a limited extent for their late crop.

English Morello.—The dark red fruits should not be mixed with the light red fruits of Montmorency in commercial processing or an irregularly colored product will result.

Ostheim.—Earlier than English Morello.

ROOTSTOCKS

Cherry growers have mostly used mazzard (*Prunus avium*) or mahaleb (*P. mahaleb*) as the rootstocks for their orchard trees. In general, mazzard seems best for sweet cherries and mahaleb for tart cherries, but, as indicated below, this is not necessarily always true.

Mazzard

This is the wild sweet cherry from which has come our cultivated sweet cherries. Its fruits show a wide range of color, type, and size, but are usually small, thin-fleshed, and dry. Wild mazzard trees often grow 12.2 m (40 ft) or more tall, with a large sturdy trunk. They grow spontaneously in some parts of the east. Results may be different if the seed is from interpollinated mazzard trees than if it has been collected from trees cross-pollinated by cultivated sweet or sour cherries.

Mahaleb

This is a bushlike tree with thick top and slender branches. It is called St. Lucie cherry in Europe. The leaves are small, smooth and glossy, resembling those of apricot more than orchard cherries. The fruits are borne in small clusters; are very small, round, black at full maturity, hard, and bitter. Mahaleb is as far removed botanically from cultivated cherries as a thorn is from an apple or pear, and not more nearly related to cherries than to plums.

Incompatibility

Mahaleb rootstocks which show graft-incompatibility with certain sweet cherry cultivars apparently differ both in phenolic composition and number of phenolics from mazzard rootstocks which do not show such incompatibility.

Montmorency may vary greatly both within and between rootstock clones (Yu and Carlson 1975).

Propagation of sweet cherry clones may be done by continuous layerage, and for tart cherry and certain hybrids by rooting softwood cuttings under mist (Westwood *et al.* 1976).

Distinguishing Mazzard from Mahaleb

If the outer layer of the roots is sliced off, exposing the inner bark, the cut surface takes on a color characteristic of the species. Typical colors of mazzard and mahaleb a few minutes after making the cuts are xanthine orange and cinnamon, respectively. After a few hours mazzard deepens in color to burnt sienna but mahaleb does not change materially.

Small pieces of bark may be cut from below the union and washed for a few seconds to remove the adhering soil. Then cut the bark into thin splinters and place them in a glass tumbler containing enough water to cover them. After 2–5 min, water containing mazzard bark becomes yellow to orange color while that containing mahaleb remains colorless. Morello stock behaves similarly to mazzard in this treatment, but growers in most cherry districts seldom need to distinguish this stock from mazzard or mahaleb.

The quantitative difference in tannic acid in mazzard and mahaleb permits distinguishing between pieces of their roots by using any of the known methods for tannic acid, viz., salts of iron (ferric chloride, ferric acetate, iron alum), acetate of copper, potassium chromate, molybdate of ammonia, canadate of sodium hydroxide, formalin, ethylnitrite, potassium carbonate, and others. For example, on a treating a water extract of dried root bark with ferric chloride, the mazzard solution turns almost black but there is only slight change in the mahaleb solution.

Usually, nursery trees on mazzard show more fine roots than those on mahaleb. Mazzard roots 1–2 years old have a trace of brown in their surface color not shown in young mahaleb roots.

Producing Rootstock Seedlings

Afterripening the Seed.—Mazzard seed requires 120–150 days, mahaleb 80–100 days, at 7.2°C (45°F) under moist conditions to break the afterripening or rest period.

The necessary afterripening conditions are usually met when the seed is sown in fall or very early in spring. Under conditions in the seedbed or in the field, a soil moisture deficiency in early spring when the seeds have become afterripened and are prepared to germinate may result in reduced ability to grow. With artificially afterripened seed, any treatment which permits the seed to partly or entirely dry prior to or following sowing may cause reduced germination.

In Oregon, the most satisfactory treatment for handling large quantities of sweet cherry (Lambert and Napoleon) seeds was to harvest the fruit when fully mature, remove the seed from the fresh fruit, stratify in sphagnum moss at alternating temperatures of 1.7°–5°C (35°–41°F) for 120 days, and plant in the field in mid-November (Zielinski 1958).

In Washington, highest germination of sweet cherry seed was obtained when no fermentation or drying out occurred and following afterripening a 2.8°C (37°F) for at least 6 months. Gibberellin (GA) seemed to be a substitute for part of this six-month afterripening. Severe rosetting, typical of plants receiving insufficient afterripening, resulted from GA treatment. However, 2–4 months afterripening followed by a GA soak seemed to offer the most promise for reducing storage losses and making possible germination during the season the seed was produced (Fogle 1958).

GA treatments (100 ppm) gave higher germination rates and appeared to partially replace the chilling requirements. Removal of the endocarp did not hasten germination or curtail the necessary chilling requirements (Pilley et al. 1965).

The following events may occur during afterripening of cherry seeds: GA may partially replace the chilling effect and bring about the stimulation of growth-substance production. With chilling fully satisfied, the balance between growth-promoting and growth-inhibiting substances may be maintained in favor of the former. With insufficient chilling this balance is disturbed, resulting in a lack of growth-promoting activity which, in turn, may delay the process of growth. Increased production of growth-promoting substances is stimulated by an adequate period of chilling during rest. A naturally GA-like compound responsible for growth process both in the development of fruit and in germination is probably lacking in the mature seed in resting condition. GA possibly stimulates respiration and protein synthesis, known to initiate growth which is in the arrested condition in these seeds (Pilley and Edgerton 1965).

Growing the Rootstock Seedlings.—Spacing affects the grade of seedlings. About 10 cm (4 in.) apart in the row is usually good spacing. A difference of 1.6 mm (1/16 in.) near the dividing line between grades makes a difference in the sale price per tree.

With a small increase in growth, many No. 2 budded nursery trees would become No. 1 trees. Improvement of nursery stock may depend as much on the growing conditions during the seedling stage as on the growing conditions or treatment during the year the bud is growing. The size of seedling budded may be important in the production of a large,

FIG. 5.2. NURSERY TREE GROWTH HABITS: (LEFT) MONT-MORENCY SHOWS NARROW-ANGLED BRANCHES AND COM-PACT HEAD; (RIGHT) EARLY RICHMOND SHOWS WIDE-ANGLED BRANCHES

1-year-old nursery tree. Irrigation plays an important role in promoting stock of good diameter.

The time of digging seedlings and the degree of maturity are important. Stocks dug too early do not start in the spring as well as do those dug when more mature.

Budding and Subsequent Nursery Practices

Time of Budding.—Budding is usually done, in the north, on mazzard stock in early to mid-August and on mahaleb stock during August or early September. Plate buds have not given better results than the T-bud, except possibly very late in the budding period.

Comparisons of 2 adjacent blocks of nearly 2000 mazzard stocks each; 1 block with vertical cut at the top in the T-method and the other block with the cut at the bottom showed no difference (Tukey and Brase 1945).

Stand and Defoliation.—A 50% stand of 2-year-old trees on mazzard stock is above average in commercial nurseries. Usually the chief factor in causing the low stand of buds is premature defoliation by leaf spot. Mazzard stock may be completely defoliated in New York by Aug. 15 in some years, and partially defoliated in most years. Since budding season is normally Aug. 1–15 to Sept. 1 on mazzard, the defoliation shortens the working period; the bark tightens and fails to slip and buds do not unite with the stock. Mahaleb stock is seldom seriously affected in this way; the bark slips well, the budding season is seldom interrupted and a good stand of buds is secured (Tukey and Brase 1945).

A site where water lies or moves slowly in the soil adversely affects cherry stocks. Mazzard is probably even more susceptible than mahaleb to "wet feet."

Winter Injury.—Low temperature in the fall, particularly after premature defoliation, often accounts for loss of buds and young trees. In both root and top, the injury may extend from the periphery back to the crown.

Storage.—An important consideration in storage of nursery stock is the treatment prior to placing in storage, such as growing, spraying, exposure to sun and wind during digging, and freezing of roots. Cording (laying the trees tightly together horizontally) in bins is commonly done. Waxing both roots and tops of cherry trees may prove harmful. Trenching (setting the trees in damp sand to the depth they stood in the field) in the nursery cellar is no improvement over cording in bins; in fact, it may result in early starting and ensuing injury to the stock in handling.

BEARING AGE AND LIFE OF CHERRY TREES

Sweet cherries begin to bear some fruit after having been in the orchard 5–6 years; tart cherries often bear 1–2 years earlier than sweet. When an orchard is 10 years old, the trees are usually in full bearing capacity for their size. Of course, as the trees become larger the yield increases.

Trees of both sweet and tart cherries occasionally live 100 years, but the profitable age of tarts is not more than 20–25 years and of sweets 25–30 years. After this time, the yield decreases and the cost of hand picking the crop from high trees becomes too great. Besides, disease and other factors may have thinned the ranks of the trees until the orchard is unprofitable.

Too many Montmorency orchards produce 5.6 tonnes per ha (2 1/2 tons per acre) that should be producing 6.7–11.2 tonnes (3–5 tons).

LOCATION AND SITE

The tart cherry is a cool-climate fruit and, in the east, does not succeed in commercial plantings south of the Potomac and Ohio rivers. Part of its production centers in areas where the mean temperatures for June, July, and August are about 15.6°C (60°F). Montmorency tart cherry is somewhat less hardy than the McIntosh apple.

The sweet cherry is more sensitive than the tart to both heat and cold. The blossoms suffer greatly from spring frosts, partly because of early bloom. As a rule, the sweet cherry cannot endure much more cold than the peach, and does not stand heat as well as the peach. Standard cherry cultivars are subject to prolonged dormancy in the south.

The cherry is not adapted to most locations in the great interior valleys of California, probably because of the high temperatures and low humidities there. Where it does thrive in the interior, the climatic conditions are usually modified by coastal influence.

Cherries are at the mercy of moisture conditions in both bloom and fruiting stages. Continued rain during bloom prevents a proper setting of the fruit. If much rain occurs at fruiting time, the firm sweet cherries often crack, and brown rot and other fungus diseases take heavy toll. Growth begins early and the crop is off the trees before the usual summer drought. However, cherries are very sensitive to high transpiration rates, contain a high percentage of water, and there are many fruits on a tree. During a hot day, with low atmospheric humidity, a water deficit may occur in the leaves and water may move from the fruits to the leaves, thus causing shriveling and a loss in volume of crop.

Young orchard trees subjected to a severe drought may come into bearing earlier than they normally would. Such trees may remain comparatively small and unprofitable for a period of years, perhaps during the life of the orchard.

Elevation

Except for locations near large bodies of water, the best sites for cherry production may be those that are elevated above the surrounding country. The blossoms of trees on such sites are less likely to be injured by spring frosts than when the orchard is on lower land. Also, these upland soils are usually better drained than those on lower and more level locations.

Soil erosion may become a problem in orchards located on elevated sites. Serious gullying and sheet erosion may occur if cultivation is practiced. A sod cover reduces erosion to a minimum but cherry trees may grow poorly because of a shortage of soil moisture and nutrients. A mulch around trees growing in sod tends to reduce moisture and nutrient problems.

Soil

Wet, sticky clay underlain with a cold, clammy subsoil is a combination which defies the best of care. An ideal cherry soil is a well-drained, warm, deep, free-working, gravelly or sandy loam. It must be able to hold moisture enough to last 2–3 weeks during the period of high water use.

Loamy soils should be more than 1.5 m (5 ft) deep. Loams and heavier soils may be less than 1.5 m (5 ft) deep; 0.61 m (2 ft) of loam, which holds a total of 10.2 cm (4 in.) of available water, is a bare minimum.

Water Drainage

If the soil is not naturally well-drained, it must be made so by artificial drainage, because cherries, like most other fruits, do not tolerate "wet feet." In fact, cherry trees are probably more susceptible than apple or pear trees to damage from poor water drainage. The life of cherry trees planted in poorly drained soil is short.

The deepest rooted cherry trees are found on soils with a medium brown to slightly mottled profile and medium to light texture. The roots show excellent distribution in the first 1.2 m (4 ft) because of adequate soil drainage.

Courtesy of Hort. Res. Inst., Vineland, Ont.

FIG. 5.3. THE EFFECT OF POOR SOIL DRAINAGE ON CHERRY TREES: (LEFT) UNDERGROUND DRAINAGE; (RIGHT) NO UNDERGROUND DRAINAGE

Shallow-rooted trees on poorly-drained soil may appear thrifty and even maintain satisfactory production as long as rainfall is normal and winter temperatures are moderate. The ill effects of the site become apparent in a dry season when the surface soil which the roots have explored dries out and the trees suffer for lack of moisture. Trees weakened by such conditions are also more subject to winter injury.

Hardpans that cause a waterlogged soil condition may make cherry trees susceptible to various ills. A temporary rise in the water table may cause damage that is only slowly overcome by improved soil drainage. Root killing below the level of the high water table causes the trees to be shallow-rooted.

In Ontario, the growth and yield of Montmorency, as with peach, were increased by blocked compared with open tile. This was probably related to moisture stress at certain critical periods. Subsoiling did not reduce growth and yield of Montmorency, or peach, as it did with sweet cherry. There was a favorable growth and yield response to planting trees directly over the tile compared with between tile lines, and so tree location in relation to tile lines may be important. Sweet cherry responded much the same as peach and tart cherry to treatments that alter soil physical conditions (normal tile drainage, blocked tile, and compaction) in relation to growth, yield, and foliar nutrient composition. Subsoiling reduced yield of sweet cherry, but not of peach or tart cherry (Cline 1967).

PLANTING THE TREES

Losses at planting time are greater with cherry, particularly sweet cherry, than with most other fruits. Reasons for this are that buds of the sweet cherry are large and are easily rubbed off in handling, the buds expand and open early in the spring, and the root system is slow in becoming established.

Tart cherry trees usually are planted 2 years from the bud. Sweet cherry trees 1-year-old usually are preferable.

When to Plant

A common practice is to plant cherry trees in early spring. They should be set very early, while the soil moisture is ample and before the buds have started to swell. Prepare the ground to the best possible condition.

Fall planting is often superior to late spring planting. Generally, the grower is not so pressed for time in late fall as in early spring. Also, the trees are freshly dug rather than having been stored over winter as may be the case with trees bought in early spring. There is often less mortality of both sweet and tart trees when set in late fall, providing the winter is not too severe.

In Ontario, fall-planted cherry trees begin root growth earlier than spring-planted ones and can, therefore, better withstand drying conditions in the spring. Storing sweet cherry trees in moist earth in a good storage house gives results inferior to either fall or spring digging with immediate planting. "Tiering" the trees with the roots in moist moss is probably even less desirable because of the danger of the moss becoming too dry.

Do not prune cherry trees planted in the fall until late winter or early spring.

Planting Distance

Good spacing for Montmorency is usually 6.1—9.1 m (20—30 ft) and sweet cherries 7.6—9.1 m (25—30 ft) apart each way. An alternative is 5.5 × 7.3 m (18 × 24 ft) for Montmorency and 7.6—8.5 × 9.8—10.7 m (25—28 × 32—35 ft) for sweet cherries. In contour planting, allow 36—45 m^2 (400—500 ft^2) for tart cherries and 63—90 m^2 (700—1000 ft^2) for sweet cherry trees.

When the trees are too close together, spraying and other orchard operations become difficult. The branches tend to be slender and upright; the lower ones are shaded and soon become unproductive. The longer life, more economical management, and higher yield per tree that can be obtained in orchards having adequate spacing offset the small gain in production during the early life of a closely planted orchard.

Filler Trees

The use of filler trees has largely been abandoned, at least in the east. The disadvantages usually outweigh the advantages.

In eastern Washington, however, it has been a fairly common practice to interplant with peaches or apricots. When the trees are set 11 × 11 m (36 × 36 ft) a double set of peaches or apricots can be used; then the initial spacing is 5.5 × 5.5 m (18 × 18 ft).

Heeling-in

Since the losses at setting time are often greater with cherries than with most other fruits, take extra care at planting time. Open the bundles of trees as soon as they arrive, and water the roots and packing material thoroughly. Unless they can be planted immediately, carefully heel-in the trees in a sandy, well-drained location.

Planting

The site to be planted should be plowed in fall or early spring and worked thoroughly prior to planting. The time and effort spent in preparing the field helps both at planting time and in the subsequent care of the trees.

Before the trees are planted, cut off broken or injured roots. Unnecessary exposure to sun and wind before planting may cause the roots to dry quickly and lessen the chances of successful transplanting.

The actual planting operation is done much the same as stated for the other tree fruits. Many failures in transplanting are due to insufficient or improper firming of the soil about the roots.

PRUNING

Pruning Sweet Cherry Trees

Sweet cherry trees grow upright, with a tendency to form a high central leader. A modified leader form is not difficult to develop because of the growth habit of the tree.

Pruning at Planting Time.—Nurserymen often trim back the tree for ease of tying into bundles. This pruning may be all that is needed at this time. However, most growers will prune to their own satisfaction. Trees that are 1 year old (1.2−1.5 m [4−5 ft] high) are commonly headed at about 0.9−1.1 m (3−3 1/2 ft). Leave somewhat longer trunks for sweet cherries than for tart cherries, but too high heads may lead to southwest injury. The head height for sweets is usually about 76 cm (30 in.) in the east and 76−91 cm (30−36 in.) in the west.

Losses following planting are usually lessened if the branches, particularly of sweet cherries, are not cut back severely. If the top must be thinned, as is usually the case with 2-year-old trees, remove whole branches rather than cut all branches severely. Probably the reason for the loss when branches are cut back is that the pruning removes the largest and best nourished buds, those which are the first to develop the leaves that give the young trees a start. A 1-year-old tree can be topped at the desired height and a good vigorous head will be formed the first summer.

Select scaffolds that have wide angles at the trunk. These scaffold branches should be 20−25 cm (8−10 in.) apart and located spirally around the trunk.

Balance the root system and top by some pruning, so that the tree thrives in the first year. Give the selected scaffold branches a minimum of heading back, consistent with a good balance between themselves and the leader and with forming a well-shaped tree.

Cherry trees on mahaleb stock often are noticeably lacking in fibrous roots. The root system may consist of 2–4 large roots only. Most new roots arise near the cut ends of the old roots after transplanting. The paucity of roots at transplanting time limits new root formation. The sweet cherry top starts into growth very early in the spring and therefore requires water very early. That the new root system may form too slowly to provide sufficient water to the top may account for the death of trees after they have leafed out. A moderate cutting back of the branches tends to reduce the water requirement of the top in two ways: (a) it removes some of the transpiring surface; and (b) it removes the terminal buds, which are first to burst, leaving buds which are slower in developing, thus retarding top growth.

Not only does heavy pruning of 2-year-old branched trees delay top growth but it induces more than one bud to form a branch, whereas from an unpruned branch the terminal bud is often the only 1 to develop a shoot.

Orchard Pruning After One Year.—Select and retain a vigorous upright-growing shoot, usually the top one, which continues the general direction of the trunk. Leave it 20–30.5 cm (8–12 in.) longer than the framework branches. Remove any other central upright branches by thinning-out cuts close to the trunk.

Young sweet cherry trees that are severely headed back tend to form framework branches too close together. Retain only 3–5 wide-angled framework branches, 20–25 cm (8–10 in.) apart, and located spirally around the tree.

Sweet cherry trees are more susceptible to winterkilling than tart cherry trees, and to avoid injury to the wood it is important, in many regions, to grow trees with well-spaced scaffolds. If several branches arise close together, winterkilling of wood may occur in the narrow crotches. Killing the wood at this point results in splitting the bark, making an opening where wood-rotting fungi may become established, thus shortening the life of the tree.

Pruning in the Formative or Prebearing Period.—Correctional thinning-out is usually essential. Give special attention to development of a strong framework since the trees are susceptible to crotch splitting. Head back the scaffolds only when the growth is exceptionally long and willowy, when it is desirable to keep the branches balanced, and to force the formation of secondary shoots.

The sweet cherry is rather slow in coming into bearing. It usually requires less pruning than other tree fruits. If pruned heavily, vigorous shoot growth may be slow in developing fruit spurs. As the tree grows larger, if the leader becomes unduly out of balance with the scaffolds, cut it back to an outward-growing branch.

Pruning Mature Sweet Cherry Trees.—Bearing trees need some pruning to keep weak and unnecessary limbs cut out, the center of the tree reasonably open, the top fairly low, and for annual production of new wood of suitable average length.

FIG. 5.4. A WELL-SHAPED MONTMORENCY CHERRY TREE IN FULL BLOOM

Sweet cherries are borne laterally on spurs which are productive for 10–12 years. Spurs are short growths that usually grow less than 2.5 cm (1 in.) in 1 year. They arise from lateral buds, usually on 1-year-old short shoots. The flower buds are borne laterally, in a cluster on the part of the spur that grew during the previous year. A spur elongates in nearly a straight line for several successive years because the terminal bud is a leaf bud.

Because of the persistent fruit-spur system, less renewal wood is needed than with most deciduous fruits. The annual terminal growth of upright wood in the top of trees in full bearing should average 20–30.5 cm (8–12 in.) long. If the trees have been properly trained and brought into bearing, little pruning is necessary to keep them productive. The pruning should renew about 10% of the fruit bearing area annually.

Trees of some cultivars, e.g., Black Tartarian, naturally grow upright. Spread them reasonably by removing upright growths and cutting to outward-growing laterals. Napoleon (Royal Ann) naturally assumes a spreading habit with age. Hence its pruning will vary from those cultivars of upright habit.

To keep bearing trees vigorous, place emphasis on proper soil conditions, including moisture, drainage, and fertility. When cherry trees are cut back into older wood, either to lower or open the tree or to stimulate growth, cut to a lateral where possible. Trees showing much dieback of the outer branches can sometimes be given rather severe pruning.

Removing one or more large branches to lower a tall cherry tree is a major operation. It may even jeopardize the life of the tree, yet it is sometimes wise to do so. In this case, cut to a large lateral branch and leave a 0.61−0.9 m (2−3 ft) stub for several years. Leaving a stub and keeping it alive helps to avoid decay that may start in the wound from going directly to the heart of the tree.

Air must circulate freely through the trees, and thorough spraying must be possible for brown rot control. To this end, remove any unnecessary or interfacing limbs and keep the center of the tree open.

Fruit on a well-pruned tree ripens more evenly than on a tree that is too thick. In a dense tree much of the fruit of the inner and shaded branches may still be immature when the rest of the fruit is mature, thus requiring extra pickings.

Pruning commonly reduces the yield of sweet cherries proportionally to the severity of the operation. Yield is influenced by the amount or number of fruit buds removed in pruning.

In the east, pruning has only a minor effect on size of the fruit. In the west, however, where sweet cherry fruits develop larger size than in the east, pruning tends to increase their size.

Pruning Tart Cherry Trees

Pruning at Planting Time.—Both 1-year and 2-year-old tart cherry nursery trees usually are branched. Some heading back of the trees is usually required at planting time in order to establish a balance between the reduced root system and the tree top. Also, trees given some heading back commonly survive the planting operation better than those that are not pruned at all, reducing the top to the point where the transpiring surfaces meet the soil water requirements of the slowly developing root system.

Leave the main stem, trunk, or strongest upright growth 13−30.5 cm (5−12 in.) longer than the selected scaffold branches. Some growers prefer to cut back the branches so as to leave 8 to 10 buds, and reduce the main leader to 5.1 to 7.6 cm (2 to 3 in.) longer than the branches.

A total of 3 to 5 thrifty, wide-angled, well-placed branches and the main leader make a suitable framework. It is often an advantage to have the lowest scaffold branch on the southwest side. This may be 38 to 61 cm (15 to 24 in.) aboveground. If narrow-angled branches are allowed to remain and become a part of the framework, they often split under loads of fruit, the weight of snow or ice, or during heavy winds; also, winter injury to narrow-angled crotches further weakens the branch.

Cut back selected branches to make them well balanced. A spacing of 10.2 to 15.2 cm (4 to 6 in.) between branches on the trunk is good, providing they are located spirally around and one of them is not directly above another on the same side of the trunk. If the branches are not well spaced around the trunk, they may choke the leader and prevent it from growing faster than they. If all are left to develop from the same height, a weak tree is inevitable.

Pruning After One and Two Years in the Orchard.—Pruning in this period should be light. It is largely corrective in nature.

The main leader is allowed to extend and is kept dominant over the side branches. But after two years in the orchard it may need to be modified by being cut back to a strong outward and upward lateral. Where possible, keep the main leader dominant until 4 to 5 scaffolds are distributed over 0.9 to 1.2 m (3 to 4 ft) of trunk above the lowest branch. If extra scaffolds were left previously, these can be removed. Thin out any superfluous branches, especially those that cause crowding, and those with weak crotches.

If some of the selected scaffold branches are more vigorous than the others, they may be suppressed by removing some of the lateral branches on the scaffold limbs, and by making a thinning-out cut to an outward and upward branching lateral.

No heading back is necessary unless the terminal growth exceeds 51 cm (20 in.), or unless some branch is outgrowing the others.

Pruning During the Prebearing Period.—During the early formative period the tree should require only light corrective training. Some pruning may be required to direct the growth of the framework branches, make a proper balance between them, eliminate inward-growing branches and those that compete with the main scaffolds, avoid weak crotches and unduly crowding branches in the perimeter of the tree, keep the center reasonably open, and remove watersprouts. The tart cherry commonly comes into fairly heavy bearing when five to seven years old. Suppress or modify the main leader about the third, fourth, or fifth year by cutting back its season's growth to an outward-growing framework branch. If some adjacent top branches are of equal length to the modified leader make them shorter than the leader or a weak crotch may result.

Pruning during the formative or prebearing period consists mostly of thinning out. If 2 branches parallel one another from the trunk eliminate 1 of them. Only a limited amount of heading back is required, chiefly to maintain balance between the main branches and to encourage desirable spread of tree.

If some of the framework branches develop more rapidly than others, the leader, as well as the weaker ones, may be choked out. But, more pruning than necessary may delay bearing excessively.

The trees are naturally symmetrical in growth and ordinarily require little shaping. It is easy to overprune the young tart cherry tree and thereby reduce yields.

If the trees are kept sufficiently thinned from the start, fruit is more likely to be borne all along the branches. Fruit buds will not develop where there is a lack of sunlight and air circulation, and tart cherry trees as ordinarily seen are fairly dense and much thicker than advisable.

Pruning Mature Tart Cherry Trees.—As the trees reach full size, thin out the weak branches in the center and top. Also, some thinning out of the side branches may need to be done. When the top and sides of the trees become thick and bushy, the lower limbs are shaded, the leaves are small and weak, and production in that area falls off rapidly.

If the tree has been properly trained and not neglected for some years, pruning will, for the most part, consist of small cuts. The tendency is to concentrate the pruning on the inside of the tree, but results are better if some of the cutting is done to thin the smaller branches from the outside. Pruning of this type, accompanied by adequate fertilization, helps maintain good terminal growth and vigorous spurs. Some heading back to laterals of the upright branches in the top helps control the height of the tree.

Growing and Bearing Habits.—For the first few years in the orchard, grow the trees so that their shoots average 30.5 to 61 cm (12−24 in.) long. While making this vigorous growth the trees bear little fruit but increase rapidly in size and develop a large bearing surface for later production. Then, until they attain full size, grow the trees so that their shoots average 15−30.5 cm (6−12 in.) long. While making growth of this type, they bear heavily and gradually increase in productivity. After the trees have attained full size and production their annual shoot growth should average 10−20 cm (4−8 in.) long. This provides for a practically indefinite maintenance of yields.

Yields of trees well into bearing may be increased by forcing them to produce shoots that average 18 cm (7 in.) or more long. They can be maintained by forcing them to produce shoots that average 13−18 cm (5−7 in.) long. It is important to maintain moderately vigorous annual growth by moderate to light pruning and especially by good soil management and an efficient spray program.

Fruit Habit.—Fruit buds of the sour cherry are produced laterally on 1-year-old terminal growth and spurs. The terminal buds of shoots and spurs are leaf buds.

Frequently, all the buds on shoots less than 15 cm (6 in.) long are flower buds. Since there are no leaf buds, lateral shoots or spurs cannot be formed. This results in a completely barren area along the shoot after it

TABLE 5.2
FRUIT BUD, FRUIT SPUR, AND LATERAL SHOOT PRODUCTION OF
MONTMORENCY SHOOTS OF DIFFERENT LENGTHS

Shoot Length (cm)	(in.)	Avg No. of Lateral Fruit Buds	Avg No. of Spurs Formed Subsequently	Avg No. of Lateral Shoots Formed Subsequently
0–2.5	0–1	5.7	0.1	0.0
2.5–5.1	1–2	6.5	0.1	0.1
5.1–7.6	2–3	7.9	0.3	0.1
7.6–10.2	3–4	8.9	0.3	0.1
10.2–12.7	4–5	9.1	0.4	0.2
12.7–15.2	5–6	9.4	0.5	0.3
15.2–17.8	6–7	9.0	1.3	0.7
17.8–20.3	7–8	8.6	1.6	1.1
20.3–22.9	8–9	7.7	2.6	1.4
22.9–25.4	9–10	7.5	3.0	2.0
25.4–30.5	10–12	7.3	3.8	1.9
30.5–71.1	12–28	4.9	4.9	3.7

Source: Tukey and Tukey (1966).

has fruited. These shoots may elongate by making short terminal leafy growth each year but are barren except for the clusters of fruit borne on the portion of their wood produced the previous year. Shoots can be made to produce some leaf buds that will develop into spurs and lateral shoots by fertilizer and pruning programs that force more vegetative growth. Trees in high-producing orchards have a high proportion of shoots over 15 cm (6 in.) long (80%) while low-producing orchards have a large proportion of terminal shoots less than 15–18 cm (6–7 in.) long (19%). Montmorency produces 35–45% of its fruit on shoots (Gardner 1930).

Fruit Spurs.—Tart cherry fruit spurs are shorter lived than those of the apple, pear, or sweet cherry. Most of the fruiting spurs occur on 2- or 3-year-old wood with some, less vigorous and productive, on 4- or 5-year-old wood.

A single crop of fruit weakens the spurs and each succeeding year they become less and less productive. Also, fruit spurs on the strong, more upright, rapidly thickening branches (such growth is encouraged by relatively heavy pruning) lose their vigor more rapidly and are shorter lived than those that grow on shoots which are less upright in habit and are less vigorous.

FIG. 5.5. SWEET CHERRY FRUIT SPUR HABIT

A moderately heavy pruning is advised, not to increase the total amount of bearing wood of all kinds (spur and shoot) but to promote a more vigorous terminal shoot growth and thereby obtain a larger percentage of spur-borne (as contrasted to shoot-borne) fruit buds. These spur-borne buds are somewhat hardier than those borne on shoots.

Most of the crop is borne within a comparatively thin rim around the outside of the tree. Theoretically, pruning should tend to invigorate the longevity of the spurs by reducing their total number and thus providing the remainder with a larger nutrient and moisture supply, and by admitting more light to them. But in reality, the spurs show just the opposite response and the pruning treatment, or its lack, which leaves a relatively large number of moderately vigorous shoots year after year, not only provides the maximum number of fruit spurs but contributes to their individual and collective efficiency and to their long life. It is futile to attempt to develop and maintain maximum production by either extensive thinning out of the top or by heading back. Maximum fruit production, spur formation, and efficiency and longevity of spurs are associated with the type of growth that accompanies very light pruning (Teskey *et al.* 1965).

In a 16-year-old Montmorency orchard, even a light pruning (chiefly the removal of some of the smaller limbs in the thick, bushy parts of the tree) reduced the yield 12% (Table 5.3). In another orchard that had not been pruned for 6–8 years, yields for a subsequent 4-year period were 46.7 and 41.3 kg (103 and 91 lb), respectively, for the unpruned and pruned trees.

Fruit Thinning.—Size of individual fruits is not a serious factor with Montmorency grown for processing. Nevertheless, reasonably large size is desirable, if for no other reason than that it makes possible higher yields. Heavily loaded trees are as likely to produce large cherries as are

TABLE 5.3
EFFECT OF PRUNING ON FRUITING RECORDS OF 16–YEAR–OLD
MONTMORENCY (AVERAGE PER TREE)

Treatment	Shoot-borne Fruits			Spur-borne Fruits			Total Fruits		
	No.	Yield (kg)	(lb)	No.	Yield (kg)	(lb)	No.	Yield (kg)	(lb)
Pruned	6,152	27.5	60.7	8,813	38.1	84.0	14,967	65.6	144.7
Unpruned	5,906	25.2	55.5	11,815	49.1	108.2	17,721	74.2	163.7

Source: Tukey and Tukey (1966).

those bearing a light crop; conversely, fruit on light yielding trees may be either large or small. Proper pruning and nutrition satisfactorily regulate fruit size.

Controlling Tree Size.—Aim to keep the tree within bounds and its top reasonably low. Hand picking fruit from very tall trees is costly and dangerous. Avoid the removal of many large limbs and the leaving of stubs. Cut out dead and broken branches and watersprouts. If the trees have been making vigorous growth and are becoming tall, head back some of the upright branches to laterals.

Overpruning, especially in the top, reduces the yield without increasing the quality of the remaining fruit. A few thinning-out cuts on mature tart cherry trees may serve to admit adequate sunlight into the tops.

Tree Renewal.—Renewing tart cherry trees by severe cutting to outside branches checks the growth of the trees the first year and decreases the yield for three years. But, when fertilizer is applied to the pruned trees, many new shoots are produced by the end of the third year which are more vigorous than those on unpruned trees. Such responses were obtained in widely separated areas in eastern North America.

PRUNING DUKE CHERRY TREES

Duke cherries, which are hybrids, are more or less intermediate in characteristics between sweets and sours. Some Duke cultivars tend toward sweets and others toward tarts, and pruning, in turn, varies with the type.

FRUIT SET IN THE CHERRY

Poor fruit set in the cherry may result from causes other than pollination. Severe winter cold or late spring frosts may kill nearly all blossoms, and cold stormy weather or strong winds at pollination time may so reduce bee activity that insufficient flowers are visited. The development of stigmatic secretion after the blossoms open may be retarded by cold weather. Rain at full bloom may destroy pollen. Pollinator trees and crop trees may not bloom at the same time.

Some basic causes of poor fruit setting are as follows: (a) pollen may fail to germinate on the stigma; (b) pollen tubes may penetrate the style for only a short distance; (c) pollen tubes may burst within the style; (d) pollen tubes may grow too slowly to reach the ovules before they disintegrate; (e) pollen tubes may reach the ovarian cavity but fail to penetrate the ovules; (f) after penetration of the ovules, the sperm and egg gametes may be unable to cooperate in forming a zygote; (g) a zygote may be formed but its development is arrested at some immature stage; (h) an embryo may be fully formed but there is no endosperm (such embryos usually die unless artificially cultured); (i) embryo sac degeneration can account for poor fruit set in the sweet cherry (Eaton 1962).

Nutritional deficiencies can result in poor fruit set. The trees may have been too low in vigor to develop large numbers of flower buds.

Sweet Cherries

Pollination.—All sweet cherry cultivars must be cross-pollinated by other suitable cultivars in order to set fruit. Self-pollination, i.e., transfer of pollen from different flowers on the same tree, or from one tree to another of the same cultivar, is not effective. Ordinarily, 25—50% of the flowers, depending on the amount of bloom, should set fruit in order to produce a full crop. Hormones may induce sweet cherry flowers to set fruit without pollination, but the percentage of flowers which set is too small commercially (Wann 1950).

Interplanting Cultivars.—In one orchard in New York, the trees 6.1 m (20 ft) apart each way, were uniform in size and growth, yet the normal blossom set in the first row of Windsor trees, adjacent to the Black Tartarian row, was 62% more than in the second row, and 127% more than in rows 3 to 11 inclusive. These differences were not due to tree individuality since hand pollinations with Black Tartarian pollen were uniformly successful in whatever section of the orchard they were tried (Tukey 1925).

It is necessary with self-unfruitful cultivars to include a pollinator in the orchard, otherwise honeybees may be of little value. Plant the suitable pollinators when the orchard is set out.

Planting alternate rows of different cultivars would be a good plan from the pollination and mechanical standpoints but it is seldom

**FIG. 5.6. TART CHERRY SHOWING CHARACTERISTIC AB-
SENCE OF LEAVES IN BEARING AREA**

practicable commercially in hand-picked orchards. The minimum proportion of pollinators to ensure good crops every year is 1 in 8–9 planted
in every third space in every third row. Where pollinators have not been
provided, it is possible to ensure cross-pollination by top grafting the
trees with other cultivars or by distributing bouquets of a suitable
pollinator in the orchard during bloom.

Inter-unfruitful Groups.—Bing, Lambert, Napoleon, and several others,
are inter-unfruitful because of a genetic incompatibility factor (Table
5.4). Deacon, Sam, and Van have been widely planted in the west as a
pollinator for Bing, Lambert, and Napoleon. Sweet cherry cultivars not
true to name sometimes affect the pollination.

TABLE 5.4
POLLEN COMPATIBILITY GROUPS OF SOME SWEET
CHERRY CULTIVARS

Group	Cultivar of Seedling
I	Black Tartarian, Early Rivers
II	Sodus, Van, Venus, Windsor
III	Bing, Emperor Francis, Lambert, Napolean, Vernon
IV	Velvet, Victor, Gold, Merton Heart, Viva, Vogue
VII	Hedelfingen, Vic
VIII	Hudson, Giant, Schmidt, Ursula
XIV	Valera
O	Seneca, Vega, Vista

Source: Bradt *et al.* (1968).

The cultivars and seedlings within each of the groups listed will not
pollinate each other and therefore should not be planted together. Any
cultivar from one compatibility group will pollinate any cultivar from
any of the other compatibility groups and therefore may be planted
together.

The incompatibility groups of 115 cultivars and seedlings have been
identified, and also for 20 additional clones (Way 1966). Knowledge of

incompatibility groups is useful in verifying synonymy of sweet cherry clones. The nomenclature of sweet cherry cultivars has been greatly confused.

Incompatibility.—East and Mangelsdorf (1925), in tobacco experiments, proposed the oppositional-factor explanation of self- and cross-incompatibility in which the responsible sterility genes (S) belong to a multiple series identified as S_1, S_2, S_3, etc. Incompatibility is governed by genetic factors, just as morphological characters are. Any two specific factors can comprise a pair of allelomorphic genes.

Cherry incompatibility is due to S genes which prevent normal growth of pollen tubes into floral styles, thus prohibiting fertilization. Incompatibility exists when the specific S gene carried by a pollen grain is the same as 1 of the 2 specific S genes carried by the somatic tissue of the receptor pistil (Crane and Brown 1937).

An antagonism may exist between the stylar tissue and the pollen grain tubes in incompatibility. One theory on gametophytic incompatibility states that S alleles are regulators. If the pollen and style S genes are identical, molecules from each unite to form a repressor molecule which prevents normal metabolism and growth by the pollen tube. If the S alleles are not identical, the repressor molecules are not formed (Ascher 1966).

Six distinct specific S genes in sweet cherry have been identified as S_1 to S_6 (Brown 1955). At least three more are known to exist. Eighteen incompatibility groups are known to exist; the specific genes are known for 11 of these groups. Cultivars which have not yet had their specific S genes determined but are known to be different from the S genes of the 18 identified groups are described as universal donors.

Overlapping of Bloom.—In Ohio in a normal year, all sweet cherries usually overlapped sufficiently in bloom for cross-pollination. Tart cherries generally bloomed too late to be good pollinators for sweet cherries. The Dukes were not good pollinators, but sweet cherries were good pollinators for early blooming Dukes (Shoemaker 1928).

Following the moderately warm winters common in California, the blossoms are delayed a little in coming out. In the east, the sweet cherry blooms as early as, or a little later than, the peach, but in most places in California the delay and the slight straggling of bloom are favorable for cross-pollination and heavy fruit set. Following especially warm winters, certain cultivars, particularly the Lambert, may be delayed so much that normal cross-pollination is prevented—a condition which probably accounts for the fact that Lambert is less desirable in California than in the Northwest where the winters are colder.

Tart Cherries

Montmorency, the main tart cherry, is self-fruitful. It produces good crops in large solid blocks. Most other tart cultivars also are self-fruitful, and cross pollination is not required.

The acclimation and deacclimation rate of wood and flower buds of tart cherries were determined over two seasons and related to the time of defoliation that occurred as the result of pathogenic infection or hand removal during the first year. Early leaf loss resulted in delayed acclimation in the fall and more rapid deacceleration in spring. The end result was reduced bud survival and decreased fruit set. Effects of early defoliation carried over into the second year (Howell and Stackhouse 1973).

Premature Fruit Abscission.—Dropping in tart cherries is not primarily the lack of pollination and fruit set. When there is an unusually light bloom due to excessive winterkilling of the blossom buds, the percentage of blossoms which sets fruit is ordinarily higher than in other years. Unfavorable nutritional conditions play a major part in the arrested development and dropping of immature fruits.

Adverse conditions during the first two weeks after pollination, resulting in a deficiency of nutrients in general, seem to be responsible for most aborting embryos.

Severe defoliation can result in a reduced fruit set the following year. The weakened fruit buds fail to set fruit, or they set but the embryos abort within 2–3 weeks.

Effect of Shade.—Cherry trees withstand considerable reduction in light intensity without materially reducing the set of fruit; therefore, cloudy weather alone is not responsible for a poor set when adequate pollination has been effected. Normally, the reduction in light intensity in the center of the tree is not sufficient to be the direct cause of a low fruit set.

Reductions in light intensity sufficient to decrease the production of total sugars and dry matter and to affect the character of the foliage do not affect the set of fruit. Also, the effect of light intensity reduction on time of bloom, pistil receptivity, bee activity, temperature, humidity, petal fall, and abscission of fruits is not significant, nor does it affect pollen germination, rate of pollen tube growth, or rate of embryo development. Very heavy reduction of light intensity, however, indirectly affects the set of fruit by reducing photosynthesis, thus causing more embryo abortion.

Sweet cherry fruits covered with aluminum foil bags at the beginning of pit hardening were visibly larger than those exposed to light when examined fourteen days later. With Bing this difference in size remained nearly constant until harvest; with Royal Ann it continued to diverge during the "final swell." Size of covered fruits was larger at harvest, but soluble solids content was much less than in exposed fruits. Alcohol insoluble substances of control mesocarp tissue of Bing decreased from 15 to 7% during the second week prior to harvest; those of dark brown fruits decreased from 13 to 10%. Trends for Royal Ann were similar (Ryugo and Intrieri 1972).

DEVELOPMENT OF FLORAL TUBE AND STYLE ABSCISSION

The interval is shorter from bloom to maturity in the cherry than in the peach. Development of floral-cup and style abscission zones in Montmorency followed the same general pattern as in the peach, with these important differences: (1) A more rapid rate of development from one stage to the next, especially beyond the appearance of the floral abscission zone as a narrow, pale greenish-yellow line around the outside of the cup, 1/3 from base to rim. (2) Nearly coincident development. In the peach the style abscission zone did not develop until the floral tube was separated from the fruit (this is a genetic difference). (3) The floral-cup abscission zone developed in a nearly straight line, rather than as a sigmoid curve; this may result from a difference in development rate. (4) Style abscission coincided with or preceded that of the floral tube instead of occurring five stages later (Lott and Simons 1966).

FRUIT DEVELOPMENT STAGES

Histological development of Montmorency tart cherry from 18 days before full bloom to fruit ripening is as follows (Tukey and Young 1939):

Three main tissues compose the ovary wall: the inner and outer epidermis, the stony pericarp, and the fleshy pericarp. The stony pericarp may be divided into an inner and an outer layer of small, thin-walled parenchyma, a middle region of large thin-walled parenchyma, and an outer or hypodermal layer of collenchyma.

Fleshy pericarp and stony pericarp are derived from distinct groups of cells, which are early separated from each other by characteristic size, shape, and periodicity of division. The inner layer of stony pericarp is derived from the inner epidermis together with a few adjacent cells bounding the inner ovary wall. The outer layer is derived from the pericarp, and the cells of which it is composed are elongated at right angles to those of the inner layer.

Cells of the fleshy pericarp increase in number during the prebloom stage and the first few days after full bloom, and enlarge during the next few days. Cell walls become progressively thicker and by the end of the period the maximum number and size of cells is attained. During the next two weeks the walls thicken and harden greatly.

Epidermal cells are elongated radially 18 days before full bloom. They increase rapidly in number during the prebloom stage and during the first 10 days after bloom; increase in size and wall thickness during the next 10 days; change but little during the following 16 days; and then greatly enlarge tangentially in the last 3 weeks.

Stomata are fully differentiated 18 days before full bloom. The guard cells increase in size as the fruit develops, but less so than epidermal cells.

LOW TEMPERATURE INJURY

Cold may injure buds and flowers, woody parts above ground, and roots. The damage may be due to extremely low temperature effects on mature tissues, but more often, probably, it is associated with immaturity of tissues.

Low Temperature Injury to Mature Tissues and Roots

Temperature in a winter in New York continued low for 3–4 days in December. For 24 hr on some of these days it remained below −17.8°C (0°F) and at times fell to −28.9° to −31.7°C (−20° to −25°F) in many places. Again in February, a period of excessive cold swept the Hudson River Valley where temperature reports ranged as low as −37.2°C (−35°F). The subsequent injury to fruit trees probably was due more to prolonged cold than to the very low temperature, which lasted but a few hours (Anderson 1935).

Relatively few cherry trees were killed outright by the cold. They produced their blossoms and small leaves normally. Near the end of bloom, after the fruit had set, however, injury began to show. Small leaves began to shrivel and die, limbs and branches appeared dry, and the bark shrunken. The trees apparently functioned on the food stored in the surrounding tissues. They died when this food was exhausted, being unable to obtain nourishment from the roots. Some trees matured their crop and died; a few lingered on until late summer. The peak of injury, however, came soon after the fruit had set.

Young trees, particularly those on mazzard roots in light sandy soils low in organic matter, suffered severely from winterkilling of the roots. In soils with a good supply of organic matter the losses were less. Older trees with deeper roots came through the winter better, but many of them sustained severe damage to the surface roots. Some such trees came back to normal; others died in a few years. Vigorous cherries on mazzard roots on fertile, well-drained soils suffered very little.

Comparatively little damage from winterkilling of roots occurred on trees on mahaleb. However, losses from other causes had occurred annually in the past to a greater extent than with trees on mazzard roots, particularly in the older orchards. Excluding orchards on light sandy soils, losses were not as great over a period of years with mazzard as with mahaleb roots.

A poorly drained cherry orchard always presents a serious problem to the grower. The trees suffer from intense cold periods regardless of the rootstock.

Injury Attributable to Premature Defoliation

If life processes are interrupted by cold early in autumn the percentage of leaves persisting gives a good indication of the amount of winter injury to expect; but when the trees mature normally before cold weather commences, then at no time in autumn does time of leaf fall show any marked relation to winter injury. The earlier that premature defoliation occurs the greater is its adverse effect on quantity of nutrients stored in the tree. If such defoliation takes place repeatedly in successive years, its effect on the crop will ultimately be apparent. If the stored nutrients are low, severe winter cold may injure the branches and cause many trees to die in the next few years (Chandler 1951).

In Pennsylvania, complete defoliation increased the susceptibility of Montmorency trees to cold injury and delayed bloom the following spring. Trees completely defoliated by Aug. 10 were more severely injured than those completely defoliated by Sept. 1; 50% defoliation by Aug. 10 did not increase the amount of injury; nitrate of soda, up to 385 kg per ha (350 lb per acre), applied on Aug. 1, did not increase the amount of injury on trees 50% defoliated by Aug. 10; the same application, however, increased cold injury on trees completely defoliated by Aug. 10 (Kennard 1949).

Defoliation due to leaf spot can be checked by proper spraying. However, defoliation may be caused by other factors such as drought, pests, and spray injury. Defoliation caused by these several factors may occur very early in the growing season and may result in increased susceptibility to winter injury.

Blackheart.—This causes the death of more cherry trees than all other factors combined and is a main factor in determining the average life of an orchard. Premature defoliation by leaf spot contributes to blackheart.

FIG. 5.7. ALFALFA MULCH IN A SWEET CHERRY ORCHARD

The wood may fail to ripen properly. Sometimes the death of trees can be traced to blackheart injury that occurred several years previously, i.e., a lingering death.

Blackheart is typically a late fall or early winter form of injury occurring chiefly in the sapwood in various parts of the tree. Cambium often is uninjured. Affected parts in mild cases may be orange to light brown in color; in more severe cases, dark brown to almost black. Frequently the affected tissue rots.

New tissue may form over the affected area and prolong life. In some cases, more commonly with young trees, at least partial recovery may be attained. Growth is usually checked through blackheart by both drought and starvation, even though the soil is normally supplied with moisture and fertility. Conduction of moisture and food materials is reduced. Spread of the trouble may keep pace with the formation of new tissue on the outside, and partial recovery is often followed by recurrence of the trouble. A dry summer or a heavy crop may hasten the death of the tree.

Application of N fertilizer and cultivation may help overcome the injury if it is not too severe. Delay pruning until it is possible to determine the injured wood so that the cuts can be made to healthy wood only.

Dieback.—Killing of the tips commonly occurs, uniformly distributed throughout the top, when severe winters follow a season of premature defoliation. It may also be associated with low temperatures following lack of maturity due to late growth. Inherent tenderness of cultivars is also involved (Bradford and Cardinell 1926).

Injury Attributable to Late Growth

Bark Splitting.—Young trees that grow in late in the fall are somewhat

FIG. 5.8. CHERRY TREE COVERED WITH LIGHT FISHNET COVERING TO PROTECT THE FRUIT FROM BIRDS

more susceptible than older trees to this form of winter injury. The injury to immature tissue usually occurs from sudden drops in temperature. Do not confuse this form of injury with normal growth cracks due to gradual increase in size, nor with southwest injury which occurs in late winter or early spring.

Bark splitting injury appears as regular, slightly ragged, deep cracks or splits on the trunk, usually near the ground. The cracks appear in late fall or early winter. The edges of the bark curl upward, thus exposing the underlying tissues. Sometimes the bark may separate from the sapwood on either side of the crack.

The cracks may be deep, or they may not extend as deep as the sapwood. Injury from bark splitting is itself a rather minor matter, but if accompanied by blackheart it may be serious.

A little bark or wood death may pass unnoticed for several years. If the area involved is small it may be covered by new growth and no real harm result. Often, however, a complication, slow in its operation but serious in its consequences, sets in. Certain fungi start growth upon the dead tissue, thus gaining entrance to the tree; and once established they may invade sound wood. As the tree grows the fungus advances, eating out the wood and making a hollow-hearted or "punky" tree or limb. If the tree slackens in growth the fungus gains. A combination of rain, wind, and weight of crop may cause the breaking of a branch or of the trunk.

To prevent bark splitting, direct cultural practices toward early maturity of wood. As a remedy, promptly tack down edges of the bark that separate from the wood, thus preventing drying out of bark and wood and promoting healing.

Crotch Injury.—Acute-angled crotches are particularly susceptible to low temperature injury.

Three distinct tissues, or groups of tissues, may be involved in crotch injury: the sapwood on the inside, the cambium in the middle, and the bark on the outside. The trouble may occur not only at the base of the crotch but may also extend up the branch. Crotch tissues mature relatively late and slowly. The injury from cold usually is due to immaturity. The darker color of the bark at the crotch may accentuate the injury. If the sapwood or all the cambium in the area, or both, are killed, severe rotting may soon begin. If the cambium is injured, healing may occur slowly from callus formation of the living cambium. Some living cambium may persist even when the bark is killed.

Wounded crotch area heals slowly, for here the movement of nutrients is somewhat restricted even under the best conditions. The main crotches may be greatly weakened and, with the heavy crops of fruit of subsequent years, may give way. The tension created by the weight of the branches is not favorable for rapid healing.

Crotch injury is commonly serious where several limbs leave the trunk at about the same level. As the tree grows older and the limbs larger, bad crotch conditions develop, not only in the interior angle but also on the sides of the limbs where it adjoins another. The condition may lead directly to rotting and splitting of the trunk into several segments. Under extreme conditions the killing may be extensive enough to girdle the branch on three sides, leading to progressive dying of the portions in a line with and above the injured area. This condition may involve subbranches coming out along this line and sooner or later lead to borer and wood-rot infection. These conditions have accelerated splitting down of many open-center trees.

Remedial measures consist chiefly of propping, wiring, or pruning. Remove the dead bark as soon as the injury becomes apparent and treat immediately with a wound dressing. If splitting is prevented, the injured crotch may function reasonably well for some time. The importance of developing a strong framework when training the young tree is emphasized.

Collar or Crown Injury.—The tree does not mature simultaneously in all its parts. The crown (part of the trunk near the ground), crotches, and tips of the branches mature last. Consequently, a sharp autumnal or early winter freeze may find part of the tree hardy because those parts are mature, and other parts tender because they are immature.

Injury Attributable to Diurnal Temperature Differences

Southwest Injury.—During the afternoon in late winter, the sun, by both the direct rays and reflection of the rays from snow, warms the tissues of the south and southwest sides of the trunk and low branches. Then with the coming of darkness the temperature falls rapidly. If the temperature change is not very marked any injury encountered will be localized to small areas that may callus over without much damage. If, however, the temperature fall is very marked the injury may be extensive, and healing slow or even impossible. In severe cases, both bark and sapwood may be killed, in effect like a partial girdling.

Preventive measures are to provide shade for the susceptible sides of the trunk by heading the trees low, by nailing boards or shingles to the trees or driving them into the ground on the southwest side of the tree, by wrapping the trunk with suitable material, or by painting the trunk with a white latex mixture. Remedial measures are to clean the affected areas and cover them with wound paint to stop drying out and rotting.

Southwest winter injury, frequently called "sunscald," should not be confused with the sunscald resulting from high and destructive temperatures in and underneath the bark, produced by the sun's rays in midsummer.

Southwest injury may afford a ready means for the entrance and work of borers.

Root Killing

Trees on dry, sandy knolls, ridges, and windy spots are very susceptible to root killing. Absence of snow covering on such sites lessens protection for the roots. Root killing is frequently associated with poor soil drainage. If a tender rootstock is used, hardy cultivars suffer equally with tender cultivars.

Although low temperature is a primary cause of root injury, secondary factors may also be involved. Excess water in the soil over a long period or a highly fluctuating water table may injure the roots directly or so interfere with normal plant functions as to contribute to winterkilling. Impervious subsoil may cause roots to grow near the surface in a position more exposed to freezing injury. Heavy cropping or early defoliation leaves the tree in a more susceptible condition than otherwise.

The following four stages indicate the nature and degree of injury from rootkilling: (1) The trees are killed without foliage appearing. This effect is associated with extremely adverse conditions during the current winter. (2) Trees die during bloom. (3) Trees produce foliage and then die at the first indication of dry, hot weather. At that time transpiration may exceed the water supplied by the roots, and wilting occurs during the day with slight recovery at night. When rains are excessive in June, the effects of winter injury may not be recognized until the hot days in early July. (4) Trees die after 1, 2, or 3 years at any of the above stages. They produce abundant fruit the first year with little terminal growth. Thereafter, growth is very slow. "Staghead" (dead branches in the top of the tree) is apparent. Sometimes trees outgrow the injury.

Large roots are sometimes frozen near the trunk, although smaller roots originating deeper in the soil are undamaged and prolong the life of the tree; however, its vigor is diminished. Some trees injured in this way ultimately recover depending largely on the number of roots left.

Deep planting may not result in less injury because the roots tend to grow upward to, or be replaced at, the same level as trees set at ordinary depth. Mounding may induce crown injury associated with immaturity.

Root injury in winter restricts the absorbing surface and may reduce the set of fruit without appreciably affecting vegetative growth, although both may be affected. Injury to the cambium in the main roots near the crown of the tree by low temperature or girdling by rodents may improve the set of fruits at first, but in the next 1–3 years may be the cause of a heavy shedding of flowers and fruits.

Provide vegetation that holds snow or that by itself is a good insulator. Induce good, but not excessive, vigor by early applications of N fertilizer. Do not expose the crown of the trees late in the year. If sod is removed, cover the crowns of the trees with grass. Occasionally, injury to roots from a very late freeze will follow very early spring cultivation.

In New York, young Montmorency trees growing in soils with a high population of the parasitic nematode *Pratylenchus penetrans* were less winter hardy than similar trees established in sites treated prior to planting with a nematocide. The small feeder roots on trees planted in soils with a high nematode population were attacked. Fungi that invaded the roots through small injuries made by the nematodes caused further damage. These trees suffered because they could not absorb adequate water and nutrients, and their metabolic processes were reduced. They did not develop maximum winter hardiness and the trees were subject to such unfavorable environmental factors as low winter temperatures. Trees on mahaleb stock seemed more winter hardy than similar trees on mazzard stock (Edgerton and Hatch 1969).

Injury to Flower Buds and Flowers

Developing buds of Montmorency may be killed by low temperatures at three different stages: while dormant; soon after growth is resumed within the bud in spring, and in the prebloom or bloom stage. Susceptibility to injury at any of these stages is not closely correlated with the degree of differentiation or development that has been attained. Loss of flower buds in any one year reduces the crop, but injury to the tree itself is more serious (Gardner 1935).

Tart cherry buds are hardier than those of sweet cherry but are less hardy than apple buds. The amount of bud injury at a given relatively low temperature cannot be predicted. In Michigan, a delayed winterkilling of flower buds, which were slightly advanced from dormancy, was responsible for the killing of six times as many Montmorency buds as were killed by much lower temperatures when the buds were dormant. A critical stage was three to four weeks before bloom. At this time the buds were one quarter to one half the size they reached before opening into flowers, and −2.2°C (28°F) temperature might have reduced the crop.

Bing sweet cherry fruit buds averaged 2.2°C (36°F) hardier than those of Elberta peach in Washington state. The difference was least in late fall and early spring, and greatest during winter. High temperature caused the minimum hardiness level of peaches to rise by midwinter; cherries not until late winter or early spring. Cherries hardened more rapidly than peaches during cold days. During dormancy cherry and peach buds hardened when the temperature was below −1.1°C (30°F) and dehardened when above that level. Hardening rates up to 1.9°C (35.4°F) per day and 2.2°C (36°F) were observed. Cherry buds hardened by cold days lost 6.1°C (43°F) of hardiness in 4 hr when exposed to 23.9°C (75°F) (Proebsting and Mills 1973).

At full bloom some damage to sweet cherries may result at −1.7°C (29°F); the lower the temperature the greater the damage. Tart cherry flowers perhaps have 0.5° to 1.1°C (1° to 2°F) more tolerance until a critical point is reached in flowers of both sweet and tart types. In both types, flower buds showing color withstand 1.1° to 2.2°C (2° to 4°F) lower temperature than open flowers.

The same general pattern of tissue injury to young developing fruits occurred in sweet cherry as that given for apricot (Chap. 7). However, injury at the base of the fruit was not as intense as that of the apricot. Stark Gold, which was three days later in development than Starking Hardy Giant, was less severely injured because the floral cup provided some protection. It also had four to six hypodermal layers compared with two layers in Starking Hardy Giant (Simons 1969).

Cultural Practices in Relation to Hardiness

Cultural practices may adversely affect hardiness, or even accentuate the effects from adverse weather, in several ways: late cultivation is associated with an immaturity form of injury; tender rootstocks, with root killing; failure to control pests or poor drainage, with a general weakening of the tree that may require several years for pronounced injury to occur. Excessively heavy dormant pruning, summer pruning, use of N fertilizer in late summer or early fall, and spraying with oil when the temperature is below 4.4°C (40°F) are other operations that may retard the ripening of tissues and result in winter injury. Cover crops tend to check growth in the fall.

Courtesy of Friday Tractor Co.

FIG. 5.9. MECHANIZED CHERRY HARVESTERS: (TOP) CONVENTIONAL LIMB SHAKER; (BOTTOM) TRUNK SHAKER SYSTEM

Weather usually is more significant than cultural practices in causing late growth. A rainy summer or fall may induce tender growth. Rain, furthermore, may favor activity of pests that destroy foliage and weaken the tree, thus increasing the possibility of winter injury. However, heavy summer rains do not of themselves promote immaturity leading to winter injury, especially if followed by a dry, late summer. Cold in early fall that causes destruction of foliage may prevent proper ripening of wood. Leaving a heavy crop unpicked may exhaust the tree and make it subject to winter injury.

Influence of Large, Deep Bodies of Water

Basic principles on this subject are discussed elsewhere in the text. Most commercial cherry plantings in the east are near the Great Lakes.

In the Hudson River Valley of New York, in order to obtain the tempering effect of the river on late spring frosts, cherry plantings are confined to a narrower belt near the river than in the case of apples. Ordinarily the river below Albany freezes over for only a short period during the middle or latter part of the winter, and often not at all. Thus, when a winter occurs with such extreme cold that the river is frozen over for a long period, its tempering effect on the fluctuating temperatures of the surrounding area is diminished.

Rest Period

Temperatures must be low enough to break the rest period, which is a factor of significance in regions with warm winters (see Prolonged Dormancy under Peaches, Chap. 4).

In many springs in California, after relatively warm winters, Lambert, Napoleon, and Bing show a marked delay and irregularity in bloom, but Black Tartarian, Burbank, Chapman, and Republican show little effect. In coastal valleys adjacent to Monterey Bay, after unusually warm winters, Lambert may be in full bloom at the same time that Black Tartarian is ripe. Under the same climatic conditions, Napoleon may show all variations from blossom to ripe fruit (Philp 1947).

SOIL MANAGEMENT

Cultivation

Most commercial cherry orchards are cultivated with a disk-harrow during spring and early summer, starting just as soon in spring as the ground can be worked to advantage. But the possibility of reducing frost damage during bloom by a temporary hard surface of the soil may warrant consideration. In a cherry tree of normal vigor, the leaf surface, the terminal growth, the set of fruit, and the initiation of fruit buds for the following year have been made or are determined by midsummer. Cultivation must be started in early spring if it is to contribute to these processes.

Cease cultivation during the latter part of early summer if rainfall has been normal. In a season with light rainfall, the orchard may well be cultivated into midsummer. Some growers feel that their orchards must be cultivated twice a week, but if the soil is thoroughly worked early, three subsequent cultivations may be as effective as more.

Cover Crops

A cover crop sown in late summer or soon after the crop is picked helps to prevent erosion, and prevents some loss of soil nutrients by leaching. By using water and nitrates in the soil it helps to mature the wood during the fall, thus making it more resistant to winter injury. By catching and holding snow it may reduce the amount of root injury.

No one cover crop is best for all conditions. In some orchards a good stand of weeds following cultivation makes a satisfactory cover crop and serves to cut production costs.

Rye (132 litres per ha [1 1/2 bu per acre]) must be disked early in the spring before it becomes too tall and old, so that it will not compete with the cherry trees for moisture and nutrients. Rye and vetch may be even better.

Sod and Mulching

In Ohio, 1/2 of a new Montmorency orchard was grown under a system of sod and mulch. The other 1/2 was cultivated each spring, after which a soybean cover crop was sown in mid-June, followed by an overwintering cover crop of rye sown in late September. A 20% N fertilizer (151 g [1/3 lb] for every year of tree age) was applied to each tree each spring.The growth and appearance indicated that the trees in both treatments were receiving adequate N for optimum development. During the first 5 years enough mulching material was secured by raking the mowed grass up around the trees. Beginning in the seventh year additional mulch was applied by hauling in straw. A sufficient layer was used to suppress the growth of grass and weeds under the branches of the trees (Judkins 1949).

Growth of the trees was similar in each treatment during the first 5 years. Then mulched trees began to be slightly larger than the cultivated trees. After 11 years the trunk circumference of the former averaged 9%

larger, but this had not affected the yield. During the 6th, 8th, and 11th years, rainfall was below average during the 5 months of the growing season from May through September. Soil moisture did not appear to be a limiting factor, however, except possibly in the 11th year when a slightly reduced growth rate occurred in all plots.

The first crop of cherries was picked when the trees were 5 years old. The orchard produced rather erratically during the next 6-year period because of spring frost damage to the blossoms. No difference in yield occurred between the trees in the sod plus mulch and the cultivated sections of the orchard. Thus, tart cherry trees may grow and produce fruit as satisfactorily under a sod plus mulch system of soil management as when land adapted to cherry production is cultivated. Cherry growers whose orchards are located on sloping sites susceptible to erosion might well consider the use of a sod plus mulch system of soil management.

FERTILIZER

As with other orchard crops, carefully note the twig growth, color and luxuriance of foliage, and fruit production. Let these factors along with soil and leaf analysis determine the need for fertilizers. Soil analysis can be very helpful in indicating the nutrient status of the soil, but leaf analysis is the most accurate method of determining the plant food requirements of the tree.

Leaf samples collected from 79 cherry orchards in Utah gave these mean values: N, 2.3%; P, 0.25%; K, 1.4%; Mn, 0.39 ppm; Cu, 11 ppm; Fe, 61 ppm; and Zn, 19 ppm. Nutritional deficiencies were not observed in any of the orchards (Christensen and Walker 1964).

Levels of nutrients considered adequate in the leaves vary from one season to another, and from year to year. A general guide used in Ontario for making fertilizer recommendations for tart cherries from leaf analysis is as follows: N, an adequate range, 2.2–2.8%; P, a minimum of 0.15–0.20%; K, minimum 1.3–2.3%; minimum levels for Ca and Mg, 1.2 and 0.35%, respectively.

Nitrogen

As a general rule, apply about 0.023 kg (0.05 lb) of actual N for each year of the tree's age, e.g., a 10-year-old tree would be given 0.23 kg (0.5 lb). Older trees often require up to 0.45 kg (1.0 lb), depending on soil conditions and tree vigor. Apply more N in sod orchards than in cultivated ones for the first 4–5 years.

The fertilizer is usually applied in early spring about 2—3 weeks before bloom. Distribute it around and under the trees to just beyond the spread of the branches.

In Washington, annual applications of N fertilizer alone, and in combination with P, to a mature sweet cherry orchard for 6 years increased the N content of leaves and fruit. Injury to the fruit buds by low temperatures had more effect on yield than did the fertilizers applied (Stanberry and Clore 1950).

In Colorado, on alkaline soils of rather low fertility, the use of N alone gave the greatest increase in yield of tart cherries.

In New York, N applied in the spring to 16-year-old Montmorency trees increased the first year the number of shoots 15 cm (6 in.) or more long, the trunk diameter, and the average size of leaf; and after 2 years, it increased the yield (Tukey 1927).

In Michigan, N applied, especially in the fall, to 10-year-old Montmorency trees growing in a rather infertile, sandy loam increased the number of fruit buds per spur and the length of shoot growth, invigorated old spurs, promoted a larger bearing surface on the trees, and increased the yields. A lack of N at bloom was not a factor in fruit setting. Application of N fertilizer is usually particularly beneficial when cherry trees reach an age where they decline markedly in vigor (Wann 1954).

Phosphorus

What has been said about P for apples and peaches also applies generally for cherries.

In Washington, application of P fertilizer increased the P content of both leaves and fruit. Increasing the P content of the fruit had no effect on maturity or keeping quality (Stanberry and Clore 1950).

Potassium

Tart cherry trees affected with leaf curling had a smaller percentage of live feeding roots and fewer large roots than trees with normal foliage in Wisconsin. Roots of trees with leaf curling were darker in color than normally. Trees fertilized with K showed less leaf curling and greater root and shoot growth than unfertilized trees (Langford 1939).

K deficiency as it developed in Wisconsin, was first noted by an upward curling of the older leaves on spurs and terminals. In the more advanced stages, the foliage became bronze on the under surface and finally the margins of leaves became necrotic. Trees which had been suffering for several years had undersized leaves, and by midsummer there was a noticeable lack of leaves in the upper portion of mature trees. Terminal growth was only slightly limited in the early stages but as the severity of deficiency increased, terminal growth practically ceased (Gilbert 1956).

Do not expect as rapid a response to K in correcting "curl leaf" symptoms as is usually obtained from N applications on N-hungry trees. Spring-applied K probably will not correct the sumptoms fully the first season. Apply corrective applications of K at any time, soon after diagnosis symptoms or leaf analyses. After correcting the K deficiency, reduce fertilizer applications to a K maintenance rate but include P since the P levels are also often on the low side. A 0-10-30 or similar fertilizer is suitable. N must not be omitted from the fertilizer program when K applications are necessary for the correction of a deficiency of K.

Recovery from marginal leaf scorching and rolling in a 6-year-old Montmorency block in New York resulted from liberal application of K. The trees responded to K when the K content of the leaves was below 0.75% on a dry weight basis; 0.9−1.1 kg (2−2 1/2 lb) of potash per tree increased the K in the leaves 300−400% (Boynton 1944).

Leaf curling has prevailed in some orchards because the grower has had little means of detecting its approach. As a result, yearly applications of sufficient N are made but without the use of K fertilizer or mulching until the trees show definite leaf curling. At least an 0.80% level in leaf analysis was necessary to be sure that the trees did not suffer from leaf curling.

It has been suspected that liberal applications of K might alter the soluble solids/titratable acids and consequently lower the sweetness in sweet cherries. A 5-year fertilizer test was conducted in New York to study the effect of K on soluble solids, titratable acids, fruit size, and leaf K content of Windsor sweet cherry. Leaf analyses showed that the K uptake was very efficient during the first year of treatment. The leaf K increased steadily in the K-treated trees as the test progressed and the increase became more significant. The titratable acids content was slightly higher with K fertilization, but K had no effect on soluble solids. Fruit size was improved when the leaf K of August samples approached the 1.25% level (Kwong 1965).

NPK and Tart Cherry Firmness

Softening of fruits during maturation is related to changes in the fruit pectin composition. This relationship involves a conversion of water-insoluble pectic substances to water-soluble forms. A loss of tissue cohesiveness, with subsequent separation of adjacent cells, is attributed to this conversion. The overall effect is a loss of firmness or fruit softening. Other alcohol-insoluble substances, e.g., starch and cellulose, may also function in fruit firmness (Curwen *et al.* 1966).

Increasing levels of N in Pennsylvania were associated with firmer Montmorency cherries having a lower juice loss on pitting and higher maturity induced by high N levels. The role of P in the pectic composition and fruit firmness relationship was not clear. Increasing levels of K resulted in softer fruit having a higher juice loss on pitting and a reduced content of water-insoluble pectic substances. A reduced fruit Ca content was associated with high K levels. High K levels induced a deficiency of fruit Ca which, in turn, resulted in a reduced water-soluble pectic content. The cumulative effect was softer fruit.

Minor or Trace Elements

Zinc.—Zn deficient symptoms in cherries resemble those described for peach. The fruit is smaller than usual. Late dormant or fall foliage sprays of zinc sulfate (29.5 g per litre [25 lb in 100 gal.] water) at 2–3 year intervals may correct this deficiency. In severe deficiency, both dormant and fall foliage applications may be required. The recommendations for peach should also be tried.

Manganese.—Mn deficiency, shown as interveinal chlorosis, may occur when leaf samples contain 17–18 ppm or less Mn. An annual early summer spray of manganese sulfate provides some control.

Boron.—B deficiency may result in hard, shriveled, blotchy fruit. Where necessary apply 113–227 g (1/4–1/2 lb) of borax on the soil under a large tree. Too heavy a rate of B may cause damage.

IRRIGATION

The water requirement is relatively heavy during spring and early summer. Most new growth is made while the fruits are developing, and an adequate supply of moisture at this period is important. Prolonged growth late in the year is not desirable in cherries, but do not let the trees suffer from a marked deficiency of water after the crop has been harvested.

Cherries are very susceptible to "wet feet." Do not overirrigate.

In the west, start the season with soil moisture at field capacity to the full depth of rooting and then irrigate only when soil samples from 0.61 to 0.91 m (2 to 3 ft) deep show dry soil. On most soils there is no need to irrigate if the soil sample can be squeezed into a ball that holds its shape when jarred slightly. If the soil crumbles after squeezing, it probably needs water. About four days after irrigating, sample again to be sure the soil is not waterlogged and adjust the irrigation schedule accordingly.

WEATHER AND CULTURAL EFFECTS

Tart cherry size, soluble solids content, and processed yield vary widely according to the year, the effect of the year exceeding the effects of both cover crop and fertilizer level (Harrington et al. 1966). Excessive moisture during the final growth stage may increase cherry size and reduce soluble solids (Tukey and Tukey 1966).

The canner seeks the maximum number of cases of canned cherries per tonne (1.1 ton) of fresh fruit. Total processed yield (dry weight/fresh weight) depends largely on sorting losses, pitting losses, and shrinkage of fruit during cooking.

CRACKING OF SWEET CHERRIES

Many sweet cherry growers have produced an excellent crop up to the picking stage only to have it completely ruined by cracking of the fruit.

In the irrigated valleys of eastern Washington, where the average annual production of sweet cherries is about 19,845 tonnes (22,050 tons), losses from cracking have exceeded 50% of the crop in some cases in the most susceptible cultivars. In individual orchards, losses of 85% or more have been experienced. Similar losses sometimes occur with certain cultivars, such as Lambert in the east.

Basic Considerations

Although early explanations of fruit cracking suggested that the condition was directly related to an excess of soil moisture likely to occur under irrigation practices, soil moisture has little influence on cracking. Cracking is commonly due to an osmotic absorption of water through the skin of the fruit, and is affected by the osmotic concentration of the fruit juice, turgor of the fruit, temperature of the water, and skin permeability. These factors may, in turn, be affected by excessive transpiration rate, inherent cultivar characteristics, and the local climatic conditions.

Anatomically, cracking of fruit has been attributed to various structural weaknesses (Sawada 1934), amount of soluble solids in the fruit, differences in colloid content of cultivars, and decrease in acidity and astringency as fruit approaches maturity (Kertesz and Nebel 1935).

Control

Mineral salts, whether injected in solution or inserted dry into tree limbs, have made no difference in cracking tendency. Temperature increases cracking in the range above 0.6°C (33°F) and sugar content of the fruit to about 21% increases the cracking index. The cations (calcium, copper, iron, aluminium, thorium, and uranium) reduce cracking when the fruit is immersed in various chemical solutions. The cracking index has been reduced as the valence of the cation increased as long as the anion was monovalent. Sprays of the sodium salt of naphthaleneacetic acid reduced cracking when applied at 0.1−1.0 ppm 3−35 days before harvest (Bullock 1952).

Cracking may be reduced to some extent by shaking the water from the tree soon after rains; by mechanically blowing water off the fruits with the downdrift from a helicopter; or by planting highly resistant cultivars.

Cultivar Differences

Different cherry cultivars show marked differences in tendency to crack during rainy periods. Individual fruits of the same cultivar also show equally great differences in this respect. In Bing, which is highly susceptible to cracking, some fruits crack during the first 1/2 hr of rain whereas other fruits, on the same tree and at a comparable stage of development, remain sound after the rain has continued for 12 hr or longer.

Cherries differ in their susceptibility to cracking because of differences in rate of absorption of water (due largely to differences in permeability of the fruit skin), and because of unequal capacities for expansion of the peripheral tissues to accommodate the increased fruit volume that results when water is absorbed. Cherries with a rapid rate of absorption and a small capacity for expansion crack readily; those with a slow rate of absorption and a large capacity for expansion tend to be immune (Verner and Blodgett 1931).

Both the degree of permeability of the fruit skin and the plasticity of the peripheral tissues may be subject to modification. Any treatment that decreases the rate of absorption of water, or increases the capacity of the fruit tissues to stretch without rupturing, should reduce the amount of cracking. But the most effective way to avoid cracking is to plant highly resistant cultivars.

Cracking, Optimum Maturity, and Birds

Records were taken on splitting (cracking) and bird damage in a sweet cherry orchard near Vineland, Ontario. In 3 out of 4 cultivars there was a net gain through leaving the fruit on the tree to the optimum maturity stage (Table 5.5). During the period in which records were taken there was relatively favorable weather for splitting (Upshall 1943).

TABLE 5.5
BALANCE SHEET FOR FINAL WEEK BEFORE OPTIMUM MATURITY
OF SWEET CHERRIES

Cultivar	Crop	Loss of Fruits by Count (%)		Size Increase by Weight (%)
Black Tartarian	Heavy			Gross 32.0
		Splitting	0	
		Birds	2.0	2.0
		Net gain from leaving on tree		30.0
Schmidt	Light			Gross 33.0
		Splitting	6.0	
		Birds	10.0	16.0
		Net gain from leaving on tree		17.0
Windsor	Heavy			Gross 19.0
		Splitting	23.5	
		Birds	3.5	27.0
		Net loss from leaving on tree		−8.0
Hedelfingen	Heavy			Gross 13.0
		Splitting	5.0	
		Birds	0.5	5.5
		Net gain from leaving on tree		7.5

Source: Upshall (1943).

The figures show that it would have been unprofitable to pick Black Tartarian, Schmidt, and Hedelfingen before the optimum maturity date. The loss incurred by deferring the picking of Windsor to the optimum stage was largely due to splitting caused by rains in the late afternoon of 2 days when the fruit was still about 5 days short of optimum maturity. The cherries in the tops of the trees were the ripest and most damaged by splitting and birds.

Robins and starlings often cause much loss by removing whole cherries, picking out portions of the riper fruits, and knocking fruit from the trees with their wings. Early cultivars and small plantings usually suffer most.

No one has yet found or developed a practical and economical method or combination of methods which suffice to prevent bird damage to the ripening fruit. Scarecrows, shot guns, acetylene contrivances, glittering and shining strings of metal and glass, or nearby mulberry trees provide only slight, if any, control. The use of various sound-effect devices appears to be no more permanent in effect on bird control than other attempts have been.

HARVESTING AND MARKETING

The small size of cherry fruits makes the cost of hand picking comparatively high. Cherry picking in New York begins about July 1 and continues 4–6 weeks if both sweet and tart cultivars are grown. To the general orchardist, this period may fill in the gap between small fruits and other tree fruits.

With Montmorency an increase in night temperatures increased the growth rate of the fruit following bloom, and shortened the period following bloom during which the embryo begins to grow rapidly. Rate of the fruit following bloom, and shortened the period following bloom during while the embryo begins to grow rapidly. Night temperatures of 15.6°–18.3°C (60°–65°F) enabled the fruits to complete the final swell in

FIG. 5.10. CHERRIES ARE BEING FLUMED FROM SOAK TANK INTO TANK OF THIS TRUCK USED BY A PROCESSOR TO MOVE CHERRIES FROM RECEIVING STATIONS TO PROCESSING PLANT

16 days—nearly twice as long as required with night temperatures of 32.8°C (91°F). Cherries produced under the former conditions had 13.4% sugar compared with 9.7% for the latter (Tukey 1953).

Hand Picking

Cherries may be picked with "stems on" when the fruit is intended for fresh fruit markets. The picker grasps the stems of several fruits in a cluster and strips them from the spur, touching only the stem with the fingers. An adept picker can do this rapidly without loosening the stems from the fruit and without damaging the spurs. If the pickers break off many fruit spurs the next year's crop may be reduced.

Cherries for processing may be "pulled," leaving the stem on the tree. Some juice exudes from the fruit when picked in this manner. This is not objectionable when the fruit goes to the processor and only a short time intervenes between picking and delivery to the plant.

If the fruit is picked by hand, payment is by piecework by the kilogram (pound). A fast picker may strip 181 kg (400 lb) a day from trees bearing a heavy crop. The Montmorency picking season in a given area lasts 2–3 weeks; a good yield is 6.8–9 tonnes per ha (3–4 tons per acre or more).

With Montmorency, 254 kg (20 lugs [about 12.7 kg or 28 lb per lug]) is a good day's work for a hand picker. A Michigan hand picker is reported to have picked 49 full lugs in 12 hr, or a total of 600 kg (1300 lb).

Do not pick the fruit until it is mature, and then remove it promptly. Cherries usually increase rapidly in size and weight during the last few days before full maturity is reached. With the fruit ripening uniformly so that the trees can be stripped at 1 picking, it may be best to have as many as 4 or more pickers per ha (10 or more per acre).

Sweet cherry trees become larger than tart cherry trees, and so more ladder work may be involved. About twice as many pickers per ha (acre) are usually needed for sweet cherries as for tart cherries. Under good cultural conditions in sweet cherry areas in the east, 15-year-old trees often average 71–95 litres (75–100 qt) per tree.

For processing purposes, sometimes cherries are loosened from the tree with the fingertips and allowed to fall into a net suspended on a frame under the tree.

Mechanized Harvesting of Tart Cherries

Over 75% of the commercial tart cherry crop in the east is now harvested mechanically. To hand pick the Michigan crop alone would require about 50,000 pickers, at a cost of about 3/5 of the price received for the fruit.

Adaptability.—The tart cherry crop is ideally suited for harvest mechanization because: (a) there is only one commercial cultivar (Montmorency); (b) it is almost 100% processed; (c) it is easy to bulk handle in water; and (d) the fruits can be moved fairly rapidly from orchard to processing line (Larsen 1969). Tart cherries, however, are subject to easy bruising, rapid oxidation and scald, which cause color and firmness to deteriorate.

Fruit Abscission.—In the tart cherry, abscission layer formation during maturation occurs between fruit and pedicel. No abscission layer is formed between the pedicel and spur.

The abscission layer is first evident 12 to 15 days before fruit maturity. This layer is composed of five to eight rows of cells in the transition zone between the fruit and pedicel and was first identified by its low affinity for hematoxylin. Cell separation occurred without rupturing of cell walls. Later cell wall collapse was apparent. Cells immediately distal and proximal to the line of separation were thin-walled and prone to separate easily. No abscission layer was formed through the vascular bundles and no cell division was noted during layer formation (Stosser *et al.* 1969).

Presence of the abscission layer in the tart cherry is important in mechanized harvesting. The force necessary to separate the fruit from the pedicel decreases during maturation, a time during which abscission layer formation occurs.

FIG. 5.11. THIS YOUNG SWEET CHERRY TREE WITH THREE MAIN BRANCHES TO PROVIDE A "HOLD" FOR SHAKER ARM IS BEING TRAINED FOR MECHANIZED HARVESTING

Pruning.—The trees should not have low branches which slow the movement of the catching units. Excessive scaffolds impede the entry of the shakers into the trees and add needlessly to the shaking time. Poorly positioned branches are often injured by improper shaker attachment. With low-hanging branches removed the fruit can be harvested in 50% of the time required to mechanically harvest unpruned trees.

Equipment.—In an early, fairly successful stage of development, harvesting machinery consisted of tractor-mounted boom shakers, hand-carried collecting frames and lug box carriers. These gave way, in part, to highly automated self-propelled units with power steering, four-wheel drive, various tilt adjustments for uneven ground or different sized trees, and lights for night harvesting. To reduce fruit bruising, the frames are usually Saran-covered, and the conveyer systems have been greatly improved. The inertia type of branch shaker, mounted on frames, is generally used.

Tree size and shape influence machinery selection. Conversely, the machinery selected influences tree size and shape. Types of equipment include wrap-around, two-piece, and roll-out. Two types of shakers are trunk and limb. All these different items have some advantages and some disadvantages over the others. They all would influence the pruning system a grower would develop. Some of the equipment is very manoeuverable, some less so.

Most catching frames used in harvesting have a self-propelled method of movement (some are hand-carried or tractor-pulled), are of the 2-wing type with 2 units which meet and seal around the tree trunk, and are equipped or optional with decelerator strips of 1–3 layers. Some types of frames feature one or more sections and have wing sloping surface, an inverted-umbrella type, a powered roller, or an inflated pillow. The wing construction is usually of plastic, nylon, mesh (fishnet), Ensolite, canvas, or foam. The catching-surface size of most units is 7.3 × 7.3 m (24 × 24 ft) or somewhat larger (occasionally considerably so).

Fruit recovery can be improved only within certain limits by altering the energy input of harvesting machines. For each crop and perhaps for each tree there is an optimum shaking time, stroke, and frequency, beyond which excessive damage to the tree results.

To convert old trees to mechanical harvesting, it may be necessary to remove 1/3–1/2 of the bearing wood. The conversion is easier in younger orchards, with the aim of narrow, upright trees. The optimum spacing of trees has not been determined but spacing of 5.5 × 6.1 m (18 × 20 ft) is one suggestion.

Shaking.—Serious damage to the trees by machine harvesting can be largely prevented in good equipment by reasonable care in operating the harvester. Long-range effects of repeated shaking of trees remain to be determined.

Do not shake for a long period of time. If the fruits will not come off in the first few seconds those left on the tree may be damaged by continued shaking. If necessary, wait until later in the season so that fruits can be removed with a milder shake even if this means fewer stems where "stems on" are required.

Handling Tart Cherries in Lugs.—Until the early 1950s, handling tart cherries in lugs by hand was the main method for processing purposes. The pickers usually put the fruit into 9.5-litre (10-qt) pails as they harvested it, and then poured the cherries into lugs which held 11.3–15.9 kg (25–35 lb) of fruit. Handling cherries in this way is inefficient because on-the-tree quality is hard to maintain, investment in lugs is high, and management is difficult (Levin and Gaston 1956).

Forklift trucks are now used for handling cherries in lugs. Pick the fruit into pails, pour into lugs, and load the filled lugs on orchard trailers, which have been provided with pallets. Stack the lugs in such a way that a pallet and the lugs piled on it are handled as a unit load. As many as 635 kg (50 lugs [12.7 kg or 28 lb per lug]) can be stacked on a pallet.

Move the loaded orchard trailer to a surfaced area in or near the orchard where a forklift truck can be used to lift the loaded pallets off the trailer. A forklift can be used at the receiving station or processing plant to remove the loaded pallets from the truck. Empty lugs can also be handled on pallets there, and again on arrival at the orchard. Handling cherries in lugs with a forklift saves time, labor, and money, and reduces bruising, congestion, spoilage, and lug breakage.

Handling Tart Cherries in Water.—At the receiving station, the processor provides truck-mounted watertight tanks. Open-top tanks are easiest to load and service. If a closed tank is used, it should be equipped with several hatches of ample size. Coat the inside surface with paint or enamel that will not react chemically with the cherries (Gaston and Levin 1956).

The tank should be large enough to carry the maximum legal load (0.03 m^3 [1 ft^3] of capacity holds about 22 kg [48 lb] of cherries and 5.4 kg [12 lb] of water). Cherries near the bottom may be crushed in a tank more than 1.5 m (5 ft) deep. The tank should be equipped with a system of pipes and valves which makes it possible to introduce water through the fruit, and let it overrun through an outlet near the top. Provide a 15 cm (6 in.) outlet near the bottom of the tank at a point that facilitates unloading the cherries.

The receiving station should have a supply of at least 76 litres (20 gal.) of cold water per minute, for filling the tanks and cooling and fluming out the fruit. Fill the tank with cold water. Then pour in the cherries from the lugs, and continue filling until the cherries are within a few centimeters (inches) of the top of the tank.

While filling the tank with cherries and after filling is completed, keep the intake valve at the bottom of the tank open so that cold water continues to circulate through the fruit and overrun at the top, carrying away some of the dirt, leaves, stems, and defective cherries. Continue the flow of water until the cherries have cooled to 12.8°−15.6°C (55°−60°F). On arrival at the processing plant, flume the cherries from the tank into a receiving hopper. Then remove them to soak tanks. Later, they are processed.

Handling the cherries in water at the receiving station helps to maintain on-the-tree quality, reduces costs, facilitates handling, and reduces the number of lugs required.

At the orchard, the grower must have truck-mounted tanks of sufficient capacity to handle the crop effectively. Besides one tank truck, he needs at least a storage tank or one or more additional trucks into which the cherries can be transferred as they come out of the orchard.

The grower also must provide a loading dock in or near the orchard, where a water supply of at least 19 litres (5 gal.) per min is available. Set up a sorting table or belt in such a way that the cherries pass across it directly into the tank of cold water.

In large orchard operations where cherries are harvested mechanically, the fruit is elevated from the catching apron directly into the water tank.

Handling cherries in water at the orchard has these advantages over other methods: maintains on-the-tree quality more effectively; is commercially feasible; eliminates the use of lugs; benefits both grower and producer. Cherries are given a cold water soak for 4−8 hr prior to sorting and pitting.

Machine Harvesting of Sweet Cherries

There are distinct markets for sweet cherries harvested with "stems off," i.e., without retaining the pedicel, as in Michigan, and with "stems on," as in Oregon, where most of the crop is brined. About 55−60% of Oregon's brine cherry crop is sold with stems for cocktail maraschino cherries.

It is simpler to machine harvest sweet cherries with "stems off" than it is with "stems on." An important reason for this is that no abscission zone occurs in the transition zone between the fruit and spur, where separation from the tree occurs in "stems on" harvested fruit. Even though the fruit removal force (FRF) decreases with maturation in the sweet cherry, considerably more force is required to effect fruit separation, and thus generally poorer fruit removal is obtained than with the tart cherry. Chemical looseners are useful to reduce the FRF through the effect at the fruit-stem zone for cherries harvested with "stems off."

Chemical Looseners.—Several chemicals have been developed which are active in promoting fruit abscission of cherries with "stems off." The most notable of these probably is ethephon which reduced the fruit removal force of sweet cherries. Concentrations of 400–800 ppm provided maximum reduction with minimum phototoxity; lower concentration required a longer time to give response similar to higher concentration; 80% of the chemical was absorbed within a 40 hr period. Fruit removed in the initial vibration by harvesting equipment was of higher quality than fruit removed after prolonged shaking. Ethephon reduces the FRF 40 to 50% and nearly doubled the amount of stemless fruit harvested in the first shake (Bukovac *et al.* 1969).

Control of the abscission process of maturing fruit provided a basis for improved hand and machine removal as well as for programmed harvesting.

Ethephon applied as a foliar spray to sweet cherry trees within two weeks of fruit maturity promoted fruit abscission at the lower (fruit pedicel) zone, as indexed by a reduction in the fruit removal force (FRF). There was no significant effect, at the concentration studied, on abscission at the upper (pedicel:peduncle) zone. Promotion of the abscission with ethephon was time and concentration dependent. Concentrations of 100 to 1000 ppm were effective, with a greater response from the higher concentration. Absorption periods of four and twenty-four hours resulted in responses equal to 73 and 94% of that observed when ethephon was present for the entire test period. Of nine sweet cherry cultivars, all responded similarly in reduction of FRF. Ethephon enhanced fruit enlargement and pigmentation when applied early in Stage III of fruit growth. The increase in weight was most pronounced in the fleshy pericarp tissue (Bukovac *et al.* 1971).

Pruning.—The sweet cherry tree is characterized normally by being larger and with longer, more willowy growth than the tart cherry. There are advocates of different shapes of trees. Some suggest a pyramidal tree with a central leader, which would encourage good exposure to light. Others suggest a vase form with 3–4 scaffold branches, which might last longer under mechanical harvesting. When the trunk becomes too large to shake, one can move up on the scaffold limbs with the shaker.

Develop a clean trunk for at least 0.9–1.2 m (3–4 ft) for ease in closing the seal and to have a good place to attach a shaker. Remove low-hanging branches, bearing in mind that the fruit-laden branches droop considerably. The dropping fruits fall into catching frames free from damage. Remove watersprouts and wood that will not shake well.

Scaffold training and cleanout of the lower scaffold area is important if a limb shaker is used. To maintain high production, the shaker operator must have good visibility and space to work the claw or clamp. This is not a problem if a trunk shaker is used.

Weighing Brined Cherries.—About 45% of the sweet cherry crop is brined. Brining of machine harvested sweet cherries should be started within 1 hr after harvest for best quality. Although some sweet cherries are now brined at the orchard, most are transported in brine from receiving stations to processing and brining plants. Whittenberger *et al.* (1969) studied the feasibility of using the weight to volume relationship, as developed earlier for tart cherries, for measuring quantities of sweet cherries and concluded that sweet cherries in brine could be bought and sold by volume.

Cherries in brine should be transported over roads of normal roughness for about 1.6 km (one mile) before volume readings are made to ensure equilibrium settling. The brine level should be above cherry level when readings are taken. Sweet cherries shrink less than 1% in 24 hr so that if they have been in brine for no longer than 18 hr, no correction factor for shrinkage is necessary. The effect of bruising on bulk density is small but will penalize a grower for severely bruised fruit. A correction factor of about 1% for every 5% of attached stems is recommended. The average bulk density of stemless sweet cherries in brine is about 675 kg per m^3 (43 lb per ft^3) (Whittenberger *et al.* 1969).

Marketable Yield and Quality.—The equipment for mechanized harvest generally costs thousands of dollars. The grower should study costs and returns for his investment in relation to the many factors involved in replacing hand picking.

Strive for removal of at least 90% or more of the fruits in a condition suitable for the market purpose intended. Early in the season machine-harvested fruit may be nearly as good as hand-picked fruit, but as the season advances the machines may greatly increase loss of quality. Also, where some fruit is left on the tree, disease control, e.g., brown rot, becomes more difficult.

Both field and processing or packing plant have some distinct items for consideration. At the brining plant, revenue may be less if there are more stems, more No. 3 and unclassified grades, less quantity suitable for cocktail purposes, more trash and dirty fruits, more plant labor, less packout yield, and less plant production. Where the intended market is for "stems off" fruit, top quality depends to a large extent on sound and well-colored fruit of acceptable size and attractive condition.

Guides to Maturity

Growers often determine when to pick sweet cherries by the aid of such factors as firmness, sweetness, size, and color. Some guides can be measured more precisely than others. Time of development of the abscission layer is a factor, as discussed earlier in mechanized harvesting.

Pressure Test.—This test is not sufficiently indicative of maturity to be a good guide to time of picking. In Napoleon, for example, the decrease in resistance for one entire season was only 17.3%, despite the fact that during this time the fruit passed from a stage of immaturity to one considerably past prime condition. During this period when the fruit could really be considered marketable, the decrease in resistance was only 5.7%. Allowing the fruit to hang a few days after the ripe stage, therefore, does not materially affect firmness.

Specific Gravity.—A rather consistent increase in sugars and other soluble solids occurs during the ripening period of sweet cherries (see also Sugar Content).

Size.—Size is not one of the main guides to maturity. However, certain sizes must be attained before cherries are acceptable for packing in the top grades. Size specifications may vary from year to year and growers should check with their packinghouse regarding sizes for the current season.

In British Columbia, size specifications have been given as follows (Atkinson 1956).

Bing and Lambert: No. 1 Large, 23.8 mm (15/16 in.) and larger, with an undersize tolerance not exceeding 10% by count and not smaller than 22.2 mm (14/16 in.); No. 1 Medium, to below 23.8 mm (15/16 in.) down to 20.6 mm (13/16 in.), with an undersize tolerance not exceeding 10% by count, and not smaller than 19.0 mm (12/16 in.).

Napoleon, Deacon, and Windsor: No. 1, 19.0 mm (12/16 in.) and larger, with an undersize tolerance not exceeding 10% by count, and not smaller than 17.5 mm (11/16 in.).

Color.—Color requirements vary with the cultivar and the purpose for which the fruit will be used. Lambert cherries of a bright tomato red are too immature for fresh fruit shipment. The absolute minimum for Lambert for fresh fruit shipment has some maroon coloring, but this maturity should be avoided. The optimum for the fresh fruit market is a still darker color and for canning is fully mature for all cultivars (Atkinson 1956).

Sugar.—A consistent and rather pronounced increase in sugars occurs during the period of maturity. In Oregon, from the first to the last pickings the juice of Napoleon increased in soluble solids from 12.1 to 22.4%, and of Lambert from 11.3 to 22.2%. Not only was the increase in sugars rapid at the beginning of the period, but it continued at a uniform rate until the fruit was past its prime (Hartman 1925).

In British Columbia, minimum desirable total solids content by refractometer has been: Lambert, 17%; Deacon, 18%; Bing, 18%; Van, 18%; Napoleon, for SO_2 processing, 15 to 17%, and for canning, 18% or more.

In Michigan, a 3-year study with Montmorency indicated that soil management practices which increased tree vigor and resulted in a lower soluble solids content of the fruit were dependent on the season, particularly in relation to soil moisture supply at or prior to harvest. A deficiency of soil moisture and N decreased the level of soluble solids in the fruit (Kenworthy and Mitchell 1952).

Also, the soluble solids and total acids content of Montmorency increased, during a three-week period, as the harvest season progressed (Table 5.6). Total solids of the fruit decreased immediately after a rain.

TABLE 5.6
TOTAL SOLIDS CONTENT OF TART CHERRIES FOR THE 20-DAY PERIOD OF COMMERCIAL HARVEST

Tree	Total Solids Percentage of Fruits Harvested on Dates Indicated					
	July 14	July 19	July 23	July 26[1]	July 30[1]	Aug. 2
A	20.3[2]	22.2	23.2	24.7	23.6	25.7
B	21.9	23.9	24.3	26.5	25.4	27.5
C	22.1	21.4	22.2	23.4	22.9	26.0
D	19.0	18.9	20.2	21.0	20.3	21.8
Avg	20.8	21.6	22.5	23.9	23.0	25.3

Source: Taylor and Mitchell (1953).
[1]3.22 cm (1.27 in.) of rain on July 27.
[2]Harvested July 13.

Then the total solids continued to increase again until the fruits shriveled on the tree. Fruits sprayed with inorganic pesticides contain a higher percentage of soluble solids than fruits sprayed with organic pesticides. The effect of fungicides on soluble solids content is greater than that of the insecticides (Taylor and Mitchell 1957).

Acid Content.—A gradual reduction of the acid content takes place during the ripening of sweet cherries. Napoleon showed a reduction from 0.64 to 0.35%, and Lambert from 0.41 to 0.24%.

Increase in Weight

Growers who sell sweet cherries for processing should take into account the loss of tonneage (tonnage) through early picking. In Ontario, it averaged 0.45–1.8% (0.5–2.0%) per day. Take the loss through early picking into account in deciding whether to sell to processing or fresh fruit markets, but against it consider the advantage of a longer total picking period for a given cultivar where fruit can be sold for both purposes (Upshall 1943).

In Oregon, Napoleon, from the time it began maturing until the fruit was fully ripe, increased in size 36.2%; Lambert increased 36.2%. This was an average daily increase of about 1.3% for each cultivar. Although the greater increase took place just before the fruit reached prime condition, the cherries continued to gain weight even after attaining full maturity (Hartman 1925).

Dessert Quality

Time of picking has a marked effect on dessert quality. Fruit of too early pickings of sweet cherries does not taste sweet. That picked very late in the season is very sweet, or may be slightly insipid. In western Oregon best quality for Napoleon was attained by the fruit picked June 20–July 6; for Lambert, July 14–19. Most of the commercial crop of these cultivars, whether for fresh consumption, canning, or drying, was gathered long before these dates. The juice of both Napoleon and Lambert became richer and more syrupy as the season progressed. Sweet cherries lose rather than gain in quality after picking.

Shipping Quality

Sweet cherries picked when mature hold up as well or better than those picked while comparatively green. They do not shrivel as badly and show less discoloration due to oxidase activity. Bruises and decay organisms have more influence than time of picking on the shipping qualities of sweet cherries.

In canning, dilution of the syrup due to osmosis occurs whenever the sugar content of the fruit itself is lower than that of the syrup in which it is canned. Since cherries picked at different times contain different amounts of sugar, the time of picking affects the amount of dilution. Cherries of early pickings show more dilution (hydrometer test) than those picked late. The dilution of the first lot of Napoleon was 43.7%; that of the last was 21.2%. Canned cherries when thoroughly ripe, therefore, require less sugar to pass the trade requirements for syrup concentration.

Firmness is a requisite of well-canned fruit. Commercial canners, in fact, sometimes reduce the length of cooking and take a chance on keeping quality in order to secure firmness. The riper cherries give a firmer product than fruit that is comparatively immature.

When processed in high degree syrup, fruits usually shrink because some of the juice of the fruit is given up to the syrup by osmosis. The greater loss in volume occurs in fruit having a low sugar content. Hence, fruit canned early in the season may show more shrinkage or loss in volume than that canned when fully ripe. Early picked cherries, in general, give a product that is small, more or less flat in taste, soft in texture, and one that displays considerable shriveling. Those picked when riper have a livelier appearance, are larger, firmer, more aromatic, and have a more pronounced cherry flavor. Those picked and canned when past their prime are of good quality and texture, but show discoloration.

Ripening of Windsor Sweet Cherries

Windsor, a late-maturing cultivar, occupies about 1/10 of the total Michigan sweet cherry hectarage (1/3 of the acreage), and is used primarily for brining. Because it is dark-fleshed, it may be harvested while relatively immature before appreciable color is developed. As a consequence, Windsor fruits are usually smaller and less desirable for maraschino cherries than light-fleshed fruits which can be grown to maximum size and quality before harvest. An alternative to brining is to process for canning. Windsor, however, does not develop sufficient flesh color for canning until late in the growing season. Fruit softening and swelling precede the intense anthocyanin accumulation required for canning. Thus, fruits must remain on the tree for one to two weeks in a relatively softened state, during which time they have a high susceptibility to brown rot and rain cracking. In view of this condition, however, growers hesitate to allow the fruits to mature for canning, even though higher crop values could be realized (Chaplin and Kenworthy 1970).

EFFECTS OF GROWTH REGULANTS

Tart Cherry

Alar (succinic acid-2,2-dimethyl hydrazide) does not advance the physiological maturity of the fruits but it induces early development of pigments. The induced earlier coloring could result in premature harvesting and possibly a reduction of 8% in weight.

Alar at 1000 to 8000 ppm two weeks before bloom increased fruit color and decreased the force required to separate Montmorency fruit from its pedicel early in the harvest season. These differences advanced the start of commercial harvesting one week. Fruit firmness was increased in both hand-picked and mechanically harvested fruit. Alar-treated fruit showed a significant ability to resist softening when mechanically harvested. Fruit color and firmness enhancement was evident in both canned and frozen processed fruit. Alar altered the fruit growth curve and contributed to a more uniform size. Fruit acidity and respiration were reduced. Alar reduced terminal growth by reducing internodal length, but increased flower-bud initiation (Unrath *et al.* 1969).

Ethephon (2-chloroethyl phosphonic acid) is a water-soluble plant growth regulator thought to be degraded within plant cells resulting in the liberation of ethylene. The ethylene released within the plant is probably responsible for the subsequently observed biological activity. Cherries on treated trees (500 ppm) were easier to harvest than those on untreated trees, appeared to be more mature, were a darker red, and were higher in soluble solids. A spaced acceleration of fruit maturity control by ethephon could extend the harvest season (Anderson 1969).

Sweet Cherry

A number of field experiments have been conducted to determine the influence of Alar on fruit ripening of Windsor and other sweet cherry cultivars. Foliar sprays of 1000, 2000, or 4000 ppm applied to mature trees promoted anthocyanin development by two weeks and sugar development by one week without reducing fruit firmness. Treated fruits, however, were smaller than nontreated fruits of comparable color and sugar content. It is thought that Alar acts directly on the enzyme systems concerned with anthocyanin and sugar biosynthesis rather than acting at the hormonal level to advance the general physiological maturity of the fruit.

STORAGE

Sweet cherries keep about 4 days at room temperatures of 18.3° to 23.9°C (65° to 75°F), 8 days at 7.2°C (45°F), 12 days at 2.2°C (36°F), and 14 days at 0°C (32°F), if promptly placed under these conditions after picking. If held longer than the period indicated, they begin to lose flavor and the bright, attractive appearance characteristic of the fresh fruit. The stems may also dry out noticeably, especially if the relative humidity is low. Color changes and decay can be retarded by the use of CO_2 in transit.

Respiration of Lambert cherries at the end of 8 days of storage was 26% greater at 2.2°C (36°F) than at −0.6°C (31°F). At 7.2°C (45°F) it was 70% greater than at 2.2°C (36°F), and 115% greater than at −0.6°C (31°F).

Condition of Stems

Condition of the stems may affect sweet cherry sales acceptance. Stems frequently become shriveled and brown; but try to keep them turgid and green. Little difference in the appearance and moisture content of the stems occurred at comparable humidities when held at 7.2°C (45°F), 2.2°C (36°F), or −0.6°C (31°F). Delayed storage of 15 hr at harvest induced shriveling, discoloration, and a large increase in loss of moisture from the stems during storage at 7.2°C (45°F). Postharvest or transit environment greatly influenced the stem condition. Moisture losses from stems were more than twice as great at 26.7°C (80°F) and 50% humidity as at 18.3°C (65°F) and 80% humidity (Gerhardt 1942).

If cherries are picked carefully and held without injury, the stems shrivel at a rate depending on temperature and humidity conditions, but do not brown. If the stems are injured, there is a rapid browning accompanied by accelerated shriveling. The injury can take place at any point during harvest, packing, or shipping. The accelerated shriveling accompanying browning is probably a result of an altered permeability of the injured cells of the stem. Shriveling of the injured stems can be stopped by a thick coating with lanolin, but browning still occurs. Therefore, stem shriveling and browning are two separate processes. Severely browned stems contain almost as much chlorophyll as fresh ones (Siegelman 1953).

Precooling

Schmidt sweet cherries and Montmorency tart cherries held in 1-qt wooden tills were precooled in refrigerated air at −0.6° to 1.7°C (31° to 35°F). In air blasts of about 770 cfm, fruits were completely precooled in 30 to 50 min, whereas more than 7 hr were required for cooling both kinds of fruits in still air. Reductions in the moisture content of the stems during precooling were smallest in the air blasts and greatest in still air. Water losses from the stems of tart cherries during 7 1/2 hr of air blast treatment were no greater than the losses from fruit precooled in still air for this period of time. The advantage of air blast precooling was retained by transferring the fruit to still air after precooling was accomplished (Dewey 1950).

Sweating

Sweating is particularly troublesome with cherries because of the dull appearance it produces on fruit displayed for sale soon after removal from iced cars. If cherries are held too long after removal from storage, especially if they are wet from sweating, brown rot, and gray, blue, or green rots may develop.

FIG. 5.12. BACTERIAL CANKER (*PSEUDOMONAS SYRINGEA*)
ON CHERRY CAUSING DEATH OF BUDS

Controlled Atmosphere

Dry ice was used for some years in carload shipments of sweet cherries from the west. This practice, however, was discontinued with the introduction of polyethylene liners which provided atmospheres of 6–8% CO_2. These liners (1.5 mil) should be perforated or opened as soon as the cherries are removed from cold storage or refrigerated cars and held at higher temperatures. This prevents the buildup of undesirable levels of CO_2 in the packs. There seems to be little advantage from CO_2 storage that cannot be achieved by sealed 1.5-mil polyethylene crate liners.

INSECTS AND DISEASES

Pest control schedules vary in different regions. Follow local schedules. Important insect and disease pests of cherries are as follows:

Insects

Cherry Fruit Fly.—The females, after emerging in late spring and early summer, lay eggs in small slits in the fruits. The eggs hatch into white maggots that burrow in the fruits.

Curculio.—The adult, a dark brown snout beetle, deposits eggs in the fruit about blossom time, making a crescent-shaped slit. The egg hatches into a whitish larva having a curved body and a brown head.

San Jose Scale and Forbes Scale.—These appear as grayish, thin, flaky masses on the bark of branches and twigs. These sucking insects decrease tree vigor. There are several other kinds of scales.

Other Insects.—Case borer, aphids, leafhoppers, and many others may cause damage to cherries.

Diseases

Leaf Spot.—This disease is probably the most destructive of tart cherries in the Great Lakes region. It overwinters in dead leaves. When the temperature nears 15.6°C (60°F) in spring the disease organism begins to spread, and may continue from before bloom until 6—7 weeks later. The spots on the leaves are first small and purplish or reddish. These enlarge and become brown, or brown surrounded by a reddish-brown. When the dead tissue falls out, the leaves assume a "shot hole" appearance and may become yellowish or chlorotic and drop prematurely.

Brown Rot.—Losses are most severe during hot, humid weather at harvest time. The fruits, especially of sweet and Duke types, rot while ripening or after being picked. Affected fruits develop small circular brown spots, which may spread over the whole fruit. Powdery, light brown or gray masses of spores appear on the rotted areas. The fungus lives over the winter in mummied fruits.

Virus Diseases.—Control consists of exclusion and removal of diseased trees from the orchard, and making new plantings with trees that have been propagated from buds of virus-free trees. Spraying is ineffective. Separate cherry orchards by at least 152 m (500 ft) from chokecherries.

The term "virus complex" has arisen to satisfy the need for a word meaning different viruses or mixed strains thought to be largely of one parent type. Originally, each disease, e.g., little-cherry, western X-disease, or red-leaf, was considered to be due to an individual virus. As research continued, similarities began to be noticed among several of these virus diseases. Research workers now attempt to relate such virus diseases as much as possible.

FIG. 5.13.   VERTICILLIUM WILT ON SWEET CHERRY

Some of the more prominent diseases that may prove to be members of the western X-little cherry complex are the "buckskin" of California cherry orchards, the X-disease of peaches in the United States and Canada, the albino of Oregon cherry orchards, the little-cherry and small-bitter cherry of British Columbia, the pink fruit of cherries in western Washington, and the yellow-leaf-roll of peach in California.

A germ plasm bank of virus-free fruit tree clones was established in Washington state with these main objectives: to assemble selections of desirable species and cultivars of deciduous fruit trees, verify their freedom from infecting viruses, maintain virus-free individuals in repositories, and distribute virus-free propagating materials to cooperating investigators for research or as foundation stock for release to industry (Fridlund 1968). Similar projects are being developed in other cherry areas.

REFERENCES

ANDERSON, J.L. 1935. Behavior of cherry trees in the Hudson River Valley with particular reference to losses from winter killing and other causes. N.Y. Bull. *653.*

ANDERSON, J. L. 1969. Effect of of Ethrel on ripening of Montmorency sour cherries. HortScience *4,* 92.

ASCHER, P.D. 1966. A gene action model to explain gametophytic self incompatibility. Euphyica *15,* 175-183.

ATKINSON, F.E. 1956. Maturity Manual. British Columbia Tree Fruits, Vancouver.

BEDFORD, C.L. 1964. Quality of hand picked and mechanically processed red cherries. Proc. Mich. State Hort. Soc. *94*, 87-90.

BOYNTON, D. 1944. Responses of young Elberta peach and Montmorency cherry trees to potassium fertilization in New York. Proc. Am. Soc. Hort. Sci. *44*, 31-33.

BRADFORD, F.C., and CARDINELL, H.A. 1926. Eighty winters in Michigan orchards. Mich. Agr. Expt. Sta. Bull. *149*.

BRADT, O.A. *et al.* 1968. Fruit varieties. Ontario Dept. Agr. Food. Publ. *430*.

BROWN, A.G. 1955. Incompatibility. John Innes Inst. Rept. (England), 7-8.

BRYANT, L.R. 1940. Sour cherry rootstocks. Proc. Am. Soc. Hort. Sci. *37*, 322-323.

BUKOVAC, M.J. *et al.* 1969. Chemical promotion of fruit abscission in cherries and plums with special reference to 2-chloroethyl phosphonic acid. Proc. Am. Soc. Hort. Sci. *94*, 226-230.

BUKOVAC, M.J. *et al.* 1971. Effects of 2-chloroethyl phosphonic acid on development of sweet cherry fruits. J. Am. Soc. Hort. Sci. *96* (67): 777-778.

BULLOCK, R.M. 1952. Inorganic compounds and growth promoting chemicals in relation to fruit cracking of Bing cherries at maturity. Proc. Am. Soc. Hort. Sci. *59*, 243-253.

CAIN, J.C. 1961. Mechanical harvesting of sour cherries: Effects of pruning, fertilizer, and maturity. Proc. N.Y. State Hort. *106*, 198-205.

CHANDLER, W.H. 1951. Deciduous Orchards. Lea and Febiger, Philadelphia.

CHAPLIN, M.H. 1970. Pros and cons of mechanical harvesting of sweet cherries. Better Fruit *64*, 17-18.

CHAPLIN, M.H., and KENWORTHY, A.L. 1970. Influence of succinic acid 2,2-dimethyl hydrazide on fruit setting of Windsor cherry. J. Am. Soc. Hort. Sci. *95*, 532-536.

CHRISTENSEN, M.D., and WALKER, D.R. 1964. Leaf analyses techniques and survey results on sweet cherries in Utah. Proc. Am. Soc. Hort. Sci. *85*, 112-117.

CLARKE, W.S., and ANTHONY, R.D. 1946. Orchard test of mazzard and mahaleb rootstocks. Proc. Am. Soc. Hort. Sci. *48*, 100-108.

CLINE, R.A. 1967. Soil drainage and compaction effects on growth, yield, and leaf composition of peaches and cherries. Ann. Rept. Hort. Res. Inst. Ontario, 28-34.

COE, F.M. 1945. Cherry rootstocks. Utah Agr. Expt. Sta. Bull. *319*.

CRANE, M.B., and BROWN, A.G. 1937. Incompatibility and sterility in sweet cherry. J. Pomol. Hort. Sci. *15*, 86-116.

CURWEN, D. *et al.* 1966. Fruit firmness and pectic composition of Montmorency cherries as influenced by differential nitrogen, phosphorus, and potassium applications. Proc. Am. Soc. Hort. Sci. *89*, 72-79.

DENNIS, F.G. 1976. Trials of ethephon and other growth regulators for delaying bloom in fruit trees. J. Am. Soc. Hort. Sci. *101* (3): 141-145.

DEWEY, D.H. 1950. Effects of air blast precooling on the moisture content of the stems of cherries and grapes. Proc. Am. Soc. Hort. Sci. *56,* 111-115.

DIENER, R.G. *et al.* 1965. Preharvest spray cooling for tart cherries. Mich. Agr. Expt. Sta. Res. Rept. *69,* 4.

EAST, E.M., and MANGELSDORF, A.J. 1925. A new interpretation of the hereditary behavior of self-sterile plants. Proc. Natl. Acad. Sci. *11,* 166-171.

EATON, G.W. 1959A. A study of the megagametophyte in *Prunus* and its relation to fruit setting. Can. J. Plant Sci. *3,* 466-476.

EATON, G.W. 1959B. Twin ovules in *Prunus avium.* Can. J. Botany *37,* 1203-1205.

EATON, G.W. 1962. Further studies on sweet cherry embryo sacs in relation to fruit setting. Ann. Rept. Hort. Res. Inst. Ontario, 26-38.

EDGERTON, L.J., and HATCH, A.H. 1969. Promotion of cherries and apples for mechanical harvesting. N. Y. State Hort. Soc. *114,* 8-12.

FOGLE, H.W. 1958. Effects of duration of afterripening, gibberellin, and other pretreatments on sweet cherry germination and seedling growth. Proc. Am. Soc. Hort. Sci. *72,* 129-133.

FRIDLEY, R.B. 1969. Tree fruit and grape harvest mechanization process and problems. HortScience *4,* 235-237.

FRIDLUND, P.R. 1968. A germ plasm bank of virus-free tree clones. HortScience *3,* 227-229.

GARDNER, V.R. 1930. Maintaining the productivity of cherry trees. Mich. Agr. Expt. Sta. Bull. *195.*

GARDNER, V.R. 1935. Susceptibility of flower buds of Montmorency cherry to injury from low temperature. J. Agr. Res. *50,* 563-572.

GASTON, H.P., and LEVIN, J.H. 1956. Weighing cherries that are transported in water. Mich. Agr. Expt. Sta. Quart. Bull. *38,* 606-611.

GASTON, H.P. *et al.* 1967. Ten years of progress in machine harvesting fruit. Proc. Mich. State Hort. Soc. *97,* 54-59.

GERHARDT, F. 1942. Respiration, internal atmosphere, and moisture studies of sweet cherries during storage. Proc. Am. Soc. Hort. Sci. *40,* 119-123.

GILBERT, F.A. 1956. Potash prevents "curl leaf" of sour cherries. Better Crops Plant Food *39,* 6-10, 46-48.

HARRINGTON, W.C. *et al.* 1966. Effect of cultural practices on processed cherry quality. Proc. Am. Soc. Hort. Sci. *38,* 184-189.

HARTMAN, H. 1925. Handling of sweet cherries. Proc. Am. Soc. Hort. Sci. *21,* 79-86.

HOWE, G.F. 1942. Cherry growing. N.Y. State Agr. Expt. Sta. Circ. *145.*

HOWELL, G.S., and STACKHOUSE, S.S. 1973. Effect of time of foliation on acclimation and dehardening in tart cherry. J. Am. Soc. Hort. Sci. *98* (2): 132-136.

JUDKINS, W.P. 1940. Sites and soil management for sour cherries. Ohio Farm Home Res. *34,* 167-169.

KENNARD, W.C. 1949. Defoliation of Montmorency sour cherry trees in relation to winter hardiness. Proc. Am. Soc. Hort. Sci. *53*, 129-133.

KENWORTHY, A.L., and MITCHELL, A.E. 1952. Soluble solids in Montmorency cherries at harvest as influenced by soil management practices. Proc. Am. Soc. Hort. Sci. *60*, 91-96.

KERTESZ, Z.I., and NEBEL, B.R. 1935. Cracking of cherries. Plant Physiol. *10*, 763-772.

KWONG, S.S. 1965. Potassium fertilization in relation to titratable acids in sweet cherries. J. Am. Soc. Hort. Sci. *86*, 115-119.

LABELLE, R.L. 1964. Recovery of Montmorency cherries from repeated bruising. Proc. Am. Soc. Hort. Sci. *84*, 103-109.

LANGFORD, L.R. 1939. Effect of potash on leaf curl of sour cherry. Proc. Am. Soc. Hort. Sci. *36*, 261-262.

LARSEN, R.P. 1969. Mechanization of fruit harvest at the eastern United States. HortScience *4*, 232-234.

LEVIN, J.H., and GASTON, H.P. 1956. Grower handling of red cherries. U.S. Dept. Agr. Circ. *981*.

LEVIN, J.H. *et al.* 1960. Mechanizing the harvest of red tart cherries. Mich. Agr. Expt. Sta. Quart. Bull. *42*, 656-685.

LOONEY, N.E., and MCMECHAN, A.D. 1970. Use of 2-chloroethylphosphonic acid and succinic acid 2,2-dimethyl hydrazide to aid in mechanical shaking of sour cherries. J. Am. Soc. Hort. Sci. *95*, 452-455.

LOTT, R.V., and SIMONS, R.K. 1966. Sequential development of floral tube and style abscission in Montmorency cherry. Proc. Am. Soc. Hort. Sci. *88*, 208-218.

MOUSTAFA, S.M. *et al.* 1968. Hydrocooling of cherries. Mich. Agr. Expt. Sta. Res. Rept. *74*.

PHILP, G.L. 1947. Cherry culture. Calif. Agr. Expt. Sta. Circ. *46*.

PILLEY, T.N., and EDGERTON, L.J. 1965. Relationships of growth substances to rest period and germination in mazzard and mahaleb seeds. Proc. Am. Soc. Hort. Sci. *86*, 108-114.

PILLEY, T.N. *et al.* 1965. Effect of pre-treatments, temperature and duration of afterripening on germination of mazzard and mahaleb cherry seeds. Proc. Am. Soc. Hort. Sci. *86*, 102-107.

POOVAIAH, B.W. *et al.* 1969. Histological and localization of enzymes in the abscission of maturing sour and sweet cherry fruit. J. Am. Soc. Hort. Sci. *98* (1): 16-18.

PROEBSTING, E.L., and KENWORTHY, A.L. 1954. Growth and leaf analysis of Montmorency cherry trees as influenced by solar radiation and intensity of nutrition. Proc. Am. Soc. Hort. Sci. *63*, 41-48.

PROEBSTING, E.L., and MILLS, H.H. 1973. A comparison of hardiness responses in fruit buds of Bing cherries and Elberta peach. J. Am. Soc. Hort. Sci. *97* (6): 802-806.

RYUGO, E. 1967. Persistence and mobility of Alar and its effect on anthocyanin metabolism in sweet cherries. Proc. Am. Soc. Hort. Sci. *80*, 160-166.

RYUGO, E., and INTRIERI, C. 1972. Effect of light on growth of sweet cherry fruits. J. Am. Soc. Hort. Sci. *9* (6): 691-694.

SAWADA, E.A. 1934. Cracking of sweet cherries. Sapporo Natl. Historical Soc. *13,* 365-376.

SHOEMAKER, J.S. 1928. Cherry pollination. Ohio Agr. Expt. Sta. Bull. *422.*

SIEGELMAN, H.W. 1953. Brown discoloration and shrivel of cherry stems. Proc. Am. Soc. Hort. Sci. *61,* 265-269.

SIMONS, R.K. 1969. Fruit tissue injury in apricots and sweet cherries as a result of a late spring freeze. J. Am. Soc. Hort. Sci. *94,* 466-470.

STANBERRY, C.O., and CLORE, W.J. 1950. Effect of fertilizers on the composition and keeping qualities of Bing cherries. Proc. Am. Soc. Hort. Sci. *56,* 40-45.

STOSSER, R. *et al.* 1969. Histological study of abscission layer formation in cherry fruits during maturation. J. Am. Soc. Hort. Sci. *94,* 239-243.

TAYLOR, O.C., and MITCHELL, A.E. 1953. Relation of time of harvest to size, firmness and chemical composition of fruit of the sour cherry. Proc. Am. Soc. Hort. Sci. *62,* 267-271.

TAYLOR, O.C., and MITCHELL, A.E. 1957. Soluble solids, total solids, sugar content, and weight of the fruit of sour cherry as affected by pesticide chemicals and time of harvest. Proc. Am. Soc. Hort. Sci. *68,* 124-130.

TENNES, B.R. *et al.* 1968. Weight to volume relationship of tart cherries. Mich. Agr. Expt. Sta. Res. Rept. *70,* 4.

TESKEY, B.J.E. *et al.* 1965. Pruning and training fruit trees. Ontario Dept. Agr. Food Publ. *392.*

TUKEY, H.B. 1925. An experience with pollinizers for cherries. Proc. Am. Soc. Hort. Sci. *21,* 69-73.

TUKEY, H.B. 1927. Responses of the sour cherry to fertilizers and to pruning. N.Y. State Agr. Expt. Sta. Bull. *541.*

TUKEY, H.B., and BRASE, K. 1945. Production of cherry trees in the nursery. Proc. Am. Soc. Hort. Sci. *27,* 88-92.

TUKEY, H.B., and YOUNG, J.O. 1939. Histological study of the developing fruit of the sour cherry. Proc. Am. Soc. Hort. Sci. *36,* 269-270.

TUKEY, L.D. 1952. Effect of night temperature on growth of the fruit of sour cherry. Botan. Gaz. *114,* 155-165.

TUKEY, L.D. 1953. Effect of night temperature on fruit development. Am. Fruit Grower. (Oct.), 32-33.

TUKEY, L.D., and TUKEY, H.B. 1966. Growth and development of Montmorency cherry from flower bud initiation to fruit maturity and some associated factors. Great Lakes Cherry Producers Market. Coop. Prog. Rept., 9-17.

UNRATH, S.R. *et al.* 1969. Effect of Alar (succinic acid 2,2-dimethyl hydrazide) on fruit maturation, quality, and vegetative growth of sour cherries. J. Am. Soc. Hort. Sci. *94,* 387-391.

UPSHALL, W.H. 1943. Fruit maturity and quality. Ontario Dept. Agr. Food Bull. *447.*

VERNER, L., and BLODGETT, E.C. 1931. Physiological studies of the cracking of sweet cherries. Idaho Agr. Expt. Sta. Bull. *184.*

WANN, F.B. 1950. Hormones fail to increase fruit set in sweet cherry. Utah Farm Home Sci. *11,* 48-49.

WANN, F.B. 1954. Cherry nutrition. *In* Fruit Nutrition. N.F. Childers (Editor). Horticultural Publications, Rutgers Univ., New Brunswick, N.J.

WAY, R.D. 1966. Identification of sterility genes in sweet cherry cultivars. Proc. 17th Intern. Hort. Congr., Maryland, Aug. 15-20. Intern. Soc. Hort. Sci. *1,* 145-146.

WAY, R.D. 1968. Pollen incompatibility groups of sweet cherry clones. Proc. Am. Soc. Hort. Sci. *92,* 119-123.

WESTWOOD, A.N. *et al.* 1976. Comparison of mazzard, mahaleb, and hybrid rootstock for Montmorency cherry. J. Am. Soc. Hort. Sci. *101* (3): 268-269.

WHITTENBERGER, E.T. *et al.* 1969. How much beating will tart cherries take and where do they get it. Proc. Mich. State Hort. Soc. *94,* 83-87.

WITTENBACH, V.A., and BUKOVAC, M.J. 1972. An anatomical and histological study of abscission in sweet cherry fruits. J. Am. Soc. Hort. Sci. *97* (2): 214-219.

YU, K., and CARLSON, R.F. 1975. Paper chromatographic determination of phenolic compounds occurring in the leaf, bark, and root of *Prunus avium* and *P. mahaleb.* J. Am. Soc. Hort. Sci. *100* (5): 536-541.

ZIELINSKI, Q.B. 1958. Factors affecting seed germination in sweet cherries. Proc. Am. Soc. Hort. Sci. *72,* 123-128.

6

Plums

The plum, like the peach and cherry, is a drupe fruit.

California, Michigan, Oregon, Washington, and Idaho account for about 90% of the total commercial crop in the United States, but plums are grown also to some extent in most of the other states. In Canada plums are grown commercially in British Columbia and Ontario.

KINDS OF PLUMS

Prunus domestica

The large-fruited European type plum is the most important species. It originated the region of the Caucasus Mountains as a result of hybridization of diploid *P. cerasifera* and tetraploid *P. spinosa* species, followed by chromosome doubling.

Prune.—This is by far the most important group commercially. Fruit is oval with bulging of central side, compressed bilaterally; blue or purple; firm, thick, meaty flesh; freestone. Prune fruits are high in sugar content, and are capable of being dried into prunes without removal of the pit. All prunes are plums but not all plums are prunes. California produces most of the prepared prunes. In the east, and in the Pacific Northwest, plums of the prune group are usually marketed as fresh fruit. Cultivars are Italian, French (Agen, Petite), Imperial Epineuse, Sugar, Giant.

Green Gage.—Fruit is greenish-yellow; medium size often round; sweet, and high in quality. Cultivars are Reine Claude, Washington, Jefferson.

Yellow Egg.—Fruit is large, long oval, more or less necked and usually with yellow skin and yellowish flesh. Cultivars are Yellow Egg, Golden Drop.

Imperatrice.—Fruit is dark blue, heavy bloom, medium to large, oval; stone usually clingy; skin thick; only fair quality. Cultivars are Arch Duke, Diamond, Monarch.

Lombard.—Fruit is purplish-red; good quality. Cultivars are Lombard, Bradshaw, Pond (Hungarian).

Prunus insititia

The small-fruited European type includes the damsons and mirabelles. Trees of this species are distinguished from those of *P. domestica* by having a more compact and dwarfed habit of growth, smaller leaves, more slender and finely divided branches, and smaller flowers. The fruits are less than 1 in. in diameter, round or oval, and uniform. They are purple (in the damsons) or yellow (in the mirabelles) with no intermediate yellows. Damsons are highly esteemed for preserves and jams, but not eaten fresh. Mirabelles have an excellent flavor.

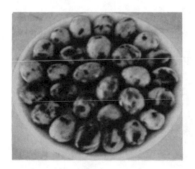

FIG. 6.1. THE ITALIAN PRUNE IS A LEADING CULTIVAR FOR DESSERT, CANNING, AND FOR MAKING PRUNES

Prunus salicina

The japanese plum is next in importance to the large-fruited European type. It blooms very early in spring and so is often injured by an early frost.

Prunus simonii

The Simon plum was introduced into the western world from China.

Crossed with *P. salicina* it gave cultivars such as Shiro and Wickson. In fact, *P. simonii* differs in few, if any, significant characters from *P. salicina*.

Methley may be a hybrid of *P. salicina* and *P. cerasifera* (myrobalan). Bruce (*P. salicina* × *P. angustifolia*) has considerable merit in Texas for marketing in May.

Native Plums

The American native plum, *P. americana*, the Canadian native *P. americana nigra*, and the Wolf, *P. americana mollis*, are highly resistant to cold. The flesh and juice have a pleasant flavor, but the astringency near the skin is objectionable. These hardy types have been used in breeding plums, e.g., in crosses with the Japanese type, for areas with severe winters.

P. Munsoniana includes Wild Goose and Downing and is an important group of native plums in the south. Another native species in the south is *P. hortulana*, as represented by Wayland.

P. angustifolia (Chickasaw plum) is a native of the Deep South and extends north to southern Ohio. The trees are thorny and thick-topped. The extreme southern forms have a low chilling requirement. This species has been successfully crossed with Japanese type plum, e.g., Terrill and McRae.

P. maritima (beach plum) is abundant on Cape Cod, and the islands of Martha's Vineyard and Nantucket, where the fruit is harvested for sale. It is native along the beaches, among the sand dunes, and on the coastal plains from Virginia to New Brunswick. It varies greatly in vigor, size, growth habit, and fruitfulness of the bush and tree-like forms, and in size, shape, color, and quality of the fruit. It is seldom eaten fresh, but is useful for jelly and jam. The fruit ripens from mid-August to early October (Bailey 1944).

P. subcordata (Pacific or western plum) grows wild in a relatively limited region of the coast range from southern Oregon to central California. When made into jams, jellies, and preserves, it is highly prized (Roberts and Hammers 1951).

CULTIVARS (VARIETIES)

Four leading cultivars in California are Santa Rosa, Casselman (five weeks later than Santa Rosa,) Laroda, and Late Santa Rosa. These three cultivars make up more than 50% of the total shipments. California's plum crop in 1974 was 128,700 tonnes (143,000 tons). Its crop is produced in a number of districts from Kern County in the south to the Sierra foothills of Placer County with emphasis on Fresno County.

In the Pacific Northwest, "prune" cultivars are marketed as fresh fruit plums.

A plum orchard in the east and midwest should include cultivars that are not only attractive in appearance and good in quality, but are also productive. For a roadside market and local trade, select a succession of good fruit suitable for all purposes from the beginning of the season to the end. Desirable plums that are attractive and pleasing to the palate, as well as productive, may be had from August to October in the east.

Japanese Type Cultivars

Japanese type cultivars are mostly used for eating fresh. Most of them ripen before the large-fruited European type and so are useful in lengthening the plum season. The fruits are fairly attractive but usually are soft and clingstone. They are round or heart-shaped, not oval as in many *P. domestica* cultivars, but some are taper-pointed, and some, e.g., Wickson, have a nipplelike apex. The trees are usually productive but bloom early and are subject to spring frosts. Almost all Japanese type cultivars are self-unfruitful.

Early Golden.—Resembles Shiro but has a reddish blush when mature, and is two weeks earlier; freestone. Tree very vigorous, and a biennial bearer. Originated in the Fonthill district of Ontario, Canada.

Beauty.—Very early; medium size; roundish conic; attractive medium red; heavy bloom; flesh yellow, tinged with red; very juicy, sweet, fair in quality.

Methley.—Blood red flesh. The mottled gray in the external skin color is not very attractive. Does fairly well in the south. New Zealand origin.

Laredo (Gaviota × Santa Rosa).—Round, large, dark red skin; flesh amber, firm, quality very good. California 1954.

Formosa.—Ripens one week after Beauty. Fruit large, oval or oblate-conic, greenish-yellow nearly overlaid with red; flesh yellow, juicy. Tree tends to biennial bearing.

Shiro.—Attractive clear yellow skin and flesh; good eating quality.

Santa Rosa.—Dark reddish-purple; large, oblong-conic; juicy; fair quality.

Abundance.—Fruit medium size; roundish ovate; red, with yellow flesh; of fairly short season.

Nubiana (Gaviota × Eldorado).—Large, flat, dark purple-red; flesh amber; quality fair. California 1954.

Wickson.—Fruit medium size; nipplelike apex; yellow flesh.

Duarte.—Large; heart-shaped; blood red flesh; rather light producer in the east.

Satsuma.—Large; nearly round; blood red flesh; rather light producer in the east.

Burbank.—Long the standard late Japanese cultivar in the east. Resembles Beauty in color, shape, and size, but has firmer flesh. Ripens unevenly, requiring 2–3 pickings. The tree is low, somewhat flat-topped, and wide-spreading.

Kelsey.—Very late, long season, good shipper; large, red, conical; from Japan in 1870; the first of the *P. salicina* plums brought to North America.

European Type Cultivars

The large-fruited cultivars are generally preferred by customers. The prune and green gage groups are canned commercially.

Queenstone.—Fruit blue, medium size, freestone, acceptable market quality. Matures several days before California Blue.

California Blue.—Ripens with Shiro. Large, nearly freestone; fair quality though tart until fully ripe; subject to preharvest drop; defoliates early in the fall; useful as an early blue plum.

Washington.—Fruit large, roundish oval, light yellow, of green gage group; flesh greenish-yellow, firm; freestone; very good quality. Useful for eating fresh and on home and local markets.

French Prune (Agen, Petite).—An important cultivar in the west for making prunes. Purplish-red, sweet, smaller than Imperial Epineuse.

Imperial Epineuse.—The west grows it particularly for drying and as a prune it brings a premium price. Fruit reddish-purple, above medium size, greenish-yellow flesh, sweet. Eagerly sought by those in the east who know it; generally less attractive appearance than the blue or black types.

Iroquois (Italian Prune × Hall).—Fruit blue, firm, greenish-yellow flesh; freestone; fair to good quality; good for canning; comes into bearing early.

Early Italian.—Shipped from the west as fresh plums. There are several earlyripening bud sports of the Italian. Except for season, they are similar, if not identical, to each other and to the Italian. Because of earliness, they may ripen better than Italian in a cool season. Useful where a prune of Italian type is wanted to precede Stanley.

Stanley (Agen × Grand Duke).—Ripens 7–10 days earlier than Italian. Flesh amber when ready to pick. Processes well. The tree is early bearing and a good pollinator for other European cultivars. Less subject to cultural difficulties than Italian. A leading cultivar in the Great Lakes region.

Grand Duke.—Fruit attractive, blue, large, only fair quality; not freestone and not as productive as Italian.

Bluefre (Stanley × President).—Large, blue, freestone, fairly good quality; some tendency to split pits.

FIG. 6.2. IN THE PLUM, THE FLOWERS APPEAR IN ADVANCE
OF THE LEAVES

Blue Bell (Stanley × President).—Larger than Stanley and about one week later. Blue, yellow flesh, freestone, high quality.

Verity (Imperial Epineuse × Grand Duke).—Fruit larger than Italian, blue, orange colored flesh, freestone; good for fresh market and processing.

Valor (Imperial Epineuse × Grand Duke).—Dark purple, somewhat larger than Italian, greenish-gold flesh, semifreestone. Good for fresh fruit.

Italian (Italian Prune, Fellenburg, York State).—Standard prune of its season and the most widely grown. Medium size, long oval, purplish-black; flesh greenish-yellow, firm, but tender; good quality; freestone; resistant to brown rot; excellent for drying and canning.

Reine Claude (Green Gage).—Matures in late September in the north. Fruit yellowish-green, medium size, roundish oval, with golden yellow flesh which is tender, sweet, mild, and very good quality; semiclingstone to freestone. The standard green gage for canning.

Lombard.—It is neither large enough nor attractive enough for fresh fruit. It can have excellent quality if left on the tree long enough to reach proper maturity. Purplish-red. Makes a good canned product.

President.—Fruit blue-black and of good size; flesh firm and of good quality; freestone.

Vision (Pacific × Albion).—Blue, large, freestone, very good eating quality.

Albion.—Ripens late (about Oct. 1 in the Great Lakes area and often hangs through several frosts in good condition through the month). Fruit large, oval, purplish-black; flesh golden yellow, firm, coarse, juicy, pleasantly flavored, and of good quality; clingstone.

Damsons

Shropshire.—The most widely grown damson cultivar, but most Shropshires grown in the United States are not the same as the Shropshire damson grown in England. Fruit small, average size for the species, oval, compressed, purplish-black with heavy bloom; flesh yellowish, juicy, tart; clingstone. Not eaten fresh. Makes excellent jams, preserves, and jellies. Well adapted to mechanical harvesting for processing.

French Damson.—Fruit comparatively large for a damson. Not as highly productive as Shropshire.

Hardy Hybrids

For regions where the European and Japanese cultivars are not hardy, several hybrids of Japanese and native plums (chiefly *P. americana*) have been developed. Cultivar examples (developed in South Dakota and Minnesota) are, in order of ripening: LaCrescent, Underwood, Pipestone, Fiebing, Redglow, Superior, Toka, Elliott, Ember, Assiniboine. Cherry plums, in which the bushlike sandcherry (*Prunus besseyi*) is one parent, include Opata, Sapa, and Sapalta. These are bushlike in habit of growth, and lacking in good character from a quality standpoint.

PROPAGATION

Plum cultivars are propagated by budding, in much the same manner as described earlier for the other tree fruits.

Seedling Rootstocks

Myrobalan (*P. cerasifera*) is the most satisfactory rootstock for plums in the east. Other stocks tested are *P. americana*, marianna, myrobalan, St. Julien, peach budded, and peach grafted.

Seed of myrobalan was obtained from several sources in the United States and from Europe and planted in the nursery. The seedlings which developed varied considerably; the leaf characteristics ranged from diverse shades of green to dark purple; the leaf serrations, texture, and

Courtesy of Hort. Res. Inst., Vineland, Ont.

FIG. 6.3. PLUMS MAY BLOOM HEAVILY ON SHORT SPURS ON
WOOD TWO TO FIVE YEARS OLD

size differed; some seedlings were thorny, some were thornless, and the vigor of the trees was not uniform. Certain seed sources may result in different degrees of compatibility when budded with standard cultivars. Seedlings of a strain of myrobalan, known as myrobalan 29, are desirable as a plum rootstock in California (Hartman 1956).

Grand Duke was grafted on trunks of peach and myrobalan plum, and on seedlings of myrobalan, peach, apricot, and almond in California. Early ripening of Grand Duke occurred in cases of uncongenial combinations, abnormal unions, and partial girdling (Day 1934).

A leaf symptom termed "chlorotic fleck" has been found on myrobalan seedlings grown from seed collected in California. Such seedlings made less growth than normal. Stanley developed a constriction at the bud union when budded to seedlings showing "chlorotic fleck" and often died after being planted in the orchard. The causal factor of chlorotic fleck was carried in the seed and symptoms were well developed on the very young seedlings but disappeared as the seedlings grew older. No source of myrobalan seed is known with certainty to be free of this trouble. Until myrobalan seed sources free of chlorotic fleck can be developed, either rogue out the young seedlings showing symptoms or at least do not use them as stocks for Stanley.

Plum cultivars can be budded on peach and marianna plum but the results are only fair. *P. americana* seedlings are used as a rootstock in the coldest fruit areas.

Certain European plums, such as Ackerman and Pershore, have some merit as a rootstock for *P. domestica* cultivars. Seedling stocks of Ackerman (also known as Marunke) must be budded earlier in the summer than myrobalan to insure good bud take. The western sandcherry (*P. besseyi*), which is budded comparatively early, is a dwarfing stock but has no place in propagating *P. domestica* and *P. salicina* cultivars for commercial fruit production.

Afterripening of Seed.—Myrobalan seed for the production of seedling rootstocks requires 100–120 days of after ripening at (40°–50°F). Ackerman plum seed requires 120–130 days at the foregoing temperatures. The artificial ripening procedure, if this method is used (see Propagation in Chap. 4), must be timed properly to avoid too early or too late starting.

Clonal Rootstocks

Much of the seed used for producing myrobalan seedlings has been imported. Inherent variability and the risk of importing viruses in the seedling rootstock are disadvantages of relying on this type of material. When a shortage occurs in availability of the usual plum rootstock, the problem is accentuated. Considerable work has been done in England on clonal rootstocks and a hedge-grown source has been suggested.

In Ontario, Canada, hardwood cuttings of clonal *Prunus* rootstocks myrobalan B, Brompton, St. Julien A. Ackerman, and marianna were successfully rooted without bottom heat. The cuttings were taken in the early fall before the rest period was completed. Treated with 0.05% IBA (ethyl alcohol plus water solution) or 0.3% commercial IBA powder, the cuttings were stored upright in portable boxes in a peat-perlite medium for a root initiation period at 7.2°C (45°F) for up to 100 days or more, depending on the cultivar. When the terminal buds began to expand, the boxes were transferred to 0°C (32°F) to suppress growth until the cuttings could be placed in the field in early spring. The cuttings made excellent budding stock the following year (Hutchinson 1967).

Of the six *Prunus* species represented by the nineteen rootstocks tested, myrobalan roots usually resulted in larger trees, heavier bloom, but lower yield efficiency than did peach roots. Trees on marianna and several *P. domestica* varied in size and yield, and most of them had greater bloom density than did trees on peach. Italian fruit firmness varied inconsistently with rootstock. Early Italian fruits were firmer on peach than on other roots, but Italian was greater on plums than on peach roots. Fewer trees on peach roots died from trunk canker than did those on several clonal plum roots, but some plum-rooted trees outgrew the canker as well as trees on peach stock (Westwood *et al.* 1973).

FLORAL TUBE AND STYLE ABSCISSION

Development of Stanley plum flowers and young fruits was characterized as having six different stages that extended from anthesis through style abscission. These stages were similar to those of peach and cherry, but growth and rapidity of development of the abscission zone differed among the three *Prunus* species. Previous research established eight stages for cherries and ten for peaches and apricots. The floral-cup base was enlarged adjacent to the ovary; in early, abortive fruit, remaining portions of the floral tube shriveled, with no evidence of abscission zone formation (Simons 1973).

SPUR-PEDICEL ABSCISSION

Parameters of spur/pedicel in Stanley plum have been recorded throughout induction and development during the early stages of fruit growth. Study of abscission included both young-persisting and "drop" fruits, each having diverse abscission potentials. Gross morphology revealed irregularity pattern and stain intensity in abscission development occurring between spur and fruit pedicel. Anatomical study indicated an effect (as far as 20 to 50 cells away from the immediate abscission) in response to safranine and fast green stains (Simons and Chu 1975).

Cultivar Identification in the Nursery

Plum cultivars can be classified into two general groups according to whether the shoots are pubescent or glabrous. The difference in this character often indicates that trueness-to-name of the trees is questionable, but it does not in itself provide cultivar identification. Another general division is that in many cultivars the growing points are green, whereas in others they are reddish to red. Various other characters must be considered before positive identification but the differences have been recorded for many cultivars (Upshall 1926A,B; Shoemaker 1927; Southwick and French 1944).

CLIMATIC REQUIREMENTS

Owing to the wide variability of the different species, plums are grown over an extensive area. Cold, wet, and very windy weather during bloom endangers the crop. Proximity to large bodies of water is desirable since water tempers the cold of winter and the heat of summer.

The standard plum cultivars of the north do not succeed in the south because of insufficient chilling.

CHOICE OF LOCATION AND SITE

Japanese type cultivars, because of their early bloom, may benefit from a northern slope. Such a slope may retard bloom and thus avoid early frost injury, but it should not be so steep as to erode. With all species, a little slope to provide both water and air drainage is better than a dead level or a pocket.

The plum does not seem hard to suit in soil requirements, provided the water drainage is good. However, most European type cultivars seem to thrive better on clay soils, whereas the Japanese favor light "peach" soils.

PLANTING THE TREES

Two-year-old trees are usually best for *P. domestica* plum, but 1-year whips are suitable for the Japanese type.

To avoid danger of winter injury, spring planting is advised. However, fall planting has merit in regions where winter cold is not too severe. Work the ground well before planting.

Best planting distance varies with the cultivar, but ordinarily 6.1 × 6.1 m (20 × 20 ft) is about right, although some strong growing cultivars may be set 7.3 m (24 ft) apart.

Trim all broken roots before setting and make the hole large enough to take the roots when spread out. Set the tree slightly deeper than it stood in the nursery, and tramp the soil firmly around the roots so that no air pockets are left. Leave a few inches of loose soil on top to act as a mulch.

POLLINATION

Pollination is an important consideration in planting a plum orchard, for most cultivars are self-unfruitful, and some are cross-unfruitful as well. Even the so-called self-fruitful cultivars probably benefit from cross-pollination.

When selecting a pollinator, choose a cultivar of good quality. Also, remember that the European and Japanese cultivars are generally incompatible. Insects are the agents of cross-pollination in both European and Japanese plums.

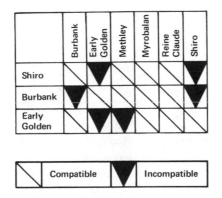

FIG. 6.4. COMPATIBILITY CHART, JAPANESE PLUM CULTI-VARS

Among the European type cultivars, Reine Claude, Yellow Egg, Sannois, and Agen are self-fruitful to a degree. Grand Duke, Jefferson, Washington, Imperial Epineuse, and Arch Duke are self-unfruitful. Italian and Stanley usually benefit from cross-pollination. Cultivars that bloom early, such as Reine Claude and Lombard, are not reliable pollinators for the later blooming Italian and Stanley.

Most Japanese cultivars are commercially self-unfruitful. A few, such as Methley, Beauty, and Santa Rosa, may set when self-pollinated. Most

Japanese type cultivars are intercompatible. Certain cultivars, such as Formosa and Kelsey, may not provide adequate amount of effective pollen to be good pollinators.

Although Burbank can be pollinated by Shiro, it will not pollinate Shiro. Therefore, include a third cultivar as a pollinator. Reine Claude, a European cultivar, sometimes pollinates Shiro, so that an orchard consisting of Burbank, Shiro, and Reine Claude may be satisfactory with respect to pollination. However, it is generally better to arrange for Japanese pollinators for Japanese cultivars.

Most plum cultivars are hybrids of two or more *Prunus* species. Both pollen germination and abortion make it seem likely that the degree of hybridity is responsible for the general level of pollen sterility and, hence, that this level is determined basically by the genotype of the cultivar (Flory and Tomes 1943).

FRUITLETS

Plum fruitlets are shed with stems attached in the first and second waves, but without stems in the third drop (June drop). Several successive abscission periods or waves commonly occur each year: (1) Flowers which reach full bloom but have aborted or defective pistils. These flowers usually fall immediately after bloom. (2) Flowers or partially developed fruits that are normal but have not been effectively pollinated. Actual transfer of pollen may have occurred but pollen-tube growth was not adequately completed. (3) Fruits of varying size in which the zygote or young embryo has aborted.

The "June drop" commonly refers chiefly to the third abscission period, but may also include the second. It also bears a relation to competition for food materials by the developing fruits, and its extent may be influenced by the extent of the earlier drops.

PRUNING

Plum trees vary more than apple, pear, peach, and cherry in growth habit because the cultivars belong to different groups or species. The large-fruited European type plums, which make up the greater number of commercial plantings, are usually upright in growth habit especially when the trees are young. Japanese cultivars are either upright or more spreading. Damson trees are very upright and less inclined to spread than the large-fruited European type.

Pruning Large-fruited European Type Trees

As far as possible, follow the natural habit of the tree in forming the leader and scaffold branches.

FIG. 6.5. THE DARK AREAS SHOW WHERE NITROGEN FER-
TILIZER WAS APPLIED AND THE RESPONSE OF THE SOD

Pruning at Planting Time.—Plum trees as obtained from the nursery may be 1-year whips, or 2-year or older branched trees. Whips are simply headed back to the desired height. If it is desired to establish the modified leader in branched trees from the nursery, select a central growth and keep it dominant over the scaffold branches.

If existing branches of 2-years-or-older trees are to be used as scaffold branches, select and retain 3–5 of them. Do not leave too many primary scaffold branches or the head will become very dense. Primary scaffold branches are commonly headed back to 41–51 cm (16–20 in. long).

The discussion on spacing of the branches, wide angles, and other factors with trees of this age, that has been given earlier for the tart cherry, applies also to the large-fruited European type plum. (See pruning information given later in the section on Mechanized Harvesting.)

Not uncommonly, the nursery practice and growth habit may result in trees that are better adapted to the open-center form of training than to the modified leader, delayed open-center form. For the latter form train the trees as described in the corresponding section for apples.

The lowest framework branch can be left a few centimeters (inches) closer to the ground than with the apple; plum trees do not grow as large as apple trees. The lowest scaffold branch in plums should originate slightly higher on the trunk than in the peach when the aim is to train to modified-leader form.

Pruning After 1 and 2 Years in the Orchard.—The main leader is kept dominant. If a good upright leader is not available, or if the one selected the year before seems poorly adapted for the modified-leader form and there is none to replace it, it may be wise to begin the open-center form.

Pruning During the Formative or Prebearing Period.—Most plum trees will form a good top even though little pruning is done. Avoid heavy heading, since it is likely to result in long upright growths and high, dense tops. Prune lightly, mostly in a corrective way, and by thinning-out cuts. Avoid heavy thinning on young, vigorous trees, since this encourages the development of excessive shoots on the tree. Thin the centers lightly to admit sunlight, to facilitate spraying, and so to develop a healthy spur system.

In the Pacific Coast states, where large fruit is required but where warm, dry summers favor heavy fruit setting and natural soil moisture is lacking, much heavier pruning of bearing trees is practiced than in other plum-growing regions. Rather heavy heading back and thinning out of small fruiting branches is often practiced to promote vigorous wood growth. Regular yields of uniformly large fruit are produced only by trees that make a good growth. Where irrigation water is limited or soils are poor, or both, relatively more severe heading back is needed to enable the tree to produce large fruits. The thinning of the fruit buds that accompanies heavy pruning reduces the crop but increases the size of fruit (Magness 1941).

Trees of prune-type cultivars are pruned very little since extra-large fruit is not required and a heavy crop is desired. Somewhat heavier pruning may be desired for canning or fresh market crops.

Pruning Mature Trees of the Large-fruited European Type

Pruning of bearing plum trees is done largely to increase the size of fruit, reduce the cost of thinning, prevent the breaking of branches by too heavy crops, and to promote new growth.

In general, the plum tree bears most of its fruit on vigorous spurs on wood 2 years old or more and on short shoots. Fruit spurs of European plums are frequently branched, longer, and more slender than those of the Japanese.

Pruning of European-type shipping plums closely resembles the pruning of Japanese type plums (see later), but it is less severe for two reasons: since the European does not tend to overproduce, a light pruning is sometimes necessary to favor fruit-bud formation, rather than a severe pruning to reduce the quantity of fruit buds; secondly, a relatively larger proportion of the fruit of the European plum is borne on spurs (Tufts 1939).

An average yearly extension (terminal growth) of the leading branches of 23–48 cm (9–19 in.) is adequate for young bearing trees. Older trees should make at least 15 cm (6 in.) of growth each summer.

Usually in old orchards that have received little pruning, the bearing area is a thin shell on the periphery of the tree. A branch 1.8 m (6 ft long), for example, often bears most of its fruit spurs on 0.61–0.91 m (2–3 ft) of the terminal end; the rest of the branch is comparatively unproductive. Also, branches tend to be brittle and break easily if allowed to become too long.

Concentration of the bearing area near the ends of the branches is generally caused by insufficient thinning out; resulting in a shading out of the interior and of the lower fruit wood. The fruit spurs, if not killed, are often long, slender and unproductive in contrast to the shorter, stockier, fruit-producing spurs on the outside of the tree. Shaded, willowy spurs produce small fruit and often are short lived.

Corrective treatment for such a condition is an adequate thinning out of the smaller branches, where the tree has been properly trained, and even in removal of a few larger branches if the framework is crowded. Exercise good judgement in thinning out the larger branches in bearing trees. With European type cultivars that are uptight in growth habit, prune by cutting back to the outgrowing branches to direct the growth outward.

Black knot is partly controlled by pruning (see Diseases).

Pruning Japanese Type Plums

Pruning at Planting Time.—Nursery trees of the Japanese type plum are more usually received by the grower as whips than as older branched trees, but both types are common in the trade. Cut back the whips to the desired height, and in older trees use the branches, where possible, in establishing the framework.

Japanese type plums are usually trained to the open-center system. They commonly assume this form even if started to the modified-leader system.

Pruning Young Trees.—Pruning of young trees should be light, and largely corrective in nature. Advice given for the other fruits applies largely, in general principle, to the Japanese-type plum.

Pruning Bearing Trees.—Most Japanese type cultivars tend to overbear. Unless pruning is moderately heavy, they may set such heavy crops that the fruit is rather small.

Japanese type plums are usually borne laterally on short, thick spurs and twigs. The spurs are found mostly on wood 2–8 years old, and live 5–8 years. Some of the crop is borne laterally on 1-year-old wood, the fruit being produced in general on the basal part of the shoot. The fruit buds are often borne in 3s, or even 4–5 at a node. Commonly, 3 flowers arise from 1 flower bud.

The Japanese plum is a prolific bearer; and, since large fruits are the most valuable commercially, prune the trees rather heavily to reduce the crop and at the same time induce sufficient new wood growth. Prune so that 25–51 cm (10–20 in.) of new wood growth annually is secured on young bearing trees and 25–30 cm (10–12 in.) on older trees.

Certain cultivars, notably Burbank, are rather low-spreading and even after the trees come into bearing, attention must be paid to diverting the growth into the more upright branches.

Wickson, Kelsey, and Santa Rosa make a comparatively narrow, upright growth. They require a type of pruning which, through thinning-out cuts to outward-growing branches, develops a more spreading tree.

SOIL MANAGEMENT

A combination of cultivation and cover crops is generally advised for the plum orchard. Clean cultivation without cover crops sooner or later exhausts the soil of humus. The plum probably thrives better than the peach or cherry under sod and mulch management systems in the east.

Fertilizer

Nitrogen, Phosphorus, and Potassium.—If the trees are making poor growth with pale green leaves, N is probably deficient. The amount of N fertilizer to apply depends on the size of the tree. Use a small amount at first and increase it as needed. A mature tree should not require more than 0.91 to 1.36 kg (2 to 3 lb) of nitrate of soda or its equivalent. The need for N or other fertilizers is usually indicated also by poor cover crop growth. P and K, if needed, for both trees and cover crop may be broadcast over the whole area at the rates of 90.7 to 136 kg (200 to 300 lb) per acre of a complete fertilizer.

Heavy applications of potash may be necessary to overcome collapse of prune trees during years of heavy crops.

Increase of the N level in leaves and fruits of 10-year old Stanley prune was significant after soil application of 0.794 kg (1.75 lb) alone or in combination with 1.02 kg (2.25 lb) K. K levels of leaf and fruit began to show increases the third year after fertilization. Ammonium nitrate favors accumulation of Mn in leaves. Higher rates of N alone or combined with K increased yields. Increasing leaf K did not affect fruit yield unless N was also increased. N had no effect on titratable acids and soluble solids of the fruit. High N increased the titratable acids in fruits in a light crop year. Leaf K should be above 2.0% and leaf N above 2.1% for optimum production of Stanley prune (Kwong 1973).

Boron.—With boron deficiency in European type cultivars, brown areas form in the fruit. The brown flesh beneath these sunken areas is firm and may extend to the pit. The fruits color up earlier than usual and drop; gum pockets may form in the flesh. The number of plums that show deficiency symptoms varies from a few fruits to the entire crop. The trouble is sometimes found on the same tree for several years in succession, but in other cases it does not recur for several years (Hansen and Proebsting 1949).

Boron injury may occur when borax is used at the rate of 454 g (1 lb) per tree, so apply about 113 to 227 g (1/4 to 1/2 lb) per mature tree in late summer or fall broadcast on the surface of the soil under the branches of the tree. This should be adequate for at least three years. If boric acid is used, apply only about 2/3 as much.

Calcium.—Ca accumulates to a higher level in prune leaves than in cherry leaves (Abdalla and Childers 1973).

Irrigation

The drying ratio of French prunes is correlated with fruit set and yields and is not changed materially by irrigation treatment, but can be changed by severe thinning of the fruit and by the ratio of leaf area to fruit. In years of large crops of prunes, the drying ratio is high.

When soil moisture is readily available throughout the growing season and crops are light to moderate, the percentage of large prunes (three largest size grades) is related to the total number of fruits per tree. When the readily available soil moisture is exhausted before the fruits are of full size, prune trees produce a smaller proportion of large fruit, even those with a light crop.

Maintaining soil moisture far above the permanent wilting percentage does not produce a higher percentage of large fruits relative to the total number than do normally irrigated trees. When the crop is light to moderate, a certain percentage of large fruits may be expected. The proportion of large fruits is not increased by the use of unnecessarily large amounts of water, but may be markedly decreased if the readily available moisture becomes exhausted before the fruits are fully grown (Hendrickson 1948, 1949).

THINNING THE FRUIT

Not all plum cultivars need thinning, but in most years some of them benefit from it. Those which tend to overbear, like most Japanese type plums, need thinning in order to harvest good sized fruit. Moreover, it is better for the health of the tree to retain only as much fruit as it can bring to normal size. Since growth of the fruit does not accelerate rapidly until relatively late in the season, comparatively late thinning is satisfactory and cheapest in hand work, but may not be as effective as early thinning. Even though thinning may be beneficial to the tree, the grower must consider whether the work will prove profitable.

Chemical thinning of the blossoms can be helpful where biennial bearing is occurring or where size is difficult to attain. The dinitros thin plums more easily than peaches. Dinitro at 62 ml per 100 litres (1/2 pint per 100 gal.) of water has given fairly satisfactory results.

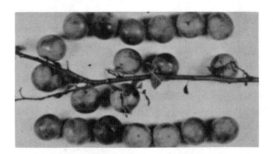

FIG. 6.6. THE TWENTY FRUITS ON THIS BRANCH WERE RE-
DUCED TO SEVEN AND THE INCREASE IN SIZE OVER FRUIT
ON UNTHINNED BRANCHES WAS ABOUT 10%

HARVESTING, HANDLING, AND STORING

Fresh Plums

The harvest season in the east may extend from July to October. A few late cultivars do not ripen until mid-October, often hanging until the end of the month. Plums usually ripen unevenly over the tree. Therefore, make 2–3 pickings, although 1–2 may suffice for cultivars where the fruit is left to become well matured before harvest. Plums which are picked for canning, such as Italian, are usually harvested at one time. Increased uniformity in maturity may be secured by harvesting full-bearing trees in 2 pickings 1 week apart.

Sugar content as measured by the refractometer is a consistent and practical indicator of harvesting maturity for prune type plums. The pressure test of prune type plums having a desirable soluble solids content varies from year to year, so this test is not a good guide for harvesting (Fisher 1940).

Mature Italian fruits for fresh use are dark purple with an amber crisp flesh, free from the pit. They have a refractometer reading of 18% or more, and are of good quality for either fresh shipment or canning. Light-colored Italian fruits may be pinkish on the outside and have green flesh; they are poor in quality. The maturity of early harvested Italian fruits will be between the two foregoing and specifications vary from year to year and with the packing house (Atkinson 1956).

Early ripening kinds of Italian are grown in the Pacific northwest mostly for fresh or canned use rather than for drying. They are subject to internal browning, a disorder that appears following storage and subsequent ripening. The use of GA at proper concentration seems promising for the production of firmer fruit, increased shelf-life, and reduction of internal browning (Proebsting and Mills 1966, 1969).

The growth regulator, ethephon (2-chloroethyl phosphonic acid), advanced color and soluble solids development by about 2 weeks, and softening by 1 week when applied at 80 ppm 3 1/2 weeks before commercial maturity of untreated fruit. Lesser effects of a similar nature were produced by rates down to 10 ppm on the same date. GA delayed softening by 1 week, color and soluble solids development by 1–2 days when applied at 50 ppm 3 1/2 weeks before commercial maturity of untreated fruit. Internal browning was reduced. Ethephon combined with GA produced better colored fruit with higher soluble solids and less internal browning than untreated checks. Ethephon stimulates ethylene production and is effective on Italian over a long period (Bukovac *et al.* 1969).

Prunes

The best dried product is obtained when prunes are harvested in the ideal condition for eating fresh, i.e., firm (no shriveling), juicy, and the flesh free of discoloration and gas pockets. Overmature prunes tend to discolor and the flesh tends to break down with the formation of gas pockets. Immature fruits form "chocolates" (brown fruits), "bloaters," "frogs," and "frog berries" (swelled fruits).

In the coastal counties of California, prunes fall from the trees naturally when mature and are gathered from the ground. The fruit should be picked up promptly, dipped in lye solution, trayed, and placed in the drying yard. If prunes are held in boxes they discolor, develop gas cavities, lose sugar, and may even ferment or mold. Culls such as bloaters, frogs and chocolates may develop from these causes. Prunes harvested from the ground are passed through a trash remover employing a strong draft of air, and then under strong sprays of clean water before they enter the dipping tank.

If harvesting of French prune is begun in the interior valleys as soon as chlorophyll has disappeared from the skin and flesh, the sacrifice of potential tonneage (tonnage) may be great and premiums for large sizes reduced, since the prunes are still growing rapidly. However, if harvest is delayed until full size is attained, quality in the latter part of the harvest season, as measured by air pockets and dark flesh color, may be greatly impaired. Best results for the entire crop occur if harvest begins about a week after the chlorophyll has disappeared. Such a practice eliminates much of the sacrifice in tonneage that results when harvest begins with chlorophyll disappearance, and usually permits completion of harvest before the flesh color has darkened seriously (exclusive of periods of excessive heat). In this way, the harvest period encompasses the period of peak quality with only a minor sacrifice of tonneage.

Flesh firmness may be used to indicate when harvest should start, and usually will be 1.4–2.3 kg (3–5 lb). The amount of soluble solids is an even more important index. When the crop is normal or less, harvest might well start when soluble solids attain 24%. If the crop is very heavy, soluble solids may not reach as high as 20% while the fruit is turgid, and thus limit the value of this index. Color of flesh and skin is perhaps the most reliable single index of maturity, but often it may be aided greatly by determining soluble solids and also by firmness measurements (Claypool and Kilbuck 1957).

IRREGULAR CROPPING

A major weakness of Italian, the chief prune cultivar of the northwest, is the tendency for irregular bearing. Initial fruit set in six Italian plum orchards for two years in Oregon was markedly influenced by post-bloom temperature. In the warm year initial sets ranged from 36 to 64%; in the cool year initial sets were 1 to 13%. Visible frost damage occurred in only one orchard and was not responsible for the general failure in the cool year. Embryo sacs were studied in pistils from the same orchards in the two years. The cause of erratic fruit setting in Italian was attributed to its genetically determined sensitivity to cool weather in the post-bloom period. Cool weather delays pollen-tube growth, fertilization, and earl y embryo and endosperm development so long that the ovule begins to degenerate. Ovule breakdown begins in the nucellus at the chalazal end, so that even if fertilization is finally accomplished, fruit set is not stimulated (Thompson and Liv 1973).

Economic stability of the 40,500 ha (100,000 acres) of the French prune industry in California is often threatened by irregular cropping. Biennial bearing results when a heavy year and associated small, poor quality fruit is followed by a small drop and large fruit. Heavy cropping of the trees also results in terminal dieback, leaf scorch, sunburn, and limb and tree breakage.

Ethephon.—Foliage sprays of ethephon were applied at 50, 100, and 150 ppm to French prune trees at 50% petal fall, and when seed length was 8.4 to 9.4 mm (0.33 to 0.37 in.). All concentrations thinned fruits within three to four weeks after treatment. The treatments increased soluble solids and fruit size, and in some instances decreased dry storage. Return bloom the next year was greater on the treated trees than on controls. Also, fall coloration patterns appeared.

Internal Browning

To minimize post-harvest development of internal browning, Early Italian has been harvested for fresh shipment while still firm, green, and astringent. Ripening for several days is required for them to become palatable.

Ripening of the fruit, harvested at a maturity suitable for fresh shipment, is enhanced by several days exposure to low temperature of 1.1°C (34°F). When cold treated, the fruits develop more color, less acidity, and are often softer than those ripened immediately at 21.1°C (70° F). No change in these characteristics occurs during the cold period. Canned whole fruit and juice reflect these changes and are readily detected by a taste panel (Proebsting *et al.* 1974).

Mechanical Shakers and Pickers

Mechanical harvesters are used in the interior valleys of California where the prunes do not fall from the trees naturally and picking by hand is costly.

To facilitate machine harvesting of prunes, train the trees with the lowest limb no less than 61 cm (24 in.) aboveground, and with a space of at least 61 cm (24 in.) on the scaffolds for easy access of the shaker head or clamp.

Equipment for mechanized harvesting of prunes includes rigid-boom, inertia type tree shakers, and self-propelled catching frames. Three men working with mechanized equipment can harvest 50−60 prune trees per hour (Bainer 1970).

Lye Dipping

Dip French prune fruits in a solution of 5.94 kg (1 lb) of lye (caustic soda) to 100 litres (20 gal.) of water at about 93.3°C (200°F), or near the boiling point. If the skins do not crack readily, use 0.9−1.4 kg (2−3 lb) of lye. Immerse for 5−15 sec. Adjust temperature and strength of solution to do the necessary checking in the shortest exposure time. Avoid large cracks or peeling of the fruits. Rinse the fruit well by immersing in running water, or by spraying.

Treat Imperial prunes to be sun-dried in a very mild lye solution for about 5 sec, or in hot water for 5−10 sec. Pretreat this fruit with care since it tends to form slabs easily.

After the lye dip, run the fruit over a needle board to make the skin more permeable to water. Then place the fruit, one layer deep, on trays for drying.

Drying

The percentage of California prunes that is dehydrated has increased so greatly that less than 20% is still sun-dried. Dehydration gives a superior dried fruit in less time and costs less. Weather does not interfere with dehydration.

The drying ratio is the number of pounds of fruit freshly picked from the ground that is required to give 0.45 kg (1 lb) of dried prunes. It usually is 0.91−1.36 kg (2−3 lb). The larger the crop and the smaller the leaf surface per fruit the more it loses in drying.

In sun drying, leave prunes in the sun for 1 week or more, until they are at least 3/4 dried. Then stack the trays and let the fruit continue to dry slowly. With large cultivars, shake the trays back and forth to turn the fruit. In case of rain, stack and cover the trays, then spread them again when weather permits. In long, continued bad weather spells, prunes may need to be finished in a dehydrator in order to avoid mold spoilage; but dehydration cannot save spoiled fruit.

When the prunes are dry enough, the flesh is firm and the pit does not slip when the fruit is squeezed between thumb and forefinger. The fruit should not be so dry as to rattle like pebbles when shaken on the trays. When a handful is squeezed the fruit will feel pliable, but when released the prunes separate.

Artificial Dehydration

Prepare a solution of lye (0.25% potassium hydroxide in water) and bring to boiling. Place the fruit in a single layer in the lye solution for 3–5 sec. Place the fruit in a single layer on trays and stack these in a cabinet where temperature and air motion are controlled. Dehydration is completed in 3–4 days in an efficient dryer, at about 65.6°C (150°F) and 60% RH. Dry the fruit to 18–19% moisture. At that point the pit in the plum has ceased to be readily movable when the fruit is squeezed with the fingers. Place the dehydrated fruit in tightly covered containers for about four weeks to let the moisture content become uniform.

Sometimes, the plum crop in the east is larger than can be absorbed by the fresh fruit and canning markets. Dehydration is not now used in the east, but the process might result in a relatively slow method of processing as compared with canning, and it would seem necessary to have refrigerated storage space to maintain supplies of ripe, raw fruit over a commercial processing season. Storage in very low relative humidity aids in controlling brown rot. An effective program of brown rot control must be maintained in an orchard supplying fruit for dehydration. Dried prunes of acceptable quality can be made from cultivars grown in the east, such as Italian, Imperial, Epineuse, and President (Truscott and Simpson 1953).

STORAGE

The storage period for fresh plums and prunes at the optimum of −0.6° to 0°C (31°–32°F) and 85–90% RH is 3–4 weeks, depending on the cultivar. After that time most cultivars become too soft for commercial handling and may suffer from breakdown, darkening of the flesh, and loss of flavor.

At 0°C (32°F), 2 weeks is about the maximum cold storage period for Italian if it must be shipped. After arrival at market, prunes shipped immediately after harvest can ordinarily be held in cold storage for about three weeks. If held longer, shriveling, mealiness, and internal browning as well as abnormal flavor may develop. Do not place too much confidence in the appearance and condition of the fruit while it is in storage since more decay, shriveling, and internal browning may take place in three days after removal from storage than during the whole storage period. Fresh prunes shipped out of storage at shipping point cannot safely be stored again after arrival at eastern markets. Storage disorders can be largely prevented by partly ripening the fruit before storage or by holding it at 4.4°–7.2°C (40°–45°F), since immediate storage at the low temperature seems to favor breakdown (Wright 1963).

A temperature of 0°C (32°F) will stop practically all ripening of plums for the approximately 12-day period which is required for shipment to some eastern markets; 4.4°C (40°F) permits only a little ripening. Temperatures of 7.2°–10°C (45°–50°F) cause appreciable coloring and softening. Hence, the plums must be picked somewhat in advance of the stage of ripeness preferred by the eastern wholesaler if transit temperatures as warm as these are used. Only early maturing or slow ripening cultivars, such as early picked Beauty and Kelsey, can withstand 12 days at 12.8°C (55°F) without becoming riper than desired for auction sale.

Breakdown is of no consequence in 12–28 days' storage of Beauty, Santa Rosa, Duarte, and President plums, but it is serious in Wickson and Kelsey, developing in early picked lots of these cultivars after 12–15 days at 4.4°C (40°F) and in some instances at 0°C (32°F). Avoid such low temperatures in transporting early picked Wickson and Kelsey plums to distant markets or in storing them for 12 days or longer. Later pickings of Kelsey are less susceptible to the trouble than early ones. Therefore, delay picking until the fruit is riper than commonly harvested.

Although 18.3°C (65°F) is generally considered a good ripening temperature for plums, 12.8°C (55°F) may cause better color and texture (Pentzer and Allen 1944).

Natural modification of atmosphere in sealed 1.5-mil polyethylene box liners provided a favorable environment for Nubiana plums during storage (0°C [32°F]). Modified atmosphere that averaged 7.8% CO_2 and 11.0% O_2, reduced fruit decay, softening, and loss of soluble solids during storage periods up to 10 weeks when compared with fruit in vented liners. The longer the storage period, the greater were the benefits attributable to modified atmosphere (Covey 1965).

Fresh prunes were packed for shipment in a study by the U.S. Dept. of Agr. (Williams 1970) in 17.6–litre (1/2-bu), ring-faced wood veneer baskets and in jumble-packed fiberboard boxes, both containing 13.6 kg

(30 lb) of fruit. Costs and charges for packing materials, packing and loading labor, equipment, cold room, and transport were 4¢ per kg (1.8¢ per lb) lower for prunes when fiberboard boxes were used instead of the other containers. The fiberboard boxes also cut prune-bruising from 21% to 7%.

INSECTS AND DISEASES

Timely and thorough applications of the proper spray materials will control the common insects and diseases. Growers should consult local authorities concerning preventive measures for pest control.

Curculio and Brown Rot (*Sclerotinia fructicola*).—Both of these pests have been discussed in earlier chapters.

Plum Pocket (*Taphrina pruni*).—Symptoms appear as greatly enlarged, distorted, hollow, bladder-like fruit. It is quite readily controlled by any good fungicide.

Plum Dwarf.—This virus disease is indicated by small, narrow, strap-like, rugose, distorted leaves; reduced terminal growth due to shortened internodes; abortion of pistils; and few fruits maturing, which are larger than normal.

Cytospora Canker.—The disease may cause serious damage to prunes in Idaho and similar areas. Symptoms are flagging or discolored terminals, dieback, streak cankers in the bark of branches, eruptive small gummy cankers, or depressed cankers. Disinfect wounds promptly.

Plum Curculio.—The adult is a rough-looking grayish to dark brown snout beetle with a hump on the middle of each wing cover. The chewing mouth parts are on the end of a long snout which projects from the head somewhat like an elephant's trunk. Soon after the petals have fallen, the beetles attack the fruit, which they injure both by feeding and by egg-laying. Injury shows on the fruit as crescent-shaped cuts, which enlarge as the fruit grows and form D-shaped russeted scars. Except in early cultivars, the plums sometimes drop when the larvae are nearly full grown, so the larvae are not always found in fruit at harvest. The adults may introduce spores of brown rot during feeding and thus indirectly cause destruction of the fruit. From Kentucky south, summer brood adults may cause damage to the fruits.

Oriental Fruit Moth.—The larvae of the first brood attack the tips of growing shoots and burrow downward for a short distance. Terminal leaves wilt, turn brown, and die. Larvae of later broods attack both shoots and fruits, but generally do not attack the fruit until it is nearly 2/3 grown. Wormy fruits result.The larvae are pinkish when mature and crawl rapidly (curculio larvae are yellowish-white when full grown and are sluggish and grublike in appearance).

Japanese Beetle.—Adults are about 1.3 cm (1/2 in.) long and have a shiny green body with reddish-bronze wing covers and a row of white dots around the hind parts. They skeletonize the foliage and eat holes in ripening fruit.

FIG. 6.7. BLACK KNOT DISEASE ON PLUM

Black Knot (*Dibotryon morbosum*).—The symptoms appear as large black cankerous-like growths on the wood. Control of this fungus disease requires the removal of twigs or branches showing the swellings. Make the cut a foot or so toward the trunk of the tree from where the knot is located with the hope of removing the mycelium of the fungus, invisible to the unaided eye, which penetrates the wood for some distance below the diseased area. Also, apply a dormant spray.

If black knot appears on the trunk, its practical control is impossible; but the tree may live for several years depending on its condition and the rapidity with which the disease develops. If other plums are growing nearby, a tree with knots on the trunk should be removed promptly to prevent spread of the disease. Burn all infected prunings immediately.

REFERENCES

ABDALLA, D.A., and CHILDERS, N.F. 1973. Calcium nutrition of peach and prune relative to growth, fruiting, and fruit quality. J. Am. Soc. Hort. Sci. *98* (5): 517-532.

ATKINSON, F.W. 1956. Maturity Manual. British Columbia Tree Fruits, Vancouver.

BAILEY, J.S. 1944. The beach plum. Mass. Agr. Expt. Sta. Bull. *422.*

BAINER, R. 1970. Mechanization in agriculture. HortScience *4*, 231-232.

BRADT, O.A. 1967. Fruit varieties. Ontario Dept. Agr. Food Publ. *430.*

BRASE, K.D. 1948. Propagating fruit trees. N.Y. State Agr. Expt. Bull. *773.*

BUKOVAC, M.J. *et al.* 1969. Chemical promotion of fruit abscission in cherries and plums with special reference to 2-chloroethyl phosphonic acid. Proc. Am. Soc. Hort. Sci. *94*, 226-230.

CHAPLIN, M.H. 1972. Effects of rootstock on leaf element content of Italian Prune. J. Am. Soc. Hort. Sci. *97* (5): 642-644.

CLAYPOOL, L.L., and KILBUCK, J. 1957. Influence of interior valley French prunes on yield and quality of the dried product. Proc. Am. Soc. Hort. Sci. *68*, 77-85.

COVEY, H.M. 1965. Modified atmosphere of Nubiana plums. Proc. Am. Soc. Hort. Sci. *86*, 166-168.

DAY, L.M. 1934. Ripening dates of Grand Duke plums on various rootstocks. Proc. Am. Soc. Hort. Sci. *63*, 150-243.

DORSEY, M.J. 1919. Sterility in the plum. Genetics *4*, 417-488.

FISHER, H.D.V. 1940. Maturity indices for harvesting Italian prunes. Proc. Am. Soc. Hort. Sci. *37*, 183-186.

FLORY, W.S., and TOMES, M.L. 1943. Plum pollen, its appearance and germination. Proc. Am. Soc. Hort. Sci. *43*, 42.

HANSEN, C.J., and PROEBSTING, E.L. 1949. Boron requirements of plums. Proc. Am. Soc. Hort. Sci. *53*, 13-20.

HARTMAN, F.O. 1956. Plum rootstock research. Ohio Agr. Expt. Sta. Spec. Circ. *93.*

HENDRICKSON, A.H. 1948. Size of prunes as influenced by differences in set and irrigation treatment. Proc. Am. Soc. Hort. Sci. *51*, 234-238.

HENDRICKSON, A.H. 1949. Effect of yields upon the apparent drying ratios of French prunes. Proc. Am. Soc. Hort. Sci. *24*, 244-249.

HUTCHINSON, A. 1967. A simple method of growing clonal *Prunus* rootstocks from hardwood cuttings. Ann. Rept. Hort. Res. Inst. Ontario, 9.

KWONG, S.S. 1973. Nitrogen and potassium fertilization effects on yield, fruit quality, and leaf composition of Stanley prunes. J. Am. Soc. Hort. Sci. *98* (1): 72-74.

MAGNESS, J.R. 1941. Pruning hardy fruit plants. U.S. Dept. Agr. Bull. *1870.*

MARTIN, G.C. 1975. Thinning French Prune with ethephon. J. Am. Soc. Hort. Sci. *100* (1): 90-93.

PENTZER, W.T., and ALLEN, F.W. 1944. Ripening and breakdown of plums as influenced by storage temperatures. Proc. Am. Soc. Hort. Sci. *44*, 148-156.

PROEBSTING, E.L., and MILLS, H.H. 1966. Effects of gibberellic acid and other growth regulators on quality of Early Italian prunes. Proc. Am. Soc. Hort. Sci. *89*, 135-139.

PROEBSTING, E.L., and MILLS, H.H. 1969. Effects of 2-chloroethane phosphonic acid and its interaction with gibberellic acid on quality of Early Italian prunes. J. Am. Soc. Hort. Sci. *94*, 443-446.

PROEBSTING, E.L. et al. 1974. Interaction of low temperature storage and maturity on quality of Early Italian fruits. J. Am. Soc. Hort. Sci. *99* (2): 117-121.

ROBERTS, A.N., and HAMMERS, L.A. 1951. The native Pacific plum. Oregon Agr. Expt. Sta. Bull. *502*.

SHOEMAKER, J.S. 1927. Eliminating variety mixtures in nursery trees. Proc. Ohio State Hort. Soc. *60*, 42-51.

SHOEMAKER, J.S. 1928. Damson plums. Ohio Agr. Expt. Sta. Bull. *426*.

SIMONS, R.K. 1973. Floral tube and style abscission in Stanley plum. J. Am. Soc. Hort. Sci. *98* (4): 393-399.

SIMONS, R.K., and CHU, M.C. 1975. Spur/pedicel abscission in plum (Stanley): morphology, and anatomy of persisting and drop fruits. J. Am. Soc. Hort. Sci. *100* (6): 656-666.

SOUTHWICK, L., and FRENCH, A.P. 1944. Identification of plum varieties from nonbearing trees. Mass. Agr. Expt. Sta. Bull. *413*.

TEHRANI, G. 1972. Pollen compatibility studies with European and Japanese plums. Fruit Var. J. *26*, (3): 63-66.

THOMPSON, M.M., and LIV, L.J. 1973. Temperature, fruit set, and embryo sac development in Italian prune. J. Am. Soc. Hort. Sci. *98* (2): 193-197.

TRUSCOTT, J.H.L., and SIMPSON, M. 1953. Quality of dried prunes from Niagara grown fruit. Ann. Rept. Hort. Res. Inst. Ontario (1952).

TUFTS, W.P. 1939. Pruning deciduous fruits. Calif. Agr. Expt. Sta. Circ. *112*.

UPSHALL, W.H. 1926A. Nursery stock identification. Ontario Dept. Agr. Food. Bull. *319*.

UPSHALL, W.H. 1926B. Government inspection of nurseries to eliminate variety mixtures. Proc. Am. Soc. Hort. Sci. *22*, 276-283.

WESTWOOD, M.N. et al. 1973. Effects of rootstock on growth, bloom, yield, maturity, and fruit quality of prune. J. Am. Soc. Hort. Sci. *98* (4): 352-357.

WILLIAMS, J.E. 1970. Quality, packaging, organization. Better Fruit-Better Vegetables *64* (4): 8.

WRIGHT, R.C. 1963. Commercial storage of fruits, vegetables, and nursery stocks. U.S. Dept. Agr., Agr. Handbook *66*.

7

Apricots and Nectarines

APRICOT

Botanically, the apricot (*Prunus armeniaca*), like the peach, cherry, and plum, is a drupe fruit. Horticulturally, the fruit is considerably smaller than that of commercial peaches, is bright orange in color, often with a red blush, has fine hair or none on the surface, and a distinctive flavor. The flesh is comparatively dry.

The apricot crop of North America has averaged about 202,500 tonnes (225,000 tons) over a period of years. Over 90% of the commercial production is in California and Washington, most in California. Utah ranks third in commercial production, followed by Colorado, Idaho, and Oregon. British Columbia leads in Canada.

Declining hectarage (acreage) in Santa Clara County, California, once a major center, has been offset to some extent in the San Joaquin Valley—mostly on the west side of Stanislaus County, which in 1976 had more than 3600 ha (9000 acres) bearing and about 800 ha (2000 acres) nonbearing. Apricots are grown in California for canning, drying, freezing, and fresh purposes. Canning is the largest market. Most of the plantings of the past decade have been of the Tilton cultivar for use in canning. Blenheim (Royal, Derby) is preferred for drying because of its flavor and color. It once dominated the industry; it now makes up about 50% of the bearing hectarage. Some newer cultivars are Modesto, Patterson, Tracy, and Westlay.

Hand-picking costs are high because of the small fruit size and uneven ripening. They may be more than double the harvest cost for cling peaches. Machine harvesting by the shake-and-catch method so far has proved practical only for the fruit for baby food processing.

The apricot contains nine times as much Vitamin A as the average of 18 other common fruits and twice as much as its nearest competitor—the nectarine. It exceeds the average of these 18 fruits in proteins, carbohydrates, phosphates, and niacin. It is slightly lower in fats and calcium.

There has been a gradual increase in test plantings for fresh fruit of some of the newer cultivars in the east. Susceptibility to frost during bloom is a main cause for its limited planting.

Besides Tilton and Blenheim, the very early Montgamet, Perfection, Stewart, and Chinese are grown commercially.

FIG. 7.1. THE APRICOT CULTIVAR MONTGAMET

In Utah, Chinese and Moorpark have long been the prominent cultivars. Production of Moorpark has declined largely for commercial planting because it is not adaptable except as puree or juice. In spite of its continued popularity, Chinese has several limitations which are: rather soft texture; high color and flavor (undesirable for baby food); susceptibility to frost; and a tendency toward small size in heavy cropping years. (Norton 1960).

A comparison of commercially canned Utah apricots (Chinese) with those from California (chiefly Blenheim and Tilton) showed that while flavor of the Utah product was somewhat comparable to California apricots, the texture (firmness and stringiness) was decidedly inferior. A few of the more promising cultivars and selections worthy of trial in Utah and perhaps elsewhere, in order of maturing are: Early Orange (patented); Wilson Delicious; Perfection; Utah 27; Sunglo (patented); Utah 18; and Utah 32 (Norton 1960).

Some cultivars for testing in the east, and their place of origin, are as follows: Alfred and Farmingdale (New York); Curtis and Goldcot (Michigan); Coffing (Indiana); and Vescot and Victory (Ontario). In 1969 HortScience listed these registrations for California: Flaming Gold; Gold Kist; Patterson; and Pinkerton.

Moongold and Sungold were developed in Minnesota for areas where hardiness is very important. Others of the Siberian type and their place of origin are: Sunshine; Anda; Manchu; Ninguta; Sing; and Tola (all from South Dakota); Scout; Robust; and Leslie (Manitoba).

Apricot cultivars are usually budded on apricot. It is resistant to nematodes. Apricot cultivars are sometimes budded on myrobalan plum for special soil, climate, and cultural conditions. But in general, results are not as satisfactory on such stocks as on apricot rootstock.

In nursery stock and young trees, apricot root is easily identified by the beet-red or blood-red color. In older trees, however, more detailed features are essential to distinguish between apricot and peach roots (Day 1934).

Age of Bearing, Tree Life, and Yield

The trees often bear a few fruits at 3–4 years of age, but more usually bear their first light crops at 4–6 years. Commercially, the life of a good orchard is 15–20 years. Yields average 3.08–3.52 kl per ha (350–400 bu per acre) of marketable fruit, and may be higher in the most suitable areas and lower in less favorable ones.

Location, Site, and Soil

The apricot thrives in soil that is suitable for the peach and Japanese plum. Because of its early bloom a northern slope may have some advantage. Poorly drained sites, for either soil or air, are not good for apricot orchards.

Suggestions on planting given earlier for peach and Japanese plum apply in general to apricot. The trees are commonly spaced 6.1–7.6 m (20–25 ft) each direction, i.e., 3–44 trees per ha (70–109 trees per acre), in leading apricot areas.

Soil Management and Fertilizer

The apricot is shallow rooted and suffers more than apples from lack of moisture when in competition with grass. Since apricots are so similar in growth and fruiting to the peach and Japanese plum, they generally respond to similar fertilizer treatments.

The effect of high N on delaying fruit maturity begins early in the cell division phase of fruit development and is cumulative until maturity (Albrigo 1966).

Apricots respond to K fertilizer with improved vigor, better bloom, and increased yields, but they fail to respond to P fertilizer except for newly transplanted trees. The P requirements of apricots are low but the trees seem able to absorb P from the soil over a long period of the year and to store it in the tissues.

Apricot trees are susceptible to Zn deficiencies, and the typical symptoms of "little leaf" or "rosette" are common in irrigated orchards of the Pacific slope. Zinc to phosphate balance, soil microorganisms, and light intensity may play a role in nutrition.

Iron malnutrition of apricot trees is associated with carbonate in the soil and, as with peaches, is commonly referred to as lime-induced chlorosis. The disorder, as indicated by yellowing of the foliage, is not the result of a simple Fe deficiency but seems to be associated with a Ca, K, and Fe interrelationship.

Saline toxicity to apricots may occur in soils containing sodium chloride in excess of 0.30%.

The chief Mn deficiency symptom of apricot is an interveinal chlorotic pattern on the foliage. The disorder has been corrected by Mn sprays, injection, and soil treatment.

Boron deficiency symptoms in apricots, especially on the fruit, have been reported frequently. In Washington, internal browning and corky tissue developing in the stone area, cracking of fruit, and shriveling, surface browning and constructions of fruit have been corrected by field applications of 227 g (0.5 lb) of borax per tree (Bullock and Benson 1948).

Irrigation

In the west, apricots are generally grown under irrigation. In California, prolonged periods of dry soil conditions during July, August, and September resulted in reduction in number of flower buds differentiated, and a slower rate of development of buds. Also, an unusually cool March prolonged the bloom period of the late-developing, later-differentiated buds, so that full bloom on trees which received no irrigation after harvest the previous year occurred about one month later than on trees receiving postharvest irrigations. Fruits set by the late-developing flowers from later-differentiated buds were characterized by long stems to which a small leaf frequently was attached. The stones of such fruit were smaller than normal. The fruits, which remained small, matured 1–3 weeks later than normal fruit (Brown 1953B).

Pollination

Most commercial apricot cultivars are self-fruitful. In Washington, Perfection and Riland proved self-incompatible, but were cross-fruitful with Blenheim, Royal, Tilton, and Wenatchee Moorpark. Pollen of

FIG. 7.2. LIKE THE PEACH, COMMERCIAL CULTIVARS OF
APRICOTS ARE SELF-FRUITFUL, BUT BENEFIT FROM BEE
ACTIVITY

Perfection and Riland was functional and normal in appearance (Schultz 1948).

The Constant cultivar contained more than one megaspore mother cell. Embryo sacs did not develop exclusively from the chalazal megaspore but from any member of the tetrad. Thus, several embryo sacs of different genetic constitution could be present in one ovule. Sometimes these embryo sacs fused or were otherwise irregular. Integuments were abbreviated and the nucellus was in direct contact with the endocarp in 26% of the ovules examined. Degeneration of embryo sacs occurred at all stages of development. Only 22% of the ovules between anthesis and petal fall contained functional egg cells. The results indicate a possible explanation for instances of poor set in apricot, even under ideal conditions of weather and pollination (Eaton 1965).

Reduction in cropping of Tilton has been noted near an aluminum reduction plant known to release fluoride. Apricots are sensitive to fluoride, with Chinese and Royal leaf tissue more susceptible to injury than either Tilton or Moorpark.

Fumigations with hydrogen fluoride (HF) decreased *in vivo* pollen tube growth of Tilton. Both tube lengths and percentage of styles with pollen tubes that reached the base of the style were adversely affected more by high HF concentration for a short time than by low concentration for a longer time (Facteau and Rowe 1977).

Cold Resistance

Apricots bloom earlier than peaches, and thus stand a greater risk of damage from spring frosts. Closed buds are highly subject to damage from low temperature and, after injury, may drop to the ground in northern regions.

About 850 hr below 7.2°C (45°F) seem necessary to break the rest period of Royal buds completely. Some of the Mediterranean types, from low altitudes, seem to have very low chilling requirements.

Increase in resistance of Royal fruits to low temperature injury was obtained by spraying 2,4,5-T about 15 hr before a frost of 0°C (32°F) for 3 hr or -0.6°C (31°F) for 1 hr that occurrred 40 days after full bloom. Also, the sprayed trees dropped 84% less fruit that was injured by frost that did the unsprayed trees. Severely frosted Tilton fruits, in which the ovules were killed and, in some cases, the endocarp tissue injured, were induced to grow to normal size by spraying with 2,4,5-T two days after the frost occurred (Crane 1954).

Fruiting Habit

Apricots tend to produce excessive crops in 1 year and, by reason of this overbearing, to develop few fruit buds for the following year. This tendency is particularly marked in British Columbia in the canning Blenheim and Tilton cultivar. The alternate bearing habit, once established, has many disadvantages over the annual bearing condition (Fisher 1951).

Fruiting habits of the apricot resemble those of Japanese plums. The tree fruits on 1-year-old shoots and on short spurs. These spurs have a life of 2 or, at most, 3 seasons.

Fruit buds are usually formed at all nodes from base to tip of the shoot. At the base of the shoot 2–3 buds form at each node and the number per node decreases toward the tip of the shoot where frequently only 1 bud is formed; 2 or more fruit buds per node are commonly found on short spurs. On biennial bearing trees in their heavy cropping year bud formation is largely confined to the shoots. Apricots usually produce more fruit on spurs than do peaches.

Thinning the Fruit

The growth stages in the apricot are much the same as in the peach, and have a similar relation to time of thinning.

In British Columbia, the time during which blossom-bud induction in apricots could be influenced was not entirely constant from year to year, and varied from 38 to 55 days from full bloom. The most effective period for influencing fruitbud formation, however, extended up to 38–41 days from full bloom. These data emphasize the importance of early thinning of biennial cultivars such as Blenheim and Tilton in order to maintain as far as possible the annual bearing condition (Fisher 1951).

Training and Pruning

Hand pruning of apricot should be much the same as that described for Japanese plum. Fruit spurs of apricot are shorter lived, however, than those of plums; and, as a result, one should practice a slightly heavier pruning with the apricot in order to keep the spur system renewed.

As far as possible train the tree to a modified leader. This is more easily accomplished than with peach trees, but not so efficiently as with apple trees.

The branches of apricot, although often heavier and larger than those of peach, are inclined to become too long and willowy unless they are headed back to desirable laterals. In like manner, the trees become too thick unless a fair amount of thinning out is practiced.

As the trees become older and especially after several heavy crops have been produced, it is desirable to increase the amount of pruning in order to stimulate growth and increase fruit size. Prune to replace branches comprised of older spurs with branches of younger spurs. Aim for 33—76 cm (13—30 in.) of shoot growth each year in young trees, and 25.4—35.6 cm (10—14 in.) in older trees. Do not allow the limbs to become unbranched and willowy.

FIG. 7.3. SOMEWHAT LIKE PEACH, THE APRICOT DOES NOT LEND ITSELF WELL TO THE MODIFIED LEADER SYSTEM OF TRAINING

Late Spring Freeze Damage

During a night in Illinois, the temperature lowered from −1.1° to −4.4°C (30° to 24°F). Fruit development of apricot fruits (Wilson Delicious) was 18−20 days after anthesis. The floral tube was absent, styles had abscissed, the ovary was enlarging rapidly, and the base was growing down around the floral-cup base. The range of equatorial diameter of the fruit at this stage of development was 0.97−1.22 cm.

Disrupted tissue appeared in the following areas: vascular bundle tissues in the mesocarp, general destruction of the endocarp, occulary tissue destruction with separation of the outer and inner integuments, and a distinct injury occurring in the chalazal end of the ovule (Simons 1969).

Harvesting

Predicting Time of Harvest.—Satisfactory predictions of the time of apricot harvest can be made in California, based on temperature data for the 42 days following bloom (Brown 1952A).

FIG. 7.4. THE WINTER INJURY TO THIS MATURE APRICOT TREE WAS DUE TO THE ACCUMULATIVE EFFECT OF THREE SEVERE WINTERS

Machine Pruning and Harvesting.—Machine pruning of the tops of apricot trees has been useful in association with mechanical harvesting in California. The tops were cut back 0.9−1.2 m (3−4 ft) in late winter. Little or no fine pruning was done. The heavy pruning did not result in a dense umbrella of new growth, though some severe pruning of the top seemed necessary to keep the umbrella-like growth under control and to promote development of lower branches. Any difference in time of fruit maturity and size depended much on the lack of fine pruning of the mechanically-pruned trees (Fridley 1969).

Minor or moderate bruises and cuts on apricots tend to cook out during canning, so that apricot injury is not as serious as with cling peaches. Thus, protective catching frames and tree shape, though important with apricots, is not as critical as with cling peaches. A primary cause of fruit loss is variable maturity both within an orchard and within a tree. At best, shakers are only slightly selective in regard to removing mature

FIG. 7.5. BROWN ROT ON APRICOT

apricots from a tree while leaving immature fruit on the tree. As a result, the best method is to harvest an orchard selectively, going over it more than once. This procedure allows for tree-to-tree variability but not for within-tree variability in maturity.

Best results, thus, call for cultural practices which will induce uniform maturity on each tree. Pruning, fertilization, crop load, hormone applications, or other factors may require adjustment. Relatively fast equipment, covering the orchard quickly, and either a slow rate of maturing or a gradient in maturity across an orchard would further add to the hectarage (acreage) which could be handled by only one machine.

Effects of 2,4,5-T

There is some evidence that sprays of 2,4,5-T on apricots may increase fruit size, hasten maturity, and stimulate color development. Besides a higher moisture content, sprayed fruits are greater in fresh and dry weights, because of a stimulation in growth of the flesh alone. Application shortly before pit hardening begins combined more of the advantages and less of the disadvantages (from fruit splitting, temporary foliage flagging, and shoot tip killing) than those made earlier or later. Besides the above responses, 2,4,5-T sprays effectively reduced preharvest fruit drop. Since the period of effectiveness of 2,4,5-T is 50 to 60 days, time of application for fruit drop is not critical (Crane 1953).

Marketing

Fresh apricots are marketed through July and August. The Brentwood lug holding 10.9 to 11.3 kg (24 to 25 lb) net had long been used in California and Washington. Different sized lug boxes holding about 11 kg (24 lb) are also used in Washington and Oregon. In Colorado, Utah, and Idaho, apricots have been shipped chiefly in 17.6 litre (1/2 bu) tub baskets. Apricots are usually packed in rows, sized, and "faced and filled." The row sizes are stamped on the box and the usual sizes are 6, 7, 8, and 9 row. Because apricots are very perishable and susceptible to bruising, they need protective packaging to reduce handling damage.

Development of high-moisture packs, suitable for eating out of hand, may open up a snack-food market outlet. Uneven ripening on one side of individual fruits is a problem requiring costly sorting expense.

Apricots are also marketed as canned, frozen, and dried products. For drying, the fruit is cut by hand or machine, exposed to the fumes of burning sulfur in a "sulfur house," and then spread out on trays in the sun. Large scale operators use dehydrators, fork lifts, other machinery, and stringent sanitation. Average dry-away is 6 kg (13.2 lb) of fresh weight for 1 kg (2.2 lb) of dried product. With cutting machines, a water spray washes sugar from the fruit to raise the ratio of fresh to dried fruit.

FIG. 7.6. VERTICILLIUM WILT ON THE TREE AT LEFT

Storage

Apricots are not stored to any important extent although they may keep for one to two weeks at $-0.6°$ to $0°C$ ($31°-32°F$) and 85 to 90% RH. When harvested at a firmness permitting storage and shipping, the fruit lacks flavor and dessert quality after ripening but is satisfactory for canning. Average freezing point of the fruit is $-2.2°C$ ($28°F$).

Canners are interested in storing apricots for a fruit salad pack in which apricots, peaches, pears, and other fruits are combined. Since apricots mature earlier than the other fruits, apricot halves are canned in relatively large containers and held until the other fruits are available and then comingled and canned in consumer containers. Two problems exist in this operation: the cost of double canning is undesirable, and double cooking is harmful to texture and appearance. Blenheim apricot, in particular, properly selected for maturity at harvest may have storage potential approaching seven weeks in CA at $0°C$ ($32°F$) (Claypool and Langborn 1972).

Insects and Diseases

Most of the insects and diseases which attack other drupe fruits also attack the apricot. In many cases, plum curculio is the chief insect pest of apricots. Apricots do not seem to be subject to leaf curl but are susceptible to rots.

NECTARINES

Botanically, the nectarine (*Prunus persica*), like the peach, cherry, plum, and apricot, is a drupe fruit. It is a seed sport of peach. Genetically, the nectarine character is due to a single recessive gene in its parent, the peach. Peaches that are heterozygous for the recessive nectarine character occasionally mutate to nectarine. Nectarine can mutate to peach as well.

Horticulturally, the trees of peaches and nectarines differ in no essential respect from each other in appearance, growth responses, bearing habits, or other general characteristics. Only the absence of pubescence, usually smaller size, greater aroma, and the distinct flavor of the nectarine fruit distinguish it from the peach. Both nectarines and peaches have freestone and clingstone forms that may be yellow, white, or even red fleshed. The flowers vary similarly in size and color; the stones and kernels are basically identical in appearance. Like peach cultivars, nectarine cultivars vary in chilling requirements. Sometimes the nectarine is called "smooth skin" peach.

**FIG. 7.7.   THE NECTARINE TREE, A SEED SPORT OF PEACH, IS
SIMILAR TO THE PEACH IN ALL RESPECTS**

Many of the newer and better nectarines have peach "blood" in the
crosses, for larger size and firmness. In peach-nectarine crosses, all the
first generation progeny are of the dominant peach, and backcrossing is
done to obtain 1 out of 4 nectarines as in Mendelian segregation.

In the east, as yet, the nectarine is grown only to a limited extent and
as a specialty fruit, seldom as large orchards. However, in the west,
particularly in California, where the production exceeds 54,000 tonnes
(60,000 tons) a year, the nectarine has assumed commercial importance
and quantities of it are shipped to eastern markets. Shipments are made
from June 10 to Oct. 1, and amount to a total of around 10 million 10-kg
(22-lb) (net) boxes a year.

Cultivars

The name nectarine is probably derived from nectar, the drink of the
gods.

In the east, much less breeding work has been done with nectarines than
with peaches, and so better cultivars of peaches than of nectarines are
available. In the west, great strides have been made in breeding
improved nectarine cultivars.

Stanwyck (from New Zealand), Gower and John Rivers (from England),
Quetta (from India), and Goldmine (from Australia) comprised most of
the first plantings on this continent. Burbank's Flaming Gold was of
early American origin.

The heaviest plantings of nectarine cultivars in California consist of
Early Sun Grand, Sun Grand, Late Grand, Merrill Sunrise, Grand River,

**FIG. 7.8. IMPROVED NECTARINE CULTIVARS MAY COMPARE
FAVORABLY WITH PEACHES**

and Merrill Princess; all of these are yellow-fleshed and all are patented. California breeders are very active in breeding work with nectarines.

Some nectarine cultivars developed in the east and state of origin include: New Jersey—Nectarose (white fleshed), Nectaheart (white), Nectacrest (white), Nectared (yellow), Garden State (yellow), Cherokee (yellow), and Red Chief (white); New York—Hunter (yellow), Morton (white), Geneva (white); Alabama—Stark Early Flame.

RedGold matures about as Elberta starts. Its fruits are large, and the tree vigorous and productive in the Pacific Northwest. EarliBlaze ripens just before Redhaven peach; fruits are red and somewhat clingy. SunGlo has fairly large, round fruits that mature just after Redhaven; it is freestone, and slow softening.

Sunred and Sungold are two low-chilling nectarines developed by R.H. Sharpe, University of Florida. The former matures in the week of May 10 and the latter in the week of June 10. Florida has a potential for a crop of 9000–13,500 tonnes (10,000–15,000 tons) but it is not likely that this state will grow nectarines that mature after June 10, when summer rains begin, because of brown rot (Sharpe 1966).

Culture

The information supplied earlier for peaches applies in general to nectarines. Nectarines are shipped from the west in the standard nectarine box, 29 cm (11 1/2 in.) wide.

Fruit Surface

Fruit surface characteristics of fifteen nectarine clones were examined by scanning electron microscopy at 200X and 2000X magnifications. The relatively unprotected surface, the apparent ridging, and minute cracks starting at the stomates have been noted. Among clones, differences were evident in prominence and coarseness of ridges, number and elevation of stomates, and cracking of the cuticle of mature fruits. The appearance of minute cracks coincided with the time of natural brown rot infection; thus the cracks seem to be the most likely entry sites for the brown rot organism. Stomates seemingly are suberized in maturing fruits. Relatively small differences in surface structures are apparently heritable, which may open the way for developing cultivars with desirable fruit surfaces (Fogle and Faust 1975).

Growth Curves

Growth curves of eastern-grown nectarines do not follow closely the three phase sigmoid curve established for peaches. All clones completed phase I at the same time, but many clones did not have a well-defined "final swell" and several showed gradual increases in growth through phases II and III. Percentages of final fruit size attained during the final few weeks of growth or percentage of calcium in leaves, fruits, and peels were not closely associated with cracking of fruits. Slower growth rate appeared linked to the nectarine character, although the linkage seems to have been partially broken in some clones. Growth rates of 41 nectarine clones were not closely associated wth field cracking or minor surface cracking (Fogle and Faust 1975).

REFERENCES

ALBRIGO, L.G. 1966. Effect of nitrogen level on development maturity of Royal apricot fruits. Proc. Am. Soc. Hort. Sci. *89*, 53-60.

BAKER, G.A., and BROOKS, R.M. 1944. Effect of temperature on number of days from full bloom to harvest of apricot and prune fruits. Proc. Am. Soc. Hort. Sci. *45*, 95-104.

BRADT, O.A. *et al.* 1968. Fruit varieties. Ontario Dept. Agr. Food Publ. *430*.

BROWN, D.S. 1952A. Use of temperature records to predict the time of harvest of apricots. Proc. Am. Soc. Hort. Sci. *60*, 197-203.

BROWN, D.S. 1952B. Relation of irrigation practice to differentiation and development of apricot fruits. Botan. Gaz. *114*, 95-102.

BROWN, D.S. 1953A. Apparent efficiencies of different temperatures for development of apricot fruits. Proc. Am. Soc. Hort. Sci. 62, 173-183.

BROWN, D.S. 1953B. Effects of irrigation on flower bud development and fruiting in the apricot. Proc. Am. Soc. Hort. Sci. 61, 119-124.

BROWN, D.S., and ABI-FADEL, J.F. 1953. Stage of development of apricot flower buds in relation to their chilling requirements. Proc. Am. Soc. Hort. Sci. 61, 110-118.

BROWN, D.S., and KOTOB, F. 1957. Growth of flower buds of apricot, peach, and pear during the rest period. Proc. Am. Soc. Hort. Sci. 69, 158-164.

BULLOCK, E.M., and BENSON, N.R. 1948. Boron deficiency of apricots. Proc. Am. Soc. Hort. Sci. 51, 199-204.

CLAYPOOL, L.L., and LANGBORN, R.M. 1972. Influence of controlled atmosphere storage on quality of canned apricots. J. Am. Soc. Hort. Sci. 97 (57): 636-638.

CRANE, J.C 1953. Responses of apricot to 2,4,5-T. Proc. Am. Soc. Hort. Sci. 61, 163-174.

CRANE, J.C. 1954. Frost resistance and reduction in drop of injured apricot fruits affected by 2,4,5-T. Proc. Am. Soc. Hort. Sci. 64, 225-231.

DAY, L.H. 1934. Rootstocks for stone fruits. Proc. Am. Soc. Hort. Sci. 30, 337-360.

DAY, L.H. 1947. Apple, quince, and pear rootstocks. Calif. Agr. Expt. Sta. Bull. 700.

EATON, G.W. 1965. Embryo sac development of apricot cv Constant. Proc. Am. Soc. Hort. Sci. 65, 95-101.

FACTEAU, T.J., and ROWE, R.E. 1977. Effect of hydrogen fluoride and hydrogen chloride on pollen germination in Tilton apricot. J. Am. Soc. Hort. Sci. 102 (1): 95-96.

FISHER, D.V. 1951. Time of blossom bud infection in apricot. Proc. Am. Soc. Hort. Sci. 58, 19-22.

FOGLE, H.W., and FAUST, M. 1975. Ultrastructure of nectarine fruit surfaces. J. Am. Soc. Hort. Sci. 100 (1): 74-77.

FRIDLEY, R.B. 1969. Tree fruit and grape harvest mechanization progress and problems. HortScience 4, 235-237.

LARSEN, R.B. 1969. Mechanization of fruit harvest in the eastern United States. HortScience 4, 232-234.

NORTON, R.A. 1960. New apricot varieties for Utah. Fruit Var. Hort. Dig. 15, 5-7.

RAMSAY, J. et al. 1970. Determination of the type of onset of set in spur and shoot buds of apricot. HortScience 5, 270-272.

SCHULTZ, J.H. 1948. Self compatibility in apricots. Proc. Am. Soc. Hort. Sci. 51, 171-174.

SHARPE, R.H. 1966. Peaches and nectarines in Florida. Florida Agr. Expt. Sta. Circ. 299.

SIMONS, R.K. 1969. Fruit tissue injury in apricots and sweet cherries as a result of a late spring freeze. J. Am. Soc. Hort. Sci. 94, 466-470.

Index